Stream-Tube Method

Jean-Robert Clermont · Amine Ammar

Stream-Tube Method

A Complex-Fluid Dynamics and Computational Approach

Springer

Jean-Robert Clermont
Laboratoire Rhéologie et Procédés
Grenoble Alpes University
Gières, France

Amine Ammar
Arts et Métiers - LAMPA
Angers, France

ISBN 978-3-030-65469-6 ISBN 978-3-030-65470-2 (eBook)
https://doi.org/10.1007/978-3-030-65470-2

This Springer imprint is published by the registered company Springer Nature Switzerland AG
The registered company address is: Gewerbestrasse 11, 6330 Cham, Switzerland

Foreword

Before the numerical simulation revolution, when models of mathematical physics were addressed, they were manipulated in very specific ways in order to obtain consistent solutions. At that time, each model had its own solution procedure. However, engineers and scientists recognized the limits and practical difficulties of those procedures.

In the second half of the last century, existing discretization techniques like finite differences, and other new formulations like finite elements or finite volumes, were massively developed and applied for solving very complex models of mathematical physics. In particular, finite elements experienced an exponential growth supported by major advances in numerical analysis (constituting its numerical foundations), computer sciences (for algorithmic aspects) and computational facilities (with the more and more powerful and cheaper computing environments).

At that time, everything seemed reachable using a single-purpose technique able to address "almost" any problem. Numerical simulation was democratized. Then, no more new thinking was needed, and only applications could simply be invoked. Computations prevailed in engineering practice, and they are today present everywhere in the technology world.

However, the need to make everything in simulations usually implies a loss of efficiency. As society is clearly moving towards a personalization (everything must be user-oriented, patient specific, etc.), the question is: how in such a new world could simulation be the exception?

Nowadays, the necessity for rationalizing efforts and for increasing efficiency is a key factor of the new fully connected society. This would enable precise and fast predictions for real-time decision-making. When possible, deploying computing platforms would also be helpful for accompanying technology in its new revolution (Industry 4.0, autonomous car, Internet of Things, etc.).

Thus, model-order reduction techniques are developing with an unprecedentedly growing pace, even when each application has its own reduced model. Artificial Intelligence is also a new protagonist where specific learning allows for optimal uses. It is in this context that traditional techniques are living a new renaissance, and the topic covered by the present work is without any doubt a perfect example.

The stream-tube method is extremely natural: when choosing a transformation that maps the different flow streamlines into a domain in which all of them become parallel, a complex geometry in 2D becomes a rectangle and in 3D a tube. In those mapped geometries, transport phenomena are naturally described and, moreover, discretization techniques work very efficiently.

That seems a very good idea. However, in the stream-tube method, determination of the unknown transformation leads to kinematic results which may involve closed streamlines. Moreover, many fluids encountered do not only provide linear or non-linear viscous properties, but they also exhibit history dependence, implying multi-scale descriptions or a series of conformational coordinates.

Of course, there is no free lunch, but these issues do have solutions, which are addressed in the present book, where complex problems are shown to lead to elegant formulations.

Professors Jean-Robert Clermont and Amine Ammar, the major protagonists of this numerical technology and both duly recognized in the computational rheology community for their major contributions in many related topics, describe in this monograph (complete, precise and very well written) all the technical ingredients for a full understanding and exploitation of the stream-tube method.

Francisco Chinesta
Professor of Computational Physics
ENSAM
Paris, France

Acknowledgements

The authors met and worked at Grenoble University, in the research team concerned by digital simulation and creation of numerical tools in Laboratory of Rheology and Processes (LRP), associated to UGA (Université Grenoble-Alpes) and Centre National de la Recherche Scientifique (CNRS). This book reflects the formative influence of the activities within the laboratory.

We thank the anonymous referees for their useful suggestions concerning this book on Stream-Tube Method.

We wish to take this opportunity for thanking Dr. N. El Kissi, Director of the Laboratory of Rheology and Processes for her support. Special thanks are addressed to Dr. A. Magnin (LRP, CNRS) who gave his time to read the book and proposed us interesting comments to improve the manuscript, D. Blésès (LRP) for his help to assist us in the realization of PC problems and Pr. F. Clermont (Australia) for suggestions. The book contents involve works related to participation of researchers among whom we would like to mention: M.-E. de La Lande, P. André, Y. Béreaux, M. Normandin, D. Grecov, A. Mahmoud, B. Mokdad and A. Chine. The book is the richer for their contributions. We also express our heartfelt thanks to Dr. I. D. Landau (Grenoble INP, CNRS) for his help.

We want to acknowledge Oliver Jackson from Springer. His remarks and professionalism have helped us to finalize the book.

Gières, France
Angers, France
May 2020

Jean-Robert Clermont
Amine Ammar

Introduction

Computational methods have been significantly developed for a great variety of situations and phenomena, for the use of more and more natural and synthetic complex materials as polymer solutions and melts. These methods are related to the field of "Rheology", the study of the flow of matter, applied in materials science and large domains of engineering such as polymer processing, agri food sectors and new technologies as microfluidics.

The present book proposes an original theoretical and numerical approach based on transformation of the physical domain, in various laminar flow conditions. The so-called Stream-Tube Method (STM) allows calculations of the kinematics and stresses in a deformed medium where streamlines are parallel and straight. Advantages of STM include the possibilities meaning fully storage and computing time. One or two mapping functions considered in respective two- and three-dimensional situations can be determined numerically, generally without secondary flow regions. A distinguishing feature of STM leads to significantly lower number of unknowns:

- In two-dimensional cases, *two unknowns* are considered in STM: one mapping function and the pressure, instead of three unknowns (two velocity components and the pressure) of classical methods.
- In three-dimensional cases, *three unknowns* are concerned: two mapping functions and the pressure, instead of four unknowns (three velocity components and the pressure) for traditional processes.

In computing operations, such properties may lead to a reduction in the number of unknowns. On the other hand, since the mapping functions are directly connected to the streamlines in STM, analytic functions with unknown coefficients (related to experiments, for example) may also be adopted with further reduction of the number of unknowns.

STM calculations are made with meshes defined in rectangles or three-dimensional rectangular parallelepiped domains. Particularly, complex materials with memory can be considered simply without particle-tracking problems. In this book, several applications (sharp contraction and free-surface flows related to polymer processing, journal bearings, solid mechanics cases, etc.) for various materials such as those concerned with kinetic theories are described along with different ways of solving problems.

Other distinguishing features of STM may be summarized as follows:

(i) The method can accommodate various non-Newtonian constitutive equations for incompressible fluids.
(ii) The incompressibility equation for the velocity vector is automatically verified from the formulation.
(iii) Computations may be carried out on successive sub-domains (the stream tubes) in the mapped computational domain. Data resulting from various numerical simulations with elementary stream tubes and additional condition equations constitute an "experimental" validation of this property.

Specific Features of the Book

In relation to our overall objectives, Chaps. 1 and 2 provide a review of the background for domain mapping studies and their applications to transformations linked with deformable bodies. Elements on vector and tensor analysis are recalled as prerequisite tools for the understanding and use of the rest of the book. We also recall elements from classic coordinate systems and conservation equations for solving continuum mechanics problems, using dyadic and tensorial notations. Chapter 2 also presents elements of kinematics, conservation laws and constitutive models widely considered for Newtonian and non-Newtonian fluid mechanics problems.

Chapter 3 depicts several domain transformation techniques proposed in the literature. The Stream-Tube Method (STM), based on kinematic and geometric concepts, is presented in laminar flows. The approach, defined *a priori* for computations of open streamlines, can be extended to closed lines under certain conditions. The primary unknowns are, together with the pressure, a transformation function which maps the physical domain into a rectangular domain. The mapped streamlines are parallel and straight, leading to simple discretized elements for the computations, such as those of Hermitian transformations. Interesting features of STM are underlined in the text. These include the automatic verification of mass conservation and the simplicity of particle tracking, which arises when materials with memory are considered for the calculations. In numerous geometries, the deformations may be determined only in, sub-domains of, the total domain by using integral forms for boundary condition equations. By doing so, the flow field can be computed in successive stream tubes of the global domain.

Chapter 4 presents STM equations in two-dimensional cases, related numerical procedures and applications to isothermal flow problems. The mapped computational domains are parallel stream bands which are constructed as rectangular or trapezoidal elements. In the simulations, mixed formulations may be adopted. Owing to the non-linearity of the governing equations where constraint conditions should be necessary, second-order solving methods such as the Levenberg–Marquardt (LM) and trust region optimization algorithms are adopted. By considering firstly the peripheral stream tube, the other sub-domains may be computed separately, under the action of the surrounding complementary domains. Flow computations for non-trivial physical examples corresponding to an abrupt contraction and to the swell problem are provided, for memory-integral and differential constitutive equations. Descriptions of three flow examples in two-dimensional cases are summarized at the end of the chapter.

Chapter 5 presents the stream-tube formulation in three-dimensional cases. Two unknown mapping functions may be computed for flows involving open streamlines. A two-dimensional section where the kinematics are known is required. An original method allowing determination of a Poiseuille flow velocity (contour values) at reference sections of different shapes is presented. Memory-integral constitutive equations can be considered without particle-tracking problems. Computing examples are provided for complex three-dimensional confined and swell flow situations in relation to processing operations. Results on the swell problem highlight consistent comparisons with experimental data. By way of illustration, an example of a three-dimensional domain is presented at the end of the chapter.

In order to study some flows with closed streamlines, corresponding to circulating flow regions, Chap. 6 presents theoretical formulations and applications of STM with domain decomposition. General transformations and basic equations in two- and three-dimensional flow conditions are presented. The approach leads to consideration of mapped sub-domains in which open and closed streamlines may be encountered. Examples are provided with two-dimensional journal bearing problems. In relation to industrial flow problems, the applications concern flows of fluids obeying different non-linear constitutive equations. Particularly, domain decomposition in STM leads to the identification of more complicated flow situations arising from inertial conditions.

Chapter 7 presents a theoretical analysis of unsteady flows in the stream-tube method. In such cases, streamlines and pathlines are not identical. The formulation and corresponding basic equations are presented for incompressible materials. The dynamic equations are written in a way that takes into account expressions of the kinematic mapping functions versus time. The computational procedure is applied to journal bearing flows under unsteady boundary conditions. This chapter provides computing examples with Newtonian and non-Newtonian constitutive equations. Comparisons of STM results with data from the experimental literature are found to be satisfactory.

In Chap. 8, different STM basic elements are applied to non-isothermal cases which involve polymer processing situations. Slip conditions at the wall are considered in two-dimensional geometries. Memory-integral constitutive equations are adopted for the materials. The mapping function and the temperatures are computed by adopting a finite element formulation. This chapter also concerns stream-tube applications to solid mechanics problems, starting from a finite element method based on energetic concepts. An example of this approach is solved for a solid beam deformed under a compressive load.

In Chap. 9, the stream-tube method is associated with a fully deterministic micro-macro model of a complex fluid flow. We found it interesting to present a deterministic approach for simulating complex fluid flows. For this, a separated representation of the fields involved in the kinetic theory is adopted. For steady conditions, the Fokker–Planck equation is used and the microscopic information is taken into account by integrating this theory along the flow streamlines in a converging geometry. In particular, STM computations provide distribution functions for active and pendant populations of fibres along the streamlines.

Examples in the Book

Several detailed examples are provided in the book. According to the original concepts of the stream-tube method, the purpose is to bring enough information and comments for solving practical and theoretical problems. Information is therefore given on the approach and is followed with STM governing equations and related boundary conditions. For applications considered in different chapters, comparisons of STM results with existing numerical and experimental literature data are found to be satisfactory. Other illustrative (non-solved) examples are given in Chaps. 4 and 5.

Expected Audience

The book presents a theoretical and computational method that is new and original. The approach may be considered as an offering of possibilities concerning flows and deformations for various materials, notably those related to kinetic theories. Many examples involve benchmark problems and industrial applications such as those encountered in polymer processing, journal bearings, rheology, microfluidics and micro-macro situations.

A good foundation in Mathematics and Physics (Calculus, Linear Algebra, Partial Differential Equations) is prerequisite for a full understanding of this book. This should be combined with basic insights from undergraduate engineering courses in Fluid Mechanics. Furthermore, detailed findings in the book will be of interest for researchers and engineers concerned with theoretical problems encountered in rheology, studies on complex materials, numerical analysis and applied mathematics.

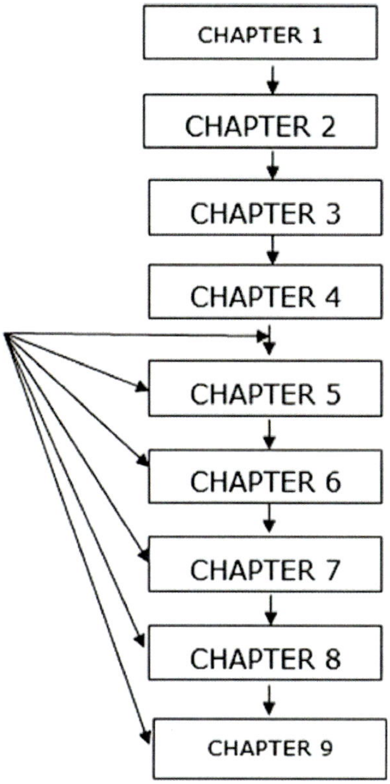

Fig. 1 Pathways through the book

Pathways Through the Book

Figure 1 provides an overview of the possible interdependence of the various chapters. For a curse on the subject, it is advisable to consider all of the nine chapters. For specific interests, Chaps. 5–9 may be considered separately, in no particular order. However, it may happen that some connections arise between the last sections in the book.

Contents

Nomenclature

Abbreviations

CAR	Carreau model
CONV	Converging geometry
DUCT	Straight domain of constant cross-section
FD	Finite differences
FE	Finite elements
FV	Finite volume
IMSL	International Mathematics and Statistics Library
K-BKZ	Kaye–Bernstein–Kearsley–Zapas
LLDPE	Low-density polyethylene
LM	Levenberg–Marquardt
NEWT	Newtonian
NR	Newton–Raphson
TR	Trust region
UCM	Upper convected Maxwell
WLF	William, Landel, Ferry
2D	Two dimensions
3D	Three dimensions

Notation

$\mathbf{a}_{\mathrm{T}}$	Scale factor in non-isothermal problem
$\underline{v}$	Velocity field
$\underline{\underline{\tau}}_p$	Extra-stress tensor
$\underline{\underline{D}}$	Strain-rate tensor
$\underline{\underline{C}}$	Cauchy tensor
$\underline{\underline{C}}^{-1}$	Finger tensor
Cp	Heat capacity

$\frac{D}{Dt}$	Material derivative
$\nabla.$	Divergence operator
<a, b>	Inner product of two functions a and b
$\underline{\equiv}$	Vector of PDF (probability distribution function) two populations
V_d, V_c	Destruction and aggregation kinetic constants
$\underline{E_R}, \underline{E_\Theta}, \underline{E_Z}$	Natural basis vectors
$\psi_1(r, \theta, \varphi), \psi_2(r, \theta, \varphi)$	Stream functions in cylindrical coordinates
$\psi_1(x, y, z), \psi_2(x, y, z)$	Stream functions in Cartesian coordinates
$\boldsymbol{H}(u)$	Hessian of a function u
$L^2(\Omega)$	Space of square-integrable functions on domain Ω
W×P×F	Space product of regular functions related to (w, p, F)
∇^2	Laplacian operator
$\tilde{N}$	Interpolation functions
$\underline{N}, \underline{M}$	Mass matrix (in configuration space and time)
$\underline{\underline{\mathcal{N}}}, \underline{\underline{\mathcal{M}}}$	Operators in configurations space and time
μ	Elasticity shearing modulus
Pe	Peclet number
Q	Flow rate
S_R	Recoverable shear rate
W_e	Weissenberg number
C_p	Heat capacity
χ	Curvilinear abscissa of a moving particle on its streamline
$\chi 0$	Abscissa of the particle at a reference section

Chapter 1
Tensor Frames

1.1 Introduction

Understanding and solving problems related to fluid and solid mechanics in research and engineering require the use of physical variables as.

- scalars (time, pressure, mass, volume, density, temperature, etc.): notation α, $\alpha \in \mathbf{R}$ (where **R** denotes the set of real numbers);
- vectors (force, velocity, acceleration, load, torque, etc.): notation $\underline{\alpha}$
- second-order tensors (stress, deformation, gradient, etc.): notation $\underline{\underline{T}}$
- higher order tensors (related to constitutive equations, etc.): for example, notation $\underline{\underline{\underline{\underline{T}}}}$ for a fourth-order tensor.

Consideration of such variables leads to define and express mathematical tools related to matrices, vectors, tensors (e.g. Borisenko and Tarapov [1], Grinfeld [2]). These basic concepts and connected topics can be associated with deformation of bodies and constitutive models used in calculations and governing equations. Different frames can also be selected to express the quantities in the most convenient conditions for the problem investigated.

1.2 Matrices

Matrices are mathematical tools related to linear and bilinear algebra. Widely employed in vector and tensor analytical problems, they are also applied in programming and numerical calculations. In this chapter, we look on some properties of these basic elements. From a practical viewpoint, we consider here a matrix as a rectangular array of $m \times n$ real or complex numbers arranged according to m lines and n columns. To define a matrix **A** (m lines, n columns), we use the following notation (index i for lines, j for columns):

J.-R. Clermont and A. Ammar, *Stream-Tube Method*,
https://doi.org/10.1007/978-3-030-65470-2_1

$$\mathbf{A} = \left[A_{ij}\right]_{1 \le i \le m, 1 \le j \le n} \quad \text{or} \quad \overset{(m,n)}{\mathbf{A}}. \tag{1.1}$$

A matrix $\overset{(m,n)}{\mathbf{A}}$ (m lines, n columns) can be considered as an element of a vector space noted $\overset{(m,n)}{E}$ that allows addition and multiplication by a scalar as follows.

- Addition:

$$\overset{(m,n)}{\mathbf{A}} + \overset{(m,n)}{\mathbf{B}} = \overset{(m,n)}{\mathbf{B}} + \overset{(m,n)}{\mathbf{A}} = \overset{(m,n)}{\mathbf{C}} \in \overset{(m,n)}{E} \tag{1.2}$$

- Multiplication by a scalar:

$$\alpha : \alpha \overset{(m,n)}{\mathbf{A}} = \overset{(m,n)}{\mathbf{D}} \in \boldsymbol{E} \tag{1.3}$$

- In this book, we will generally refer to square matrices (m, m) of real numbers that involve the same number of lines and columns. In such cases, we thus simplify as follows our previous notations for the operations and properties concerning matrices.
- Matrix transposition:

$$\mathbf{A} = \left[\mathrm{A}_{ij}\right]_{I \le i \le m, 1 \le j \le m} : \mathbf{A}^{\mathrm{T}} = \left[\mathrm{A}_{ji}\right]_{I \le j \le m, 1 \le i \le m} \tag{1.4}$$

- A matrix $\mathbf{A}$ is symmetric if it is unchanged by transposition of its elements

$$\mathbf{A} = \mathbf{A}^{\mathbf{T}}, \quad \left[\mathrm{A}_{ij}\right] = \left[\mathrm{A}_{ji}\right]. \tag{1.5}$$

- A matrix is antisymmetric (or skew) if

$$\mathbf{A} = -\mathbf{A}^{\mathbf{T}}, \quad \left[A_{ij}\right] = \left[-\mathrm{A}_{ji}\right], \tag{1.6}$$

- which implies that the diagonal terms of $\mathbf{A}$ are zero.
- The inversed matrix $\mathbf{A}$, noted $\mathbf{A}^{-1}$ is defined such that

$$\mathbf{A} \cdot \mathbf{A}^{-1} = 1, \tag{1.7}$$

- where $\mathbf{I}$ denotes the identity (or unit) matrix.
- A matrix $\mathbf{A}$ is orthogonal when its inversed and transposed matrices are identical

$$\mathbf{A}^{-1} = \mathbf{A}^{\mathbf{T}}. \tag{1.8}$$

We find it useful to recall the following typical operations with matrices:

- Product of square (m, m) matrices:

$$\begin{aligned}&\mathbf{A} = \left[\mathrm{A_{ij}}\right]_{1 \le i \le m, 1 \le j \le m}, \quad \mathbf{B} = \left[\mathrm{B_{ij}}\right]_{1 \le i \le m, 1 \le j \le m} \\ &\mathbf{A} \cdot \mathbf{B} = \left[\mathrm{A_{ik} B_{kj}}\right]_{I \le i \le m, I \le j \le m}\end{aligned} \quad (1.9)$$

- with an implicit summation on the indices $k = 1, \ldots,$ m
- Decomposition of a square matrix
- A square matrix **A** can be decomposed into symmetric and antisymmetric matrices, according to the following relationships (the first contribution is the symmetric part and the second one is the antisymmetric part):

$$\mathbf{A} = \frac{1}{2}\left(\mathbf{A} + \mathbf{A}^{\mathrm{T}}\right) + \frac{1}{2}\left(\mathbf{A} - \mathbf{A}^{\mathrm{T}}\right) \quad (1.10)$$

- $A_{ij} = \frac{1}{2}\left(A_{ij} + A_{ji}\right) + \frac{1}{2}\left(A_{ij} - A_{ji}\right)$.

1.3 Vectors and Basis

Vectors are elements of vector spaces. Operations among them are defined according to general theory on vector sets, independently of components. The use of basis vectors allows to provide expressions of vectors using coordinate systems. Therefore, they can be considered as column (m, 1) or line (1, n) matrices (respectively $\overset{m,1}{\mathbf{A}}$ or $\overset{1,n}{\mathbf{A}}$). In space position, displacement, velocity, acceleration, force and torque, among many others, are examples of vectors encountered in different sections of this book.

1.3.1 Vectors in Cartesian Coordinates

Firstly, we consider vectors in a three-dimensional space. To refer spatial points in such domains, a coordinate system is required. The Euclidian three-dimensional space E of vectors of order 3 is referred with *orthogonal Cartesian unit vectors* $\left(\underline{\varepsilon_1}, \underline{\varepsilon_2}, \underline{\varepsilon_3}\right)$ such that any vector $\underline{a}$ can be expressed by

$$\underline{a} = \alpha^1 \underline{\varepsilon_1} + \alpha^2 \underline{\varepsilon_2} + \alpha^3 \underline{\varepsilon_3} = \sum_{i=1}^{i=3} \alpha^i \underline{\varepsilon_i} \quad (1.11)$$

using operations of addition and multiplication by a scalar. Referring to the basis $\left(\underline{\varepsilon_1}, \underline{\varepsilon_2}, \underline{\varepsilon_3}\right)$, the decomposition of vector $\underline{a}$ by means of scalars $\alpha^1, \alpha^2, \alpha^3$ called scalar components defined uniquely. The components of a vector with respect to the Cartesian basis are called *contravariant*.

To express a vector in terms of a given basis, we will frequently use the Einstein convention notation. Accordingly, a monomial expression involving a subscript symbol and the same as superscript is considered as a summation, shown for the following example (1.11):

$$\underline{\mathrm{a}} = \alpha^i \underline{\varepsilon}_{\mathrm{i}}.$$

The index "i", called "dummy index", is only attached to the monomial expression, using the Einstein convention notation for a compact relationship.

1.3.2 *Basis Vectors*

A set of three spatial vectors $\left(\underline{b_1}, \underline{b_2}, \underline{b_3}\right) \equiv \left(\underline{b_i}\right)$ is said to be a basis if these vectors are not parallel to one plane, which means that the following relation

$$\alpha^1 \underline{e_1} + \alpha^2 \underline{e_2} + \alpha^3\ \underline{e_3} = \underline{0} \tag{1.12}$$

can hold only when all the numbers $\alpha^i\,(i = 1, 2, 3)$ are zero. In this equation, the scalar quantities $\alpha^i\,(i = 1, 2, 3)$ are uniquely defined and are called *contravariant components* with respect to the basis vectors.

To refer spatial points in a three-dimensional space, a coordi*nate system is required*. This consists of *three scalar nu*mbers, as made with Cartesian coordinates, such that a vector element $\underline{X}$ of the vector space E, of contravariant coordinates $\left(x^1, x^2, x^3\right)$, is expressed by

$$\underline{X} = x^1 \underline{b_1} + x^2 \underline{b_2} + x^3 \underline{b_3} = x^i \underline{b_i}. \tag{1.13}$$

The spatial vectors $\left(b_1, \underline{b_2}, \underline{b_3}\right) \equiv (b_i)$ define a basis of the space E provided the condition associated to Eq. (2.12) is satisfied.

1.3.3 *Natural Basis: Dual of Natural Basis*

It may be of interest to adopt, for every point of the space, a vector basis $(b_1, b_2, b_3) \equiv (b_i)$ called the natural basis defined, related to coordinates $\left(x^1, x^2, x^3\right)$ by the following equations:

$$\underline{b}_1 = \frac{\partial \underline{X}}{\partial x^1}\ ;\ \ \underline{b}_2 = \frac{\partial \underline{X}}{\partial x^2}\ ;\ \ \underline{b}_3 = \frac{\partial \underline{X}}{\partial x^3}. \tag{1.14}$$

The natural basis $\left(\underline{b_i}\right)$ is generally non-orthonormal. With Cartesian coordinates, the natural basis is unchanged for all the space points and identical to unit vectors

$(\underline{\varepsilon_1}, \underline{\varepsilon_2}, \underline{\varepsilon_3})$. The vector components with respect to the natural basis are called *covariant*.

The dual $\underline{(B^i)}$ of the natural basis (b_j) is defined by vectors $(\underline{B}^1, \underline{B}^2, \underline{B}^3) \equiv (\underline{B}^i)$ such that

$$\underline{B}^i \underline{b}_j \begin{cases} = 0 & \text{if } i \neq j \\ = 1 & \text{if } i = j \end{cases}. \tag{1.15}$$

It can be easily verified that, for the Cartesian coordinate system, the natural basis is identical to basis unit vectors $(\underline{\varepsilon_1}, \underline{\varepsilon_2}, \underline{\varepsilon_3})$ and unchanged for all points.

1.3.4 Contravariant and Covariant Components

Given a basis and its coordinate system, a vector can be expressed by using contravariant or covariant components.

The vector components related to a natural basis are called contravariant. The corresponding symbol is defined by a superscript symbol, as written for example in Eq. (1.13).

The components of a vector related to the dual basis $\underline{(B^i)}$ a of the natural basis $\underline{(b_j)}$ are called covariant, as illustrated by subscript symbols for components expressed by the following example:

$$\underline{X} = x_1 \underline{B}^1 + x_2 \underline{B}^2 + x_3 \underline{B}^3 = x_j \underline{B}^j. \tag{1.16}$$

For a natural basis $(\underline{b}^i)$, the relation between the respective contravariant and covariant components c^i and C_j of a vector $\underline{\mathrm{X}}$ can be expressed by the transformation matrix $\boldsymbol{Q}$ such that

$$c^i = Q^{ik} C_k \tag{1.17}$$

and the direct transformation of the covariant components C_j by using the inverse matrix $\boldsymbol{Q}^{-1}$ according to the relation

$$C_j = Q_{kj}^{-1} c^k. \tag{1.18}$$

When using Cartesian coordinates, we get the respective following relations for the sum and the scalar product of two vectors $\underline{a}$ and $\underline{b}$:

$$(\underline{a} + \underline{b})_i = a_i + b_i \quad ; \quad (\underline{a} + \underline{b})^j = a^j + b^j \tag{1.19}$$

$$\underline{a} \cdot \underline{b} = \underline{b} \cdot \underline{a} = a^i b^i = b^i a^i \quad \text{(with implicit summation).} \tag{1.20}$$

Thus, the scalar product of a vector $\underline{a}$ by itself is expressed by

$$\underline{a} \cdot \underline{a} = \underline{a}^2 = a^i a^i, \tag{1.21}$$

and the magnitude of $\underline{a}$ is defined as

$$\|\underline{a}\| = \sqrt{a^2}. \tag{1.22}$$

According to Eq. (1.7), contravariant and covariant components can be expressed by using scalar products according to the following relations:

$$\underline{A} = \alpha^1 \underline{b_1} + \alpha^2 \underline{b_2} + \alpha^3 \underline{b_3}; \alpha^i = \underline{A} \cdot b^i \quad , i = l, 2, 3 \tag{1.23}$$

$$\underline{A} = \alpha_1 \underline{b^1} + \alpha_2 \underline{b^2} + \alpha_3 \underline{b^3}; \alpha_j = \underline{A} \cdot \underline{b_j} \quad , i = 1, 2, 3. \tag{1.24}$$

1.3.5 Change of Coordinates

Let us consider the following transformation E $\rightarrow$ E defined by

$$i = 1, 2, 3: \quad \tilde{x}^i = \alpha^i\left(x^1, x^2, x^3\right) \text{ or } \tilde{x}^i = \alpha^i\left(x^j\right), \tag{1.25}$$

where α^i are three scalar continuously differentiable functions of the coordinates $\left(x^i\right)$ and $\left(\tilde{x}^i\right)$ the new coordinates of a point of E.

Given a vector $\underline{A}$, its contravariant and covariant components in terms of the new coordinates $\left(\tilde{x}^i\right)$ are expressed by the respective relationship:

$$i = 1, 2, 3: \quad \tilde{a}^i = \frac{\partial \tilde{x}^i}{\partial x^j} a^j, \tag{1.26}$$

$$i = 1, 2, 3: \quad \tilde{a}_j = \frac{\partial \tilde{x}_j}{\partial x_k} a^k. \tag{1.27}$$

1.3.6 Vector Matrix

For analytical and numerical purposes, the use of matrix expressions for vectors is generally required. Given a basis $\left(\underline{b_j}\right)$, a vector $\underline{A}$ can be identified by its components written on a line or column matrix form as follows:

$$\underline{A}_{[b_i]} = [a_i] \quad : \quad \underline{A}_{[b_i]} = \begin{bmatrix} a_1 \\ a_2 \\ a_3 \end{bmatrix} \text{ or } \underline{A}_{[b_i]} = \begin{bmatrix} a_1 \; a_2 \; a_3 \end{bmatrix}. \tag{1.28}$$

1.3.7 Gradient of a Scalar Function

The gradient of a scalar function φ, assumed to be continuously differentiable, is a vector noted $\underline{\nabla}\varphi$ given by the following relation in terms of a vector $\underline{x}$:

$$\underline{\nabla}\varphi = \partial\varphi/\partial\underline{x}. \tag{1.29}$$

In a Cartesian basis $(\underline{\varepsilon_1}, \underline{\varepsilon_2}, \underline{\varepsilon_3})$, the gradient vector can be expressed by

$$\underline{\nabla}\varphi = \left(\frac{\partial\varphi}{\partial x^{\mathrm{i}}}\right)\underline{\varepsilon_{\mathrm{i}}}. \tag{1.30}$$

1.4 Tensors

In this section, we will consider more specifically second-order tensors and related operations, in a three-dimensional space $\mathrm{E} = \mathrm{R}^3$. In this case, a second-order tensor $\underline{\underline{\mathrm{T}}}$ can be considered as a linear form transforming points of a vector space $\tilde{E} = R^9$ into points of the same vector space $\tilde{E}$

$$\underline{\underline{\mathrm{T}}} : (x \in \tilde{E}) \to \left(f_i(x)_{\mathrm{i}=1,2,..9}\right) \in \tilde{E}. \tag{1.31}$$

Given a basis $\left(\underline{b_{\mathrm{j}}}\right)$ in the three-dimensional space E, we can express the components of the second-order tensor $\underline{\underline{\mathrm{T}}}$ as

$$\underline{\underline{\mathrm{T}}} = \mathrm{T_{ij}}\underline{\mathrm{b_i}} \otimes \underline{\mathrm{b_j}}. \tag{1.32}$$

This equation corresponds to the so-called dyadic notation. The quantity $\underline{b_i} \otimes \underline{b_j}$ is defined as the tensorial basis of $\underline{\underline{\mathrm{T}}}$ in E and the components of the tensor $\underline{\underline{\mathrm{T}}}$ can be expressed by the following matrix:

$$\mathbf{T} = \mathbf{T}|_{(\underline{\mathrm{b}_i})} = \mathbf{T}_{ij} = \begin{bmatrix} \mathrm{T}_{11} & \mathrm{T}_{12} & \mathrm{T}_{13} \\ \mathrm{T}_{21} & \mathrm{T}_{22} & \mathrm{T}_{23} \\ \mathrm{T}_{31} & \mathrm{T}_{32} & \mathrm{T}_{33} \end{bmatrix}. \tag{1.33}$$

In other words, given a basis $\left(\underline{b_j}\right)$, a tensor can be expressed by components reported in a matrix for the basis $\left(\underline{b_j}\right)$.

We now consider tensor operations in Cartesian coordinates, using the Cartesian orthonormal basis $\left(\underline{\varepsilon_1}, \underline{\varepsilon_2}, \underline{\varepsilon_3}\right)$. The scalar product of these basis vectors leads to the following relations:

$$\underline{\varepsilon_i} \cdot \underline{\varepsilon_i} = 0 \ \text{ if } i \neq j \ , \ \underline{\varepsilon_i}\, \underline{\varepsilon_i} = 1 \ \text{ if } \ i = j. \tag{1.34}$$

These relations can be associated to the unit second-order tensor which corresponds to the Kronecker delta tensor $\underline{\underline{\delta}}$ given by the following matrix:

$$\delta = \delta|_{(\underline{b_i})} = \delta_{ij} = \begin{bmatrix} 1\,0\,0 \\ 0\,1\,0 \\ 0\,0\,1 \end{bmatrix}. \tag{1.35}$$

According to the definition of the tensor on the vector space $\tilde{E}$, classic operations in vector spaces as addition and multiplication by a scalar can be made. When expressing second-order tensors in the orthonormal basis $\left(\underline{\varepsilon_1}, \underline{\varepsilon_2}, \underline{\varepsilon_3}\right)$, we can obtain similar relations to those of square matrices. Thus, we now consider different useful tensor operations defined for use in the three-dimensional space E.

Many second-order tensors are used in continuum mechanics as, for example, stress and strain tensors, as also those related to velocity and deformation gradients.

- The Levi–Civita symbol L allows to express the determinant of a square matrix and the cross product of two vectors in three-dimensional Euclidean space, by using an index notation, such that

$$L_{ij} = \begin{bmatrix} 1 \text{ if } (i, j) = (1, 2) \\ -1 \text{ if } (i, j) = (2, 1) \\ 0 \ \ \text{if } i = j \end{bmatrix} \tag{1.36}$$

in two dimensions, and in three dimensions,

$$L_{ijk} \begin{bmatrix} 1 \text{ if}(i, j, k) = (1, 2, 3), (2, 3, 1) \text{ or } (3, 1, 2) \\ -1 \text{ if}(i, j, k) = (3, 2, 1), (2, 1, 3) \text{ or } (1, 3, 2) \\ 0 \text{ if } i = j, j = k \text{ or } k = i \end{bmatrix}. \tag{1.37}$$

1.4.1 Tensor Operations

We now consider some operations for second-order tensors:

- Transposition

For a second-order tensor $\underline{\underline{T}}$, the transpose operation is written as for a matrix transposition:

$$\underline{\underline{T}}^T = \left(T_{ij}\ \underline{b}_i \otimes \underline{b}_j\right)^T = T_{ij}\ \underline{b}_j \otimes \underline{b}_i \quad \text{or} \quad \left[T^T_{ij}\right] = T_{ji}. \tag{1.38}$$

- Contraction of a tensor $\underline{\underline{T}}$

Using the Einstein notation, we may define the following contraction operation:

$$T_{ii} = \sum_{i=1}^{i=3} T_{ii} = T_{11} + T_{22} + T_{33}. \tag{1.39}$$

This scalar, noted tr $(\underline{\underline{T}})$, remains unchanged in another frame resulting from a rotation and is called the trace of $\underline{\underline{C}}$. This invariant is denoted "first invariant" of $\underline{\underline{T}}$ and satisfies the following properties:

$$\mathrm{tr}(\underline{\underline{C}} + \underline{\underline{D}}) = \mathrm{tr}(\underline{\underline{C}}) + \mathrm{tr}(\underline{\underline{D}}), \tag{1.40}$$

$$\mathrm{tr}(\alpha\,\underline{\underline{C}}) = \alpha\ \mathrm{tr}(\underline{\underline{C}}). \tag{1.41}$$

- Dot product of two tensors $\underline{\underline{C}}$ and $\underline{\underline{D}}$

This operation of symbol "." leads to a second-order tensor by the contraction product (the dot is the contraction operator):

$$\underline{\underline{C}} \cdot \underline{\underline{D}} = \left(C_{ij}\underline{\varepsilon}_i \otimes \varepsilon_j\right) \cdot \left(D_{kl}\underline{\varepsilon}_k \otimes \underline{\varepsilon}_l\right). \tag{1.42}$$

By using properties of the Kronecker unit tensor $\underline{\delta}$, we can write

$$\underline{\underline{C}} \cdot \underline{\underline{D}} = C_{ij}\ \underline{\varepsilon}_i \underline{\delta}_{jk} \mathbf{D}_{kl}\ \underline{\varepsilon}_l = C_{ij} D_{jl} \underline{\varepsilon}_i\ \underline{\varepsilon}_l. \tag{1.43}$$

The components of the product tensor are those of the product of the two matrices **C** and **D**.

A tensor $\underline{\underline{C}}$ is invertible if there exists a so-called inverse tensor $\underline{\underline{C^{-1}}}$ such that

$$\underline{\underline{C}} \cdot \underline{\underline{C}}^{-1} = \underline{\underline{C}}^{-1} \cdot \underline{\underline{C}} = \underline{\underline{I}}. \tag{1.44}$$

- Double dot (or scalar) product of two tensors $\underline{\underline{C}}$ and $\underline{\underline{D}}$

The double dot operation of symbol ":" leads to a scalar according to the following relationships:

$$\underline{\underline{C}} \cdot \underline{\underline{D}} = \left(C_{ij}\underline{\varepsilon}_i \otimes \varepsilon_j\right) \cdot \left(D_{k1}\underline{\varepsilon}_k \otimes \underline{\varepsilon}_1\right) = \left(C_{ij}\ \underline{\varepsilon}_i\right)\delta_{jk} \cdot \left(D_{k1}\underline{\varepsilon}_1\right) = C_{ij}D_{k1}\delta_{jk}\delta_{i1} = C_{ij}\ D_{ji}. \tag{1.45}$$

- Tensor product of two tensors $\underline{\underline{C}}$ and $\underline{\underline{D}}$

The tensor product is a fourth-order tensor defined as

$$\underline{\underline{\underline{\underline{T}}}} = \underline{\underline{C}} \otimes \underline{\underline{D}} = \left(C_{ij}\ \underline{\varepsilon}_i \otimes \underline{\varepsilon}_j\right) \otimes \left(D_{kl}\underline{\varepsilon}_k \otimes \underline{\varepsilon}_l\right) = C_{ij}D_{kl}\underline{\varepsilon}_i \otimes \underline{\varepsilon}_j \otimes \underline{\varepsilon}_k \otimes \varepsilon_l. \tag{1.46}$$

1.4.2 *Invariants of Second-Order Tensors*

Scalar functions can be associated to tensors. They are independent of the basis where the tensor components are expressed. These are called invariants. One important invariant is the "trace" of a tensor, which is the sum of the diagonal elements of its diagonal matrix.

The trace of a tensor $\underline{\underline{C}}$ in a basis $\left(\underline{b}_1, \underline{b}_2, \underline{b}_3\right)$ of components C_{ij} is expressed by

$$\mathrm{tr}(\underline{\underline{C}}) = C_{11} + C_{22} + C_{33} \tag{1.47}$$

or using the following notation:

$$I = \mathrm{tr}(\underline{\underline{C}}) = C_{ii}, \tag{1.48}$$

according to the Einstein notation. It is the so-called "first invariant".

We may also define the second and third scalar invariants I_2 and I_3 of a tensor $\underline{\underline{C}}$, based on operations on trace. We thus write the relations:

$$II = \mathrm{tr}\left(\underline{\underline{C}}^2\right) \quad ; \quad III = \mathrm{tr}\left(\underline{\underline{C}}^3\right). \tag{1.49}$$

In the literature, notably for the writing of constitutive equations, the following invariants are usually used:

$$I_l = I, \quad I_2 = \frac{1}{2}\left(I^2 - II\right) \quad, I_3 = \frac{1}{6}\left(I^3 - 3I \times II + 2III\right) = \det(\underline{\underline{C}}). \tag{1.50}$$

1.5 Operations with Derivatives

1.5.1 Gradients

- Gradient of a scalar:

The gradient of a scalar φ is a vector noted $\underline{\nabla}\varphi$ given by the following relation in terms of a vector $\underline{x}$:

$$\underline{\nabla}\varphi = \partial\varphi/\partial\underline{x}. \tag{1.51}$$

In a Cartesian basis $\left(\underline{\varepsilon}_1, \underline{\varepsilon}_2, \underline{\varepsilon}_3\right)$, the gradient vector can be expressed by

$$\underline{\nabla}\varphi = \left(\frac{\partial\varphi}{\partial x^i}\right)\underline{\varepsilon}_i. \tag{1.52}$$

According to this relation, we can define the gradient operator $\underline{\nabla}$ as follows:

$$\underline{\nabla} = \underline{\varepsilon}_i\frac{\partial}{\partial x^i} = \underline{\varepsilon}_1\frac{\partial}{\partial x^1} + \underline{\varepsilon}_2\frac{\partial}{\partial x^2} + \underline{\varepsilon}_3\frac{\partial}{\partial x^3}. \tag{1.53}$$

- Gradient of a vector

The gradient of a vector $\underline{a} = \alpha^i\underline{\varepsilon}_i$ is expressed by the following relation:

$$\underline{\nabla}(\underline{a}) = \underline{\varepsilon}_i\frac{\partial}{\partial \mathbf{x}^i}\left(\alpha^j\underline{\varepsilon}_j\right). \tag{1.54}$$

This corresponds to a second-order tensor related to the tensorial basis $\underline{\varepsilon}_i \otimes \underline{\varepsilon}_j$ with components $\partial\alpha^i/\partial \mathrm{x}^j$ that can be expressed in the Cartesian basis according to the following matrix:

$$\underline{\nabla}(\underline{a})|_{(\underline{b}_i)} = \underline{\underline{G}}|_{(\underline{b}_i)}\begin{bmatrix} \partial\alpha^1/\partial\mathbf{x}^1 & \partial\alpha^1/\partial\mathbf{x}^2 & \partial\alpha^1/\partial\mathbf{x}^3 \\ \partial\alpha^2/\partial\mathbf{x}^1 & \partial\alpha^2/\partial\mathbf{x}^2 & \partial\alpha^2/\partial\mathbf{x}^3 \\ \partial\alpha^3/\partial\mathbf{x}^1 & \partial\alpha^3/\partial\mathbf{x}^2 & \partial\alpha^3/\partial\mathbf{x}^3 \end{bmatrix}. \tag{1.55}$$

1.5.2 Divergence

- The divergence of a vector $\underline{a} = \alpha^i\underline{\varepsilon}_i$ is a scalar (invariant) expressed by

$$\underline{\nabla} \cdot \underline{a} = \underline{\varepsilon}_i \frac{\partial}{\partial x^i}\left(\alpha^j \underline{\varepsilon}_j\right) = \underline{\varepsilon}_1 \cdot \underline{\varepsilon}_1 \delta^{ij} \frac{\partial}{\partial x^i}\left(\alpha^j \underline{\varepsilon}_j\right) = \delta^{ij} \frac{\partial \alpha^j}{\partial x^i} = \frac{\partial \alpha^i}{\partial x^i} = \frac{\partial \alpha^1}{\partial x^1} + \frac{\partial \alpha^2}{\partial x^2} + \frac{\partial \alpha^3}{\partial x^3}. \tag{1.56}$$

- The divergence of a tensor $\underline{\underline{T}}$ is a vector defined by the relation

$$\underline{\nabla}.\underline{\underline{T}} = \left(\underline{\varepsilon}_k \frac{\partial}{\partial x_k}\right) \cdot \left(\mathrm{T}_{ij} \varepsilon_i \varepsilon_j\right) = \underline{\varepsilon}_j \cdot \frac{\partial}{\partial x^i}\left(\mathrm{T}_{ij}\right). \tag{1.57}$$

1.5.3 *Curl of a Vector*

- The curl (also named "rot") of a vector $\underline{a}$ is a vector given by the following relation, using the symbol "$\nabla \times$":

$$\begin{aligned} \nabla \times \underline{a} &= \left(\underline{\varepsilon}_i \frac{\partial}{\partial x_i}\right) \times \left(\mathrm{a}_{ji} \varepsilon_j\right) = \underline{\varepsilon}_i \times \underline{\varepsilon}_j \frac{\partial \mathrm{a}_j}{\partial x_i} = L_{kij} \underline{\varepsilon}_k \frac{\partial \mathbf{a}_j}{\partial x_i} \\ &= \left(\frac{\partial \mathrm{a}_3}{\partial x_2} - \frac{\partial \mathrm{a}_2}{\partial x_3}\right)\underline{\varepsilon}_1 + \left(\frac{\partial \mathrm{a}_1}{\partial x_3} - \frac{\partial \mathrm{a}_3}{\partial x_1}\right)\underline{\varepsilon}_2 + \left(\frac{\partial \mathrm{a}_2}{\partial x_1} - \frac{\partial \mathrm{a}_1}{\partial x_2}\right)\underline{\varepsilon}_3. \end{aligned} \tag{1.58}$$

1.6 Special Non-Cartesian Coordinate Systems

In Cartesian coordinates $\left(x^l, x^2, x^3\right)$, as pointed out in Sect. 1.3.3, the basis $\left(\underline{\varepsilon}_i\right)$ and its dual all coincide and are independent of the point referred as $\underline{X} = x^1\underline{\varepsilon}_1 + x^2\underline{\varepsilon}_2 + x^3\underline{\varepsilon}_3 = x^{\mathrm{i}}\underline{\varepsilon}_{\mathrm{i}}$. The orthonormal basis unit vectors have the same directions than those of the coordinate axes. We consider here two important frames corresponding to cylindrical and spherical coordinate systems, starting from a general set of coordinates $q^i (i = 1, 2, 3)$. In this case, we find it convenient to write $q^i = q^i\left(x^1, x^2, x^3\right)$.

1.6.1 *Cylindrical Coordinates*

The cylindrical coordinates (r, θ, z) (Fig. 1.1a) are defined such that a vector $\underline{X}$ is expressed by the equation

$$\underline{X} = r \cos\, \theta \underline{\varepsilon}_1 + r \sin\, \theta \underline{\varepsilon}_2 + z\, \underline{\varepsilon}_3, \tag{1.59}$$

according to the following relationships:

$$x = r \cos\, \theta;\ y = r \sin\, \theta, \tag{1.60}$$

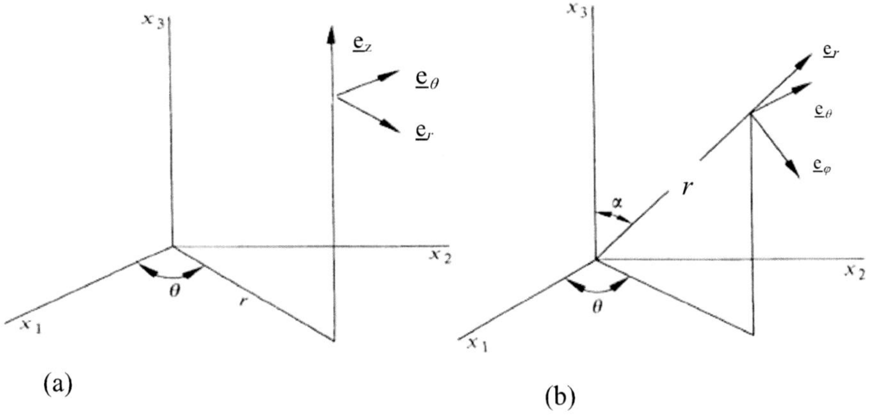

Fig. 1.1 **a** Cylindrical coordinates; **b** spherical coordinates, with corresponding frames of reference

$$r = \sqrt{x^2 + y^2};\ \theta = \tan^{-1}\left(\frac{y}{x}\right). \tag{1.61}$$

In the Cartesian frame, the basis vectors $\left(\underline{e_r}, \underline{e_\theta}, \underline{e_z}\right)$ related to cylindrical coordinates are given by

$$\underline{e_r} = \frac{\partial \underline{X}}{\partial r} = \cos\ \theta\underline{\varepsilon_1} + \sin\ \theta\underline{\varepsilon_2}, \underline{e_\theta} = \frac{\partial \underline{X}}{\partial \theta} = r\left(-\sin\theta\underline{\varepsilon}_1 + \cos\theta\underline{\varepsilon}_2\right),\ \ \underline{e_z} = \underline{\varepsilon_3} \tag{1.62}$$

and, versus the cylindrical basis vectors, the corresponding Cartesian basis vectors are expressed as

$$\underline{\varepsilon_1} = \cos\theta\ \underline{e_r} - \frac{1}{r}\sin\theta\ \underline{e_\theta},\quad \underline{\varepsilon_2} = \sin\theta\ \underline{e_r} + \frac{1}{r}\cos\theta\ \underline{e\theta},\quad \underline{\varepsilon_3} = \underline{e_z}. \tag{1.63}$$

Accordingly, the norms of the basis vectors $\left(\underline{e_r}, \underline{e_\theta}, \underline{e_z}\right)$ are expressed by

$$\left\|\underline{e}_r\right\| = 1;\ \left\|\underline{e}_\theta\right\| = \mathrm{r};\ \left\|\underline{e}_z\right\| = 1. \tag{1.64}$$

We now consider the orthonormal frame $\left(\underline{e_r'}, \underline{e_\theta'}, \underline{e_z'}\right)$ defined by the relations

$$\underline{e}'_r = \underline{e}_r\ \ ;\underline{e}'_\theta = \frac{1}{r}\underline{e}_\theta\ \ ;\underline{e}'_z = \underline{e}_z. \tag{1.65}$$

In this orthonormal basis $\left(\underline{e}'_r, \underline{e}'_\theta, \underline{e}'_z\right)$, vector and tensor components, called physical components, can be written.

1.6.2 *Spherical Coordinates*

The spherical coordinates (r, θ, φ) (Fig. 1.1b) are defined such that a vector $\underline{X}$ is expressed by the following equation:

$$\underline{X} = r\cos\varphi\ \sin\theta\ \underline{\varepsilon}_1 + r\sin\varphi\ \sin\theta\underline{\varepsilon}_2 + r\ \cos\theta\ \underline{\varepsilon}_3 \tag{1.66}$$

with the following relations:

$$x = r\cos\varphi\ \sin\theta,\quad y = r\sin\varphi\ \sin\theta,\quad z = r\ \cos\theta. \tag{1.67}$$

The basis vectors are given by

$$\begin{aligned} \underline{e}_r &= \sin\theta\ \cos\varphi\,\underline{\varepsilon}_1 + \sin\theta\ \sin\varphi\,\underline{\varepsilon}_2 + \cos\theta\ \underline{\varepsilon}_3 \\ \underline{e}_\theta &= \cos\theta\ \cos\varphi\,\underline{\varepsilon}_1 + \cos\theta\ \sin\varphi\,\underline{\varepsilon}_2 - \sin\theta\ \underline{\varepsilon}_3 \\ \underline{e}_\varphi &= -\sin\ \varphi\ \underline{\varepsilon}_l + \cos\ \varphi\ \underline{\varepsilon}_2 \end{aligned} \tag{1.68}$$

and

$$\begin{aligned} \varepsilon_1 &= \sin\theta\ \cos\varphi\,\underline{e}_r + \cos\theta\ \cos\varphi\,\underline{e}_\theta - \sin\varphi\,\underline{e}_\varphi \\ \underline{\varepsilon}_2 &= \sin\theta\ \sin\varphi\,\underline{e}_r\underline{1} + \cos\theta\ \sin\varphi\,\underline{e}_\theta + \cos\varphi\,\underline{e}_\varphi \\ \underline{\varepsilon}_3 &= \cos\theta\ \underline{e}_r - \sin\theta\ \underline{e}_\theta. \end{aligned} \tag{1.69}$$

References

1. Borisenko AI, Tarapov IE (1979) Vector and tensor analysis with applications. Dover Publications, New-York
2. Grinfeld P (2013) Introduction to tensor analysis and the calculus of moving surfaces. Springer

Chapter 2
Kinematics–Conservation Laws: Constitutive Equations

2.1 Introduction

This chapter presents the kinematics in a deformable body and the conservation laws to be written in problems of continuum mechanics. Different types of constitutive equations that can be associated to theoretical or practical problems are presented.

Movements of material points may be steady or unsteady, uniform or non-uniform, laminar or turbulent (occurring in fluid flows). Turbulent flow conditions are not considered in this book. Furthermore, though some special cases could be treated as being one-dimensional, deformations and flows investigated in the present book are considered in two- or three-dimensional situations, depending on the geometrical and applied boundary conditions.

Before going further in topics presented in this paper, let us refer to some classic definitions.

- In a fluid body deformed versus time, streamlines are curves drawn through the medium to indicate the direction of motion of a material point. At any point on the curve and at every time t, the tangent gives the direction of the velocity of the particle of the medium. Thus, the velocity vector is tangent to the streamline and has a zero component normal to the streamline (Fig. 2.1a).
- A stream tube is a part of the domain under deformation limited by a set of streamlines (Fig. 2.1b). This concept will be widely considered in large parts of this book.

In two-dimensional (planar or axisymmetric) situations, the particles are assumed to move in reference planes.

Streaklines are sets of points of all the fluid particles that have passed continuously through a particular spatial position in the past. Pathlines (or trajectories) are lines that individual fluid particles follow. The direction of the pathlines is determined by the streamlines of the fluid at each moment.

J.-R. Clermont and A. Ammar, *Stream-Tube Method*,
https://doi.org/10.1007/978-3-030-65470-2_2

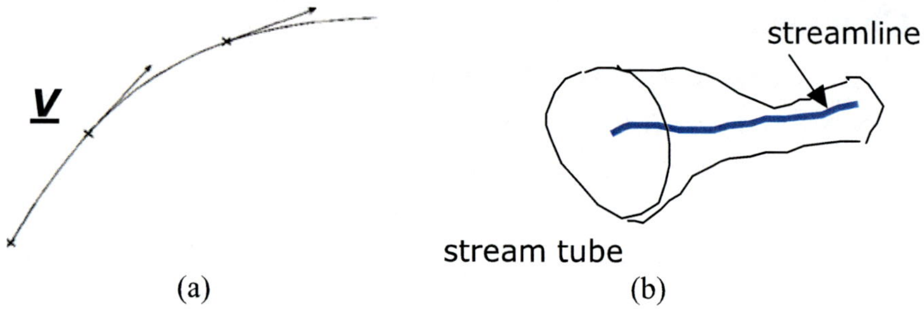

Fig. 2.1 **a** Streamline and velocity vectors $\underline{\mathbf{V}}$; **b** stream tube

2.2 Kinematics

2.2.1 *Basic Elements: Eulerian and Lagrangian Descriptions: Material Derivative*

In a given Euclidean space, the basic elements related to movement of particles of a material domain Ω under deformation are considered, using a reference frame (e.g. Coleman et al. [1]). A material point M occupies a position $\underline{X}$ at a given time $t = t_0$, corresponding to a reference configuration. The particle moves in the three-dimensional space. One denotes by $\underline{x}$ its current vector position, function of time and related to the reference configuration. The relations may be written as follows:

$$\underline{x} = \underline{M}(\underline{X}, t) \text{ or } x_i = M_i\left(X_j, t\right), \tag{2.1}$$

with

$$\underline{X} = \underline{M}\left(\underline{X}, t_0\right) \text{ or } X_i = M_i(X_i, t_0). \tag{2.2}$$

The functions defined in Eq. (2.1) are assumed to be twice differentiable and invertible. Such writing corresponds to the *Lagrangian description.*

In the spatial description of the motion, the reference position can be expressed by the following relation, in terms of the current configuration, using a transformation function ϕ such that

$$\underline{X} = \phi^1(\underline{x}, t) \text{ or } X_i = \phi^I\left(x_j, t\right). \tag{2.3}$$

The velocity vector $\underline{V}$ and the acceleration vector $\underline{\Gamma}$ are defined by the vector and components forms, in Cartesian coordinates

$$\underline{V} = \frac{\partial}{\partial t}\underline{M}\left(\underline{X},\ t\right) \text{or } \underline{V}^i = \frac{\partial}{\partial t}M^i\left(\underline{X^j},\ t\right) \tag{2.4}$$

and

$$\underline{\Gamma} = \frac{\partial^2}{\partial t^2}\underline{M}(\underline{X}, t) \text{ or } \Gamma^i = \frac{\partial^2}{\partial t^2}M^i\left(\underline{X^j}, t\right). \tag{2.5}$$

When the velocity and acceleration vectors are expressed in terms of the current position $\underline{x}$, this formalism is called *Eulerian description*. Thus, given the Eulerian velocity vector $\underline{V}$, the acceleration vector $\underline{\Gamma}$ can be expressed by the following relation:

$$\underline{\Gamma} = \frac{\partial}{\partial t}\underline{V}(\underline{X}, t) \mid_{\underline{x}} = \frac{\partial \underline{V}}{\partial t} + \frac{\partial \underline{V}}{\partial \underline{x}} \cdot \left.\frac{\partial \underline{x}}{\partial t}\right|_{\underline{x}}. \tag{2.6}$$

By introducing the *velocity gradient tensor*

$$\underline{\underline{L}} = \left(\frac{\partial \underline{V}}{\partial \underline{x}}\right)^{\mathrm{T}} = \left(\underline{\underline{\nabla}} \cdot \underline{V}\right)^{\mathrm{T}} \text{ or } L^{ij} = \left(\frac{\partial V^i}{\partial x^j}\right), \tag{2.7}$$

Equation (2.6) becomes

$$\underline{\Gamma} = \frac{\partial \underline{V}}{\partial t} + \underline{V} \cdot \underline{\underline{\nabla}} \cdot \underline{V} = \frac{\partial \underline{V}}{\partial t} + \underline{\underline{L}} \cdot \underline{V}. \tag{2.8}$$

Equation (2.8) allows to define the *material derivative* operator $\frac{D}{Dt}$ as follows:

$$\frac{D}{Dt} = \frac{\partial}{\partial t} + \underline{V} \cdot \underline{\nabla}. \tag{2.9}$$

If the operator concerns a scalar and

$$\frac{D}{Dt} = \frac{\partial}{\partial t} + \underline{V} \cdot \underline{\underline{\nabla}}, \tag{2.10}$$

then the derivative is related to a vector.

The deformation of a continuous medium is *steady* or *stationary* if the Eulerian velocity vector is independent of the time t, thus $\underline{V} = \underline{V}(\underline{x})$.

2.2.2 *Kinematic Tensors*

Kinematic tensors are defined from velocity or displacement of material points in a medium. According to the previous sub-section, two tensors can be associated to the velocity gradient tensor $\underline{\underline{L}}$:

(1) The *strain-rate tensor* $\underline{\underline{D}}$, given by the symmetric part of $\underline{\underline{L}}$:

$$\underline{\underline{D}} = \frac{1}{2}\left(\underline{\underline{L}} + \underline{\underline{L}}^T\right). \tag{2.11}$$

(2) The *vorticity tensor* $\underline{\underline{\Omega}}$, related to the antisymmetric part of $\underline{\underline{L}}$:

$$\underline{\underline{\Omega}} = \frac{1}{2}\left(\underline{\underline{L}} - \underline{\underline{L}}^T\right). \tag{2.12}$$

According to this relation, the diagonal terms corresponding to the vorticity tensor matrix are zero. The corresponding non-diagonal quantities of the matrix are generally non-zero.

Among the kinematic tensors, one often refers to the deformation gradient tensor $\underline{\underline{F}}$, defined by the gradient of $\underline{x}$ versus $\underline{X}$ such that

$$\underline{\underline{F}}(t) = \left(\frac{\partial \underline{x}}{\partial \underline{X}}\right)^T (t) \text{ or } F^i_j = \frac{\partial x^i}{\partial X^j}. \tag{2.13}$$

At the initial time $t = t_0$, $\underline{\underline{F}}$ becomes the identity tensor:

$$\underline{\underline{F}}(t_0) = \underline{\underline{I}}. \tag{2.14}$$

Let ρ be the density of the medium, and ω and ω_0 be the respective volumes of a region of the body at times t and t_0. The corresponding mass of the domain is given by

$$\int_{\omega} \rho \, \mathrm{d}\underline{x} = \int_{\omega_0} \rho \,|\Delta|\, \mathrm{d}\underline{X}. \tag{2.15}$$

In this equation, Δ denotes the determinant of the velocity gradient tensor

$$\Delta = \det \underline{\underline{F}}, \tag{2.16}$$

required to be positive and non-zero. For an incompressible material, we have the following property:

$$\det \underline{\underline{F}} = 1. \tag{2.17}$$

According to the correspondence between vectors $\underline{x}$ and $\underline{X}$, the inverse tensor of $\underline{\underline{F}}$ is written as follows:

$$\underline{\underline{F}}^{-1}(t) = \left(\frac{\partial \underline{X}}{\partial \underline{x}}\right)^T (t) \text{ or } F^{-1}{}^i_j = \frac{\partial X^i}{\partial x^j}. \tag{2.18}$$

Various kinematic tensors have been defined in the literature in order to describe the rheological behaviour of materials. These are related to the velocity gradient tensor or are derived from the deformation gradient tensor, as, for example, the Cauchy and the Finger tensors, denoted by $\underline{\underline{C}}$ and $\underline{\underline{B}}$, respectively (Crochet et al. [2]). We get the following relations:

$$\underline{\underline{C}} = \underline{\underline{F^T}}.\underline{\underline{F}} \tag{2.19}$$

for the Cauchy tensor, and

$$\underline{\underline{B}} = \underline{\underline{F}}.\underline{\underline{F^T}} \tag{2.20}$$

for the Finger tensor.

2.2.3 *Stress Tensor–Stress Vector*

The action of forces across a surface is not only due to the pressure and neither necessarily orthogonal to the surface. In relation to the surface normal vector, the stress tensor allows evaluation of the force acting on this surface, in addition to the pressure. Its behaviour is described by the so-called *constitutive equation*, which provides a relation between the stress and the deformation of the material, related to the kinematics and also to a thermodynamic equation of state. Referring to the constitutive equation for constant density materials, one may define by $\underline{\underline{\tau}}$ the second-order stress tensor, such that the corresponding stress vector $\underline{S}$ at a point M of the material is expressed by the following relation:

$$\underline{S} = \underline{\underline{\tau}} \cdot \underline{n}, \tag{2.21}$$

where $\underline{n}$ denotes the unit vector normal to the surface and $\underline{\underline{\tau}}$ is a symmetric tensor (e.g. [1]).

2.3 Laws of Conservation

The conservation laws concern the writing of equations related to mass, linear and angular momentum and energy. The Reynolds transport theorem allows to provide mathematical expressions leading to write the conservation equations, by using tools of differential analysis. This theorem starts from the following relation where the volume integral is given in the reference configuration:

$$\frac{D}{Dt}\int_{G_0} \Phi \, \mathrm{d}\omega = \frac{D}{Dt}\int_{G_0} \Phi \left| \frac{\partial \underline{x}}{\partial \underline{X}} \right| \mathrm{d}\omega, \tag{2.22}$$

where $\left|\frac{\partial x}{\partial \underline{X}}\right| = \det\left(\underline{\underline{F}}\right)$ is the Jacobian of the transformation ($\mathrm{d}\underline{x}$ can be exchanged with $\mathrm{d}V$). G_0 denotes the region occupied by the reference configuration.

The Reynolds transport theorem may be expressed as follows.

When considering a scalar, vector or tensor field noted as $\Phi(\underline{x}, t)$, defined a region **G** occupied by a body at time t, one may write the following equation:

$$\frac{D}{Dt}\int_{G} \Phi \, \mathrm{d}g = \int_{G} \left(\frac{D}{Dt}\Phi + \Phi \, \nabla \cdot \underline{V}\right) \mathrm{d}g = \int_{G} \left(\frac{D}{Dt}\Phi + \Gamma \, \nabla \cdot (\Phi \, \underline{V})\right) \mathrm{d}g. \tag{2.23}$$

In this equation, $\underline{V}$ denotes the velocity field and $\frac{D}{Dt}$ the material derivative.

2.3.1 *Mass Conservation: Incompressible Materials*

2.3.1.1 The Mass Conservation Equation

Consider the mass in an arbitrary volume ω is constant at all times in a domain of densityρ. The principle of conservation of mass of a given system is assumed to be of finite size. For this system, integral balance equations are written and, with any specified system, the mass conservation is given by

$$\begin{bmatrix} \text{net flow of mass} \\ \text{in} \\ \text{the system} \end{bmatrix} = \begin{bmatrix} \text{increase of} \\ \text{mass within} \\ \text{the system} \end{bmatrix}. \tag{2.24}$$

This relation, involving density and velocity, should be involved in the mathematical expressions of this equation. Thus, we use the material derivative operator

$$\frac{D}{Dt}\left(\int_{\omega} \rho\left(\underline{x},\, t\right) \mathrm{d}v\right) = 0, \tag{2.25}$$

and according to the Reynolds theorem, Eq. (2.25) becomes

$$\int_{\omega} \frac{D}{Dt}\left(\rho\left(\underline{x},\, t\right)\right)\, \mathrm{d}v \;=\; \int_{\omega} \left(\frac{\partial \rho}{\partial t} \;+\; \nabla \cdot \left(\rho\, \underline{V}\right)\right) \mathrm{d}v \;=\; 0. \tag{2.26}$$

Since this relation is verified for any volume ω, a necessary and sufficient condition for mass conservation can be locally written as

$$\frac{\partial \rho}{\partial t} \;+\; \nabla \cdot \left(\rho\, \underline{V}\right) = \frac{\mathrm{d}\rho}{\mathrm{d}t} \;+\; \nabla \cdot \left(\rho\, \underline{V}\right) \;=\; 0. \tag{2.27}$$

2.3.1.2 Incompressible Materials: Stream Functions in Two-Dimensional Cases

For an incompressible medium, the density is a constant and the mass conservation Eq. (2.25) thus becomes

$$\nabla \cdot\, \underline{V} \;=\; \frac{\partial V^{i}}{\partial x^{i}} \;=\; \mathrm{tr}(\underline{\underline{L}}) \;=\; \mathrm{tr}(\underline{\underline{D}}) \;=\; 0, \tag{2.28}$$

where "$\nabla\cdot$" denotes the divergence operator.

Using the deformation gradient tensor $\underline{\underline{F}}\,(t)$, the mass conservation can be written as follows, for every time t (no volume variation):

$$\det\; \underline{\underline{F}}\,(t) \;=\; 1. \tag{2.29}$$

Two-dimensional situations are related to planar or axisymmetric cases. Thus, the velocity vector $\underline{V}$ is given by the following equation, in a Cartesian frame:

$$\underline{V} \;=\; u\,(x,\, z)\, \underline{\varepsilon}_{x} \;+\; w\,(x,\, z)\, \underline{\varepsilon}_{z} \tag{2.30}$$

for a planar flow, and

$$\underline{V} \;=\; u\,(r,\, z)\, \underline{e}_{r} \;+\; w\,(r,\, z)\, \underline{e}_{z} \tag{2.31}$$

in axisymmetric situations.

The incompressibility condition (2.28) gives us, for a planar flow,

$$\nabla \cdot \underline{V} = \frac{\partial u}{\partial x} + \frac{\partial w}{\partial z} = 0. \tag{2.32}$$

This equation leads to introduce a function $\Psi(x, z)$ such that

$$u = \frac{\partial \Psi}{\partial z}(x, z), \; w = -\frac{\partial \Psi}{\partial x}(x, z), \tag{2.33}$$

where $\Psi(x, z)$ is the stream function in the planar case.

In two-dimensional axisymmetric situations, the incompressibility condition is written as follows:

$$\frac{\partial u}{\partial r} + \frac{u}{r} + \frac{\partial w}{\partial z} = 0, \tag{2.34}$$

where

$$u = \frac{1}{r}\frac{\partial \Psi}{\partial z}(r, z), \; v = -\frac{1}{r}\frac{\partial \Psi}{\partial z}(r, z), \tag{2.35}$$

in terms of the stream function Ψ. In both cases, we get, given a streamline (L), the following property:

$\forall M \in (\mathcal{L})$, the stream function :

- $\Psi(x,z)$ in the planar case (Cartesian variables),

or

- $\Psi(r,z)$ in the axisymmetric case (cylindrical coordinates),

is a constant on the streamline (2.36)

Such considerations are taken into account for practical problems involved in the next chapters.

2.3.1.3 Incompressible Materials: Stream Functions in Three-Dimensional Cases

Consider now the case of a three-dimensional flow in a domain Ω where the velocity vector $\underline{V}$ is expressed as follows:

$$\underline{V} = u(x, y, z)\,\underline{\varepsilon}_x + v(x, y, z)\,\underline{\varepsilon}_y + w(x, y, z)\,\underline{\varepsilon}_z, \tag{2.37}$$

in a Cartesian orthonormal basis $(\underline{\varepsilon}_x, \underline{\varepsilon}_y, \underline{\varepsilon}_z)$.

The z-axis is assumed to be the mean flow direction.

From the incompressibility condition $\nabla \cdot \underline{V} = 0$, a pair of stream functions $\Psi_1(x, y, z)$ and $\Psi_2(x, y, z)$ can be defined such that $\underline{V}$ is given by the following vector product:

$$\underline{V} = \underline{\nabla}\Psi_1 \times \underline{\nabla}\Psi_2. \tag{2.38}$$

According to Eq. (2.28), a streamline (L) of the physical domain Ω appears as the line of intersection of the two surfaces $\Psi_1\ (x,\ y,\ z)$ and $\Psi_2\ (x,\ y,\ z)$, with the following equations for the velocities:

$$u = \frac{\partial \Psi_1}{\partial y}\frac{\partial \Psi_2}{\partial z} - \frac{\partial \Psi_1}{\partial z}\frac{\partial \Psi_2}{\partial y}, \tag{2.39}$$

$$v = \frac{\partial \Psi_1}{\partial z}\frac{\partial \Psi_2}{\partial x} - \frac{\partial \Psi_1}{\partial x}\frac{\partial \Psi_2}{\partial z}, \tag{2.40}$$

$$w = \frac{\partial \Psi_1}{\partial x}\frac{\partial \Psi_2}{\partial y} - \frac{\partial \Psi_1}{\partial y}\frac{\partial \Psi_2}{\partial x}. \tag{2.41}$$

2.4 Momentum Conservation Equations

Momentum conservation equations, also called *dynamic equations*, concern equilibrium of internal forces in the medium and of external forces, body forces and surface forces at the boundary. The forces in the medium can be evaluated by means of the concept of the *Cauchy stress tensor*. The principle of conservation of momentum may be written as follows:

$$\begin{bmatrix}\text{Rate of increase}\\ \text{of momentum}\\ \text{of the system}\end{bmatrix} = \begin{bmatrix}\text{Net flow of all}\\ \text{surface forces}\\ \text{acting on the system}\end{bmatrix} + \begin{bmatrix}\text{Sum of all}\\ \text{body forces acting}\\ \text{acting on the system}\end{bmatrix} + \begin{bmatrix}\text{Rate of increase}\\ \text{of momentum of}\\ \text{the system}\end{bmatrix}. \tag{2.42}$$

Equation (2.42) involves four variables:

- density,
- pressure,
- velocity and
- stress.

The writing of stresses requires the knowledge of the physical behaviour of the material: the constitutive equation. A thermodynamic equation of state is also necessary in non-isothermal problems.

The pressure and the stress should be written in Eq. (2.42), for any arbitrary surface in the material.

2.4.1 Linear Momentum Equation

Given a time t, let ω be a volume of a body, limited by a bounding surface $\partial\omega$. The momentum of the material occupying the volume $\omega(t)$ is expressed as

$$\int_{\omega} \rho\, \underline{V}\, \mathrm{d}v. \tag{2.43}$$

According to the *Euler principle of linear momentum*, we may write

$$\frac{D}{Dt}\left(\int_{\omega} \rho\, \underline{V}\, \mathrm{d}v\right) = \int_{\partial\omega} \underline{T}\, \mathrm{d}s + \int_{\omega} \rho\, \underline{b}\, \mathrm{d}v, \tag{2.44}$$

where $\underline{b}$ denotes the body force per unit mass.

Then, using the Reynolds transport theorem and Eq. (2.27), under the assumption of incompressible materials, the relation can be expressed as follows:

$$\int_{\omega} \rho \frac{D}{Dt}\underline{V}\, \mathrm{d}v = \int_{\partial\omega} \underline{T}\, \mathrm{d}s + \int_{\omega} \rho\, \underline{b}\, \mathrm{d}v. \tag{2.45}$$

2.4.2 Conservation of Angular Momentum

The principle of moment of momentum, also provided by Euler, is considered here. At time t, one considers a material volume $\omega(t)$ where $\underline{V}(t)$ denotes the velocity vector. The net torque is assumed to be only referred to the stress vector $\underline{S}$ and the body force $\underline{b}$. Thus, it may be written

$$\frac{D}{Dt}\int_{\omega(t)} \left(\underline{x} \times \rho\, \underline{V}\right) \mathrm{d}v = \int_{\partial\omega(t)} \left(\underline{x} \times \underline{S}\right) \mathrm{d}s + \int_{\omega(t)} \left(\underline{x} \times \rho\, \underline{b}\right) \mathrm{d}v. \tag{2.46}$$

According to the Reynolds transport theorem, this equation can be expressed as follows:

$$\int_{\omega(t)} \left(\underline{x} \times \rho \frac{\mathrm{D}}{\mathrm{D}t}\underline{V}\right) \mathrm{d}v = \int_{\partial\omega(t)} \left(\underline{x} \times \underline{S}\right) \mathrm{d}s + \int_{\omega(t)} \left(\underline{x} \times \rho \underline{b}\right) \mathrm{d}v. \tag{2.47}$$

This equation corresponds to the conservation of angular momentum.

2.5 Conservation of Energy

This topic is related to the first law of thermodynamics. The rate of change of energy for a system is the contribution of the rate of work with this system and the rate corresponding to the evolution of the energy in the system. The conservation of energy, for a fluid in a volume $\omega(t)$, is expressed by the following equation:

$$\frac{\mathrm{d}}{\mathrm{d}t}\int_{\omega(t)} \rho\left(\frac{1}{2}V^2+E\right)\mathrm{d}v=\int_{\omega(t)} \rho\,\underline{b}\cdot\underline{V}\,\mathrm{d}v+\int_{\partial\omega(t)}\left(\underline{n}\cdot\underline{T}\right)\cdot\underline{V}\,\mathrm{d}S+\int_{\partial\omega(t)}\left(\underline{n}\cdot\underline{q}\right)\mathrm{d}S. \tag{2.48}$$

In this equation, E denotes the internal energy per unit mass and $\underline{q}$ the heat flow vector per unit area. The left-hand expression corresponds to the rate of change and internal energy of the fluid in $\omega(t)$. In the second side of this equation, the first and second integrals involve the contribution of the rate of working for the fluid in the volume $\omega(t)$ and outside, respectively. The third term is the contribution of the rate of energy to the fluid element, related to heat flow.

According to Fourier's law of heat conduction that links the heat transfer rate $\underline{q}$ to the temperature gradient, one may write the following equation:

$$q=-k\underline{\nabla}T, \tag{2.49}$$

where k denotes the thermal conductivity of the material and T the temperature. By applying the Reynolds transport theorem and the divergence theorem to the last quantities of the right side of Eq. (2.48), the following expression is written, under the assumption of continuity of all the functions, where $\underline{S}$ denotes the stress vector:

$$\int_{\omega(t)}\left[\frac{\mathrm{d}}{\mathrm{d}t}\left(\rho\left(\frac{1}{2}V^2+E\right)\right)+\rho\nabla\cdot\underline{V}\left(\frac{1}{2}V^2+E\right)\right]\mathrm{d}V$$
$$=\int_{\omega(t)}\left[\left(\rho\,\underline{b}\cdot\underline{V}\right)+\left(\nabla\cdot\underline{\tau}\cdot\right)\cdot\underline{V}\cdot+\underline{S}:\nabla\cdot\underline{V}+\nabla\cdot(k\nabla\cdot T)\right]\mathrm{d}v. \tag{2.50}$$

The use of the continuity condition (2.27) and the equation of motion (2.45), under the assumption of incompressibility of the material, leads to the following internal energy equation:

$$\rho\,C_p\frac{\mathrm{d}T}{\mathrm{d}t}=\underline{S}:\nabla\cdot\underline{V}+\nabla\cdot(k\nabla\cdot T). \tag{2.51}$$

In this equation, C_p denotes the heat capacity at constant pressure, for unit mass of material. In isothermal conditions, the mass conservation and the dynamic equations must be written. Thus, the primary unknowns are the pressure and velocity components. Since the stress tensor components should be involved to evaluate the equilibrium equations, a constitutive model of the material should be written together with the governing equations.

2.6 Constitutive Equations

From a macroscopic viewpoint, the constitutive equation of a given material is a tensorial explicit or implicit relation between the stress tensor and the kinematics, under deformation. The relation, generally expressed in two or three dimensions, may also be written in the one-dimensional case, without tensorial expressions. A universal law on behaviour of materials cannot be written explicitly, according to the very large variety of their properties. Researchers and scientists have contributed to define relations which may be realistic for materials in some strain conditions. Those can be given in terms of different deformation tensors. In a second step, the constitutive rheological laws may be inserted in the dynamical equations, leading to compute numerically—in general—the velocities, stresses, pressure and temperature variations related to the deformed medium.

In this section, some mathematical equations for materials (without the perfect fluid) that have been already defined in the field of rheology are written, in relation to theoretical and engineering problems. In this book, the materials considered are assumed to be isotropic, for the sake of simplicity.

Formulations of constitutive equations also require some specific assumptions which have been largely provided in the literature [2–4]. These principles are summarized as follows:

(1) Principle of objectivity: The constitutive equation must be invariant under a change of coordinate system.
(2) Principle of local action: The behaviour of a moving infinitesimal element of a medium depends only on the history of this element, not on those in its vicinity.
(3) Principle of determinism: The thermodynamic process in the medium under consideration is completely determined by the history of the transfer of energy.
(4) Principle of fading memory: For a material with elastic properties, the recent kinematic events are more influent on the strain history than events long past.

A constitutive equation must also obey the laws of thermodynamics.

Many materials exhibit a complex behaviour. However, to facilitate their classification, two groups of constitutive equations are considered here for so-called *simple fluids* [1]:

- inelastic laws and
- viscoelastic laws.

Materials considered in industrial applications are increasingly concerned by a complex microstructure. To product realistic results, numerical simulations involving such products must adopt significant theoretical models in the governing equations.

2.6.1 *Inelastic Models: Newtonian Fluid*

For these models, the stress tensor $\underline{\underline{\tau}}$ is expressed at time t as a function of the kinematic tensors $\underline{\underline{K}}_i(t)$, according to the general relation:

$$\underline{\underline{\tau}}(t) = \underline{\underline{F}}\left[\underline{\underline{K}}_i(t)\right], i = 1, \ldots, n, \tag{2.52}$$

where $\underline{\underline{K}}_i(t)$ denote kinematic tensors.

2.6.1.1 Newtonian Fluid

A simple example corresponds to the Newtonian fluid law:

$$\underline{\underline{\tau}}(t) = 2\eta \underline{\underline{D}}(t), \tag{2.53}$$

where η is the constant fluid viscosity and $\underline{\underline{D}}(t)$ denotes the symmetric rate-of-deformation tensor. The following simple constitutive equation

$$\underline{\underline{D}} = \frac{1}{2}\left(\nabla\underline{V} + \nabla\underline{V}^T\right) \tag{2.54}$$

is often used in the calculations for testing the numerical codes. Its results are considered as reference data for those obtained by using more complicated constitutive equations.

2.6.1.2 Purely Viscous Non-Newtonian Fluids

If the viscosity is not constant in Eq. (2.28) and does not involve parameters as those related to elasticity, the fluid is a non-Newtonian purely viscous fluid. The stress tensor is written as follows:

$$\underline{\underline{\tau}} = 2\eta\left(\underline{\underline{D}}\right)\underline{\underline{D}}. \tag{2.55}$$

The viscosity function η is generally expressed as depending on the second invariant of $\underline{\underline{D}}$. More often, the viscosity is a decreasing function of the shear gradient $\dot{\gamma}$ (shear-thinning or "pseudo-plastic" fluid case). On the contrary, the material is called shear-thickening or "dilatant" fluid, as for concentrated suspensions of solids.

Particular equations for purely viscous materials have been proposed in the literature, as, for example, power-law fluids. The stress tensor is expressed by a relation of the type of Eq. (2.53), with a viscosity function written as dependent on the velocity gradient $\dot{\gamma}$:

$$\eta(\dot{\gamma}) = \eta_0 \cdot II_{\underline{\underline{D}}}^{n-1}, \tag{2.56}$$

where $II_{\underline{\underline{D}}}$ denotes the second invariant of $\underline{\underline{D}}$, expressed by

$$II_{\underline{\underline{D}}} = (1/2)\sqrt{\underline{\underline{D}} : \underline{\underline{D}}}. \tag{2.57}$$

This invariant can be considered as a generalized velocity gradient in two- and three-dimensional situations. For planar (variables (x, y, z)) and axisymmetric (variables (r, θ, z)) cases, the second invariant is expressed, respectively, by the following relations:

$$II_{\underline{\underline{D}}} = \frac{1}{2}\sqrt{(\underline{\underline{D}} : \underline{\underline{D}})} = \begin{cases} \frac{1}{2}\sqrt{\left[(2\cdot\frac{\partial u}{\partial x})^2 + (2\cdot\frac{\partial v}{\partial y})^2 + 2\cdot(\frac{\partial u}{\partial y} + \frac{\partial v}{\partial x})^2\right]} & \text{(2D planar case)} \\ \frac{1}{2}\sqrt{\left[(2\cdot\frac{\partial u}{\partial r})^2 + (2\cdot\frac{u}{r})^2 + (2\cdot\frac{\partial w}{\partial z})^2 + 2\cdot(\frac{\partial u}{\partial z} + \frac{\partial w}{\partial r})^2\right]} & \text{(2D axisymmetic case).} \end{cases} \tag{2.58}$$

Figure 2.2 provides a viscosity curve versus the velocity gradient. The Carreau model [4] is an example of a purely viscous fluid, where the viscosity $\eta(\dot{\gamma})$ is written by using the following equation:

$$\frac{\eta\,(\dot{\gamma}) - \eta_\infty}{\eta_0 - \eta_\infty} = [1 + (\lambda \cdot \dot{\gamma})^2]^{\frac{n-1}{2}}, \tag{2.59}$$

where $\dot{\gamma}$ is the shear rate, and η_0 and η_∞ denote the viscosity at $\dot{\gamma} = 0$ and $\dot{\gamma} = \infty$, respectively.

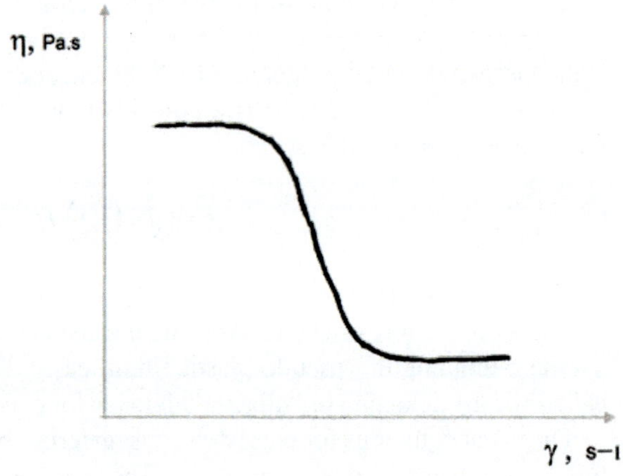

Fig. 2.2 An example of a viscosity curve $\eta(\dot{\gamma})$ versus the velocity gradient $\dot{\gamma}$ corresponding to a polymer material

2.6.2 *Viscoelastic Constitutive Equations*

Such equations concern materials with memory. The stress tensor at time t may be written as an implicit or explicit function of different kinematic tensors. In a general explicit form, the stress tensor $\underline{\underline{T}}\,(\underline{X},\ t)$, related to a position $\underline{X}$ at time t, is expressed as follows:

$$\underline{\underline{T}}\,(\underline{X},\ t) = \underline{\underline{\mathcal{H}}}_{t'=-\infty}^{t'=t}\left[\underline{\underline{K}}_i\,(t, t')\right],\ i = 1, \ldots, n. \tag{2.60}$$

In this equation, $\underline{\underline{K}}_i\,(i = 1,\ \ldots,\ n)$ denotes kinematic tensors and $\underline{\underline{\mathcal{H}}}_{t'=-\infty}^{t'=t}$ is a functional expressed as a simple or a multiple integral. Generally, the tensors $\underline{\underline{K}}_i$ may be written according two possibilities:

(i) The kinematic tensors are related to the rate-of-deformation tensor, based on the velocity, which is not a priori an objective physical quantity. The velocity of a given particle should be made objective under the condition that it is defined with respect to a given frame. Otherwise, if not referred, this velocity can be zero if the material point moves versus time at the same velocity of this particle. Obviously, for small velocities, this concept may be ignored.
Generally speaking, when a material is considered in large deformation problems as those encountered for fluids, the constitutive equation should be required to obey the objectivity principle. Thus, the constitutive equation should be written in a corotational frame (Bird et al. [4]). To express the equation of state concerning the stress tensor coordinates at a given time t in a fixed reference basis, the kinematic quantities given by the original expression in a corotational frame should be computed, at every time t. The rotation matrix between the corotational frame and the reference frame should be considered. The reader may refer to [5]. Differential and integral constitutive equations are initially written in a corotational frame for reasons of objectivity. Differential or integral forms can be expressed for corotational models.

(ii) When using the codeformational formalism, the kinematic tensors are related to strain. In these cases, the deformation of a material is the same whatever the frame selected for its expression. According to the use of kinematic tensors related to strain, the codeformational kinematic tensors obey the objectivity principle and are associated to the deformation tensor. As in the corotational case, codeformational constitutive equations may be written with differential or integral tensorial equations.

Some examples of corotational and codeformational constitutive equations are presented. They illustrate differences resulting from the models when solving practical problems, in relation to the differential or integral forms of expressions for the stress tensor.

2.6.2.1 Codeformational Models

Such models are referred to deformation tensors based on the use of the objective deformation gradient tensor $\underline{\underline{F}}$ defined by Eq. (2.12), related to strain. For differential models, specific material derivative operators are used for the writing of the constitutive equations.

Differential Codeformational Models

In these constitutive equations, material derivative operators are defined in order to write the derivatives in the differential tensor expressions. The material derivative operator for a tensor $\underline{\underline{A}}$ is generally defined as follows:

$$\frac{D}{Dt}\underline{\underline{A}} = \frac{D_J}{Dt}\underline{\underline{A}} + a\left(\underline{\underline{A}}\,\underline{\underline{D}} + \underline{\underline{D}}\,\underline{\underline{A}}\right) + b\,\mathrm{tr}\left(\underline{\underline{A}}\,\underline{\underline{D}}\right) + c\,\underline{\underline{D}}\,\mathrm{tr}\left(\underline{\underline{A}}\right), \tag{2.61}$$

where a, b and c are constant numbers. $\frac{D_J}{Dt}$ denotes the Jaumann derivative operator given by

$$\frac{D_J}{Dt}\underline{\underline{A}} = \frac{D}{Dt}\underline{\underline{A}} + \frac{1}{2}\left(\underline{\underline{\mathcal{R}}}\,\underline{\underline{A}} - \underline{\underline{A}}\,\underline{\underline{\mathcal{R}}}\right) = \frac{D}{Dt}\underline{\underline{A}} + \underline{\underline{\mathcal{R}}}\,\underline{\underline{A}} - \underline{\underline{A}}\,\underline{\underline{\mathcal{R}}}. \tag{2.62}$$

In Eq. (2.62), $\underline{\underline{\Omega}}$ denotes the rotation tensor previously defined in Eq. (2.12), according to the following relation:

$$(1/2)\underline{\underline{R}} = \underline{\underline{\Omega}}. \tag{2.63}$$

Thus, the tensor $\underline{\underline{\Omega}}$ is given by the antisymmetric part of the velocity gradient tensor:

$$\underline{\underline{\Omega}} = \frac{1}{2}(\underline{\nabla}\,\underline{V} - \underline{\nabla}\,\underline{V}^T). \tag{2.64}$$

For $a = 1$ and $b = c = 0$, the model corresponds to the so-called lower convected derivative of tensor A, expressed by one of the following symbols:

$$\frac{D_C}{Dt}\underline{\underline{A}} \text{ or } \overset{\triangle}{\underline{\underline{A}}}. \tag{2.65}$$

The case $a = -1$ and $b = c = 0$ corresponds to the upper convected derivative of $\underline{\underline{A}}$, noted as

$$\frac{D^c}{Dt}\underline{\underline{A}} \text{ or } \overset{\nabla}{\underline{\underline{A}}}. \tag{2.66}$$

(1) Oldroyd-B and Maxwell differential models

Such models are expressed by the following equation:

$$\underline{\underline{\sigma}} = -p\underline{\underline{I}} + 2\eta_s \underline{\underline{D}} + \underline{\underline{\tau}}. \tag{2.67}$$

This relation is associated to the following tensorial differential equation:

$$\lambda \frac{D\underline{\underline{\tau}}}{Dt} + \underline{\underline{\tau}} = 2\eta_m \underline{\underline{D}}, \tag{2.68}$$

where λ denotes a relaxation time. The related concept to these fluids concerns a polymer dissolved in a solvent, with elastic properties and assumed to be of constant viscosity η. Two coefficients η_{m} and η_{s}, which correspond to the respective viscosities of the polymer and of the solvent, are considered in the following relation:

$$\eta = \eta_m + \eta_s. \tag{2.69}$$

A retardation time λ_r of the fluid may be defined as follows:

$$\lambda_r = \lambda \frac{\eta_m}{\eta_s}. \tag{2.70}$$

When writing the following relation,

$$\underline{\underline{\tau_e}} = 2\eta_s \underline{\underline{D}} + \underline{\underline{\tau}}, \tag{2.71}$$

where $\underline{\underline{\tau_e}}$ is called the extra-stress tensor, Eq. (2.71) becomes

$$\underline{\underline{\tau_e}} + \lambda \frac{D\underline{\underline{\tau_e}}}{Dt} = 2\eta\, [\underline{\underline{D}} + \lambda_r \frac{D}{Dt} \underline{\underline{D}}]. \tag{2.72}$$

In the literature, for numerical simulations, notably for benchmark problems involving the Oldroyd-B equation (Crochet al. [2]), the authors have generally adopted the following relation:

$$\frac{\eta_m}{\eta_m + \eta_s} = 0.89. \tag{2.73}$$

The viscoelastic differential Maxwell model [4] may be considered as a particular case of the Oldroyd B fluid, where

$$\eta_s = 0, \quad \eta_m = 1. \tag{2.74}$$

Such values point out the assumption that the material does not contain solvent. Only the polymer is considered. The Upper Convected Maxwell model (UCM) is the most popular version of this constitutive equation, according to the use of the operator (2.61) applied to the elastic stress tensor.

(2) Other differential constitutive laws

The Phan–Thien–Tanner model is defined by the two following equations:

$$\begin{aligned} &\underline{\underline{\sigma}} = -p\underline{\underline{I}} + 2\eta_s\underline{\underline{D}} + \underline{\underline{\tau}} \\ &\lambda\frac{D\underline{\underline{\tau}}}{Dt} + \exp\left[\frac{\varepsilon\lambda}{\eta_m}\mathrm{tr}\left(\underline{\underline{\tau}}\right)\right]\underline{\underline{\tau}} = 2\eta_s\underline{\underline{D}}, \end{aligned} \tag{2.75}$$

where η_m and η_s have the same meanings than those previously defined for the Oldroyd fluids, λ is a relaxation time and ε a material constant. This model shows a decreasing viscosity function versus the shear rate in shearing flows (and stress hardening in elongational situations), which allows the model to be more representative of real materials.

The White–Metzner model is given by equations of the type of Eq. (2.75), where the quantities η_m, η_s and λ are functions of the second invariant $II_{\underline{\underline{D}}}$ of the strain-rate tensor $\underline{\underline{D}}$.

Other constitutive equations as Leonov, Giesekus and FENE models have also been used for numerical simulations in the literature (Bird et al. [4]).

Integral Codeformational Models

Three memory-integral viscoelastic equations expressed in terms of codeformational tensors are selected in this section, where parameters are fitted by experimental data.

(1) Lodge model

For this constitutive equation, considering microstructural features (macromolecular configurations (Lodge [6]), the stress tensor $\underline{\underline{\tau}}$ is given by

$$\underline{\underline{\tau}} = \int_{-\infty}^{t}\sum_{i=1}^{i=n}\alpha_i \exp\left(-\frac{t-t'}{\beta_i}\right)\underline{\underline{B}}_t(t')\mathrm{d}t'. \tag{2.76}$$

(2) Wagner model

This model also involves the Finger tensor. The stress tensor is written as

$$\underline{\underline{\tau}} = \int_{-\infty}^{t} m(t - t')\, h(I_1, I_2)\, \underline{\underline{B}}_t(t')\, \mathrm{d}t'. \tag{2.77}$$

The relaxation function is written as the product of a function of time t by a damping function h depending on the first and second invariants I_1 and I_2 of the strain Finger tensor $\underline{\underline{B}}_t$ and of the Cauchy tensor $\underline{\underline{C}}_t$, respectively, written as follows:

$$I_1 = \mathrm{tr}(\underline{\underline{B}}_t) \tag{2.78}$$

$$I_2 = \mathrm{tr}(\underline{\underline{C}}_t). \tag{2.79}$$

$m(t)$ is the memory function of the material.

(3) K-BKZ model

In this constitutive equation, widely considered in numerical simulations, the stress tensor is expressed as follows, using the respective Finger and Cauchy tensors $\underline{\underline{B}}_t$ and $\underline{\underline{C}}_t$:

$$\underline{\underline{\tau}} = \int_{-\infty}^{t} \left(\frac{\partial U}{\partial I_1} \underline{\underline{B}}_t(t') - \frac{\partial U}{\partial I_2} \underline{\underline{C}}_t(t') \right) \mathrm{d}t', \tag{2.80}$$

where U denotes a potential function which depends only on the two invariants I_1 and I_2 of the deformation Finger and Cauchy tensors $\underline{\underline{B}}_t$ and $\underline{\underline{C}}_t$, respectively.

2.6.2.2 Corotational Models

The Corotational Frame

In this chapter, the strain-rate (or rate-of-deformation tensor) $\underline{\underline{D}}$ and the vorticity tensor $\underline{\underline{\Omega}}$ have been considered previously (Sect. 2.2.2). For a material point M of the moving material at time t, the corotational frame corresponds to the orthonormal basis where the vorticity vector related to $\underline{\underline{\Omega}}$ is zero. This means that, in this frame noted ($\mathbf{c}*_i$), the strain rate is not associated to a rotation in the medium and is objective when expressed in the corotational frame. For small deformations, the corotational frame may be considered as identical to the reference frame but, in large strain velocities, the objectivity principle should be required. At time t, the constitutive equation is written in the corotational frame and the vorticity vector may then lead to write the constitutive law in the reference frame considered for the study. Thus, when considering large deformations as those corresponding to fluid flows, a corotational model should be written in a corotational frame to be determined, according to the kinematic strains or flow conditions.

Differential Corotational Models

Differential derivatives for second-order tensors have been proposed in the literature. Given a second-order tensor $\underline{\underline{S}}$, the corotational Jaumann derivative $\frac{D_J}{Dt}$ is defined by [5]

$$\frac{D_J}{Dt}\left(\underline{\underline{S}}(t)\right) = \frac{D}{Dt}\left(\underline{\underline{S}}(t)\right) + \underline{\underline{S}} \cdot \underline{\underline{R}} - \underline{\underline{\Omega}} \cdot \underline{\underline{R}}, \tag{2.81}$$

where $\underline{\underline{\Omega}}$ denotes the vorticity tensor. The Jaumann derivative can be used when defining corotational differential models as the example given by the corotational Jeffrey model, written in a corotational frame. The rheological stress tensor is defined by the following equation, involving the Jaumann derivative:

$$\underline{\underline{\tau}} + \lambda_1 \frac{\partial}{\partial t}\underline{\underline{\tau}} = \eta_0 + \lambda_2 \frac{D_J}{Dt}\underline{\underline{\tau}}, \tag{2.82}$$

where λ_1, λ_2 ($\lambda_1 > \lambda_2 > 0$) are constants and η_0 is a viscosity.

Integral Corotational Models

An example of a non-linear integral model, the Le Roy–Pierrard constitutive equation [5] expressed in a corotational frame, is presented here. This constitutive equation involves functions of the invariants of the rate-of-deformation tensor $\underline{\underline{D}}$. The first invariant $B_1 = \mathrm{tr}(\underline{\underline{D}})$ of the rate-of-deformation tensor is zero, under the incompressibility assumption of the material. The non-zero second and third invariants of the rate-of-deformation tensor are given by the following relationships:

$$B_2 = D^{ij}\, D^{ji},\ B_3 = D^{ih}\, D^{hj}\, D^{jh}. \tag{2.83}$$

The integral equation of the model provides the extra-stress tensor $\underline{\underline{\tau}}$ to be expressed as follows:

$$\underline{\underline{\tau}}(t) = \int_{-\infty}^{t} \chi_1(t-\tau)\frac{1}{B_2}\Phi_1(B_2, B_3)\underline{\underline{D}}(\tau)\,\mathrm{d}\tau$$

$$\int_{-\infty}^{t} \chi_2(t-\tau)\frac{1}{B_2}\Phi_2(B_2, B_3)\left[B_3\,\underline{\underline{D}}(\tau) - B_2\left(\underline{\underline{D}} : \underline{\underline{D}} - \frac{B_2}{3}\right)\underline{\underline{I}}\right]\mathrm{d}\tau, \tag{2.84}$$

where χ_1 and χ_2 are memory functions. For $\chi_2 = 0$, this equation corresponds to a corotational Maxwell model [10], written as follows:

$$\underline{\underline{\tau}}(t) = \int_{-\infty}^{t} \chi(t-\tau)\, \underline{\underline{D}}(\tau)\, \mathrm{d}\tau. \tag{2.85}$$

2.7 Concluding Remarks

The basic conservation equations to be considered with the kinematics required in flow studies have been presented in this chapter.

- Materials with complex physical properties often require to set up constitutive models that allow fitting of experimental results and consistency with industrial data.
- Some differential and integral constitutive equations which can be realistic for engineering applications and research may be retained.
- When using computational codes, differential models may allow more simplicity: no particle-tracking problems. At the present time, from a mathematical point of view, there exists in the literature few results on the existence and unicity of solutions [7].
- Memory-integral models involve consideration of the particle history in a deformed medium, leading to more complex expressions for the governing equations. These laws provide additional coefficients, easily written and adopted for a better agreement with experimental data. Such property may be of particular importance in problems encountered for research results to innovation and practical achievements.

References

1. Coleman BD, Markowitz H, Noll W (1966) Viscometric flows of non-Newtonian fluids. Springer
2. Crochet MJ, Davies AR, Walters K (1984) Numerical simulation of non-Newtonian flow. Elsevier, Amsterdam
3. Astarita G, Marrucci G (1974) Principles of non-Newtonian fluid mechanics. McGraw-Hill
4. Bird RB, Armstrong RC, Hassager O (1987) Dynamics of polymer liquids. Wiley
5. Le Roy P, Pierrard J-M (1973) Fluides viscoélastiques non linéaires satisfaisant à un principe de superposition. Rheol Acta 12:449–454
6. Lodge AS (1964) Elastic liquids. Academic Press, New-York
7. Joseph DD, Renardy M, Saut JC (1985) Hyperbolicity and change of type in the flow of viscoelastic fluids. Arch Rat Mech Anal 87:212–251

Chapter 3
Domain Transformations: Stream-Tube Method in Two-Dimensional Cases

3.1 Introduction

Before considering domain transformations as a tool for computing deformation and flow problems, discretization methods for local domains of small size are briefly recalled in a given geometry. Computing the primary unknowns is generally based on *velocity–pressure* formulations, vorticity-stream function in the physical domain. The discretization of the governing equations is thus carried out with techniques as Finite Differences (FD), Finite Elements (FE) or Finite Volumes (FV). Particularly, the finite element method, introduced many years ago (e.g. Zienkiewicz [1]), allows to consider very complex geometries and use tools related to numerical analysis. Sometimes, a theoretical environment allows to set up robust and efficient algorithms for computing solutions, notably in the non-linear case [2].

The basic idea of these methods is decomposition of the physical domain (of a solid or fluid material) under deformation into polygonal elements in two or three dimensions. The unknowns are approximated by parametric functions. For all the numerical approaches, discretization in space requires refinement procedures in singularity zones for the unknowns, in order to reduce the numerical diffusion. Structured and unstructured grids, formed of straight or curved segments, are considered. The solution is obtained by minimizing an associated error function, using variational methods.

Discretization in space always requires refinement procedures in singularity zones for the unknowns, in order to reduce the numerical diffusion. Structured and unstructured grids, formed of straight or curved segments, may be considered. However, the numerous applications of the FE method to solve numerous problems in physics bear witness to its success.

Computational mechanics often involve numerical difficulties, as those encountered when integral model fluids are involved in the calculations. Such problems come out from the non-linearity of the governing equations. In the other hand, when particle

J.-R. Clermont and A. Ammar, *Stream-Tube Method*,
https://doi.org/10.1007/978-3-030-65470-2_3

tracking on its trajectory is required, these points do not correspond necessarily to the mesh points. Approximations may lead to significant errors in the computations in two-dimensional cases that are emphasized in three-dimensional problems where the pathlines are warping curves.

3.2 Global Transformations for Physical Domains

In this section, global transformation functions allowing to map a global domain or a limited number of sub-domains into one or several regions are considered. Computations are done in simple transformed global domain (Fig. 3.1).

The idea of transforming a complicated two- or three-dimensional geometry into a simple region in order to solve boundary value problems of partial differential equations has been developed in numerous cases. This has concerned particularly grid generation techniques in computational fluid dynamics, notably for solving Navier–Stokes equations (e.g. Thompson et al. [3], Eiseman [4]). The set of transformations includes conformal and non-conformal mappings. In this chapter and the following ones, the basic elements of the so-called *Stream-Tube Method* (STM) will be extensively presented and developed, in relation to domain transformations.

Generally, the objectives particularity concern:

- the accurate mapping of a given complex geometry,
- the easy representation of the solution in the whole domain,
- the determination of local effects.

In some cases, the transformations may be applied to a limited number of sub-domains. It should also be pointed out that, for computations, consideration of a simple transformed domain Ω^* of a physical domain leads to expect writing of more complicated governing equations.

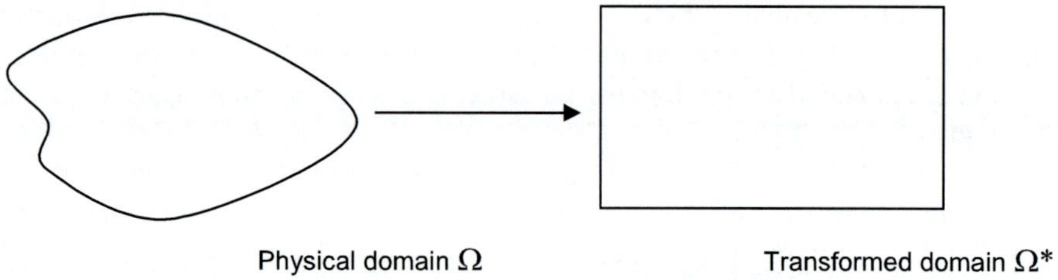

Fig. 3.1 Mapping of a physical domain into a rectangular computational domain

3.2.1 Conformal Mappings—Grid Generation Techniques

Some methods considered in the literature for numerical computations are now recalled here.

3.2.1.1 Transformations

Let us consider the following set of equations in $\mathbb{R}^2$, related to a transformation T between two sub-domains A and B:

$$X = \alpha(x, y), Y = \beta(x, y). \tag{3.1}$$

If to each point (x, y) ∈ A in the plane there corresponds one and only one point (X, Y) ∈ B and conversely, the transformation T is a one-to-one transformation or mapping. Accordingly, the functions α and β are assumed to be continuously differentiable, and the Jacobian J of the transformation is expressed as follows:

$$\mathrm{J} = \frac{\partial \alpha}{\partial \mathrm{x}}\frac{\partial \beta}{\partial \mathrm{y}} - \frac{\partial \alpha}{\partial \mathrm{y}}\frac{\partial \beta}{\partial \mathrm{x}} = \frac{\partial(\alpha, \beta)}{\partial(\mathrm{x}, \mathrm{y})}. \tag{3.2}$$

The transformation T is one-to-one provided J ≠ 0. Its inverse is given by

$$\mathrm{J}^{-1} = \frac{\partial \mathrm{x}}{\partial \alpha}\frac{\partial \mathrm{y}}{\partial \beta} - \frac{\partial \mathrm{y}}{\partial \alpha}\frac{\partial \mathrm{x}}{\partial \beta} = \frac{\partial(\mathrm{x}, \mathrm{y})}{\partial(\alpha, \beta)}. \tag{3.3}$$

3.2.1.2 Complex Mapping Functions and Conformal Mappings

When x and y are real and imaginary parts in an analytic function of a complex variable, one can write

$$z = x + iy, \quad f(z) = f(x + iy) = \alpha + i\beta = w. \tag{3.4}$$

According to properties of an analytic function, the JacobianJof the transformation is expressed by

$$\mathrm{J} = \frac{\partial(\alpha, \beta)}{\partial(x, y)} = \left|f'(z)\right|^2. \tag{3.5}$$

Hence, the transformation is one-to-one in regions where $f'(z) \neq 0$. The property of conformal mapping can be expressed as follows.

If $f(z)$ is analytic such that $\left|f'(z)\right| \neq 0$ in a domain Ω, then the mapping $w = f(z)$ is conformal for all points of Ω. Conformal mapping preserves angles both in

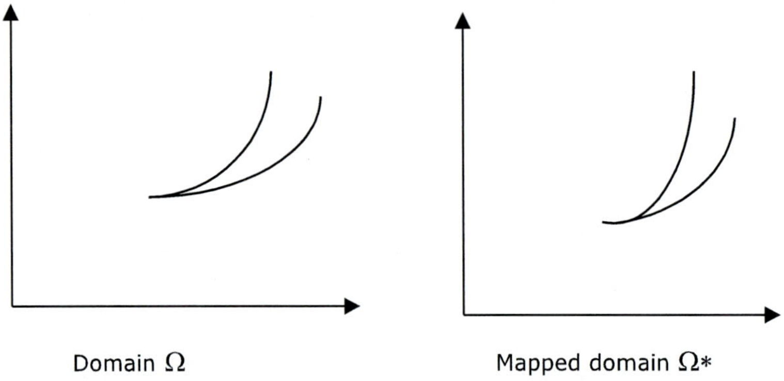

Fig. 3.2 Conservation of angles for a conformal mapping

both magnitude and sense (Fig. 3.2) in the mapped domain Ω^*. This method can be applied only for two-dimensional domains.

3.2.1.3 Complex Mappings in the Literature

Conformal mappings have been often applied to flow computations, as, for example, the numerical work by Lee and Fung [5] who considered an axisymmetric viscous fluid flow involving a local constriction (Fig. 3.3). In cylindrical coordinates, the physical domain is mapped into a rectangle by using the following conformal transformation:

$$\zeta + i\,\xi = \mathsf{F}(Z - i\,R). \tag{3.6}$$

Figure 3.3 shows a regular rectangular mesh in the mapped domain where computations are done. A finite-difference technique has been applied for solving the equations in the transformed domain.

Ryan and Dutta [6] have studied numerically a free surface flow for a Newtonian incompressible fluid at the exit of a circular tube, for the so-called swelling problem. A constant solid flow is obtained at approximately one diameter downstream the exit section (Fig. 3.4). The authors adopted a conformal transformation to simplify the computational domain, when compared to the physical one. The latter area involves an unknown non-rectilinear free surface (Fig. 3.4a). The transformed rectangular domain Ω^* allows to define a new system of orthogonal coordinates. A finite-difference scheme is associated to the stream function $\Psi(r, z)$ and the vorticity ω. The numerical results have found to be in agreement with those resulting from other finite element computations (Nickell et al. [7]).

The study of theoretical and practical features of mesh adaptivity has been largely provided in the book of Huang and Russel [8].

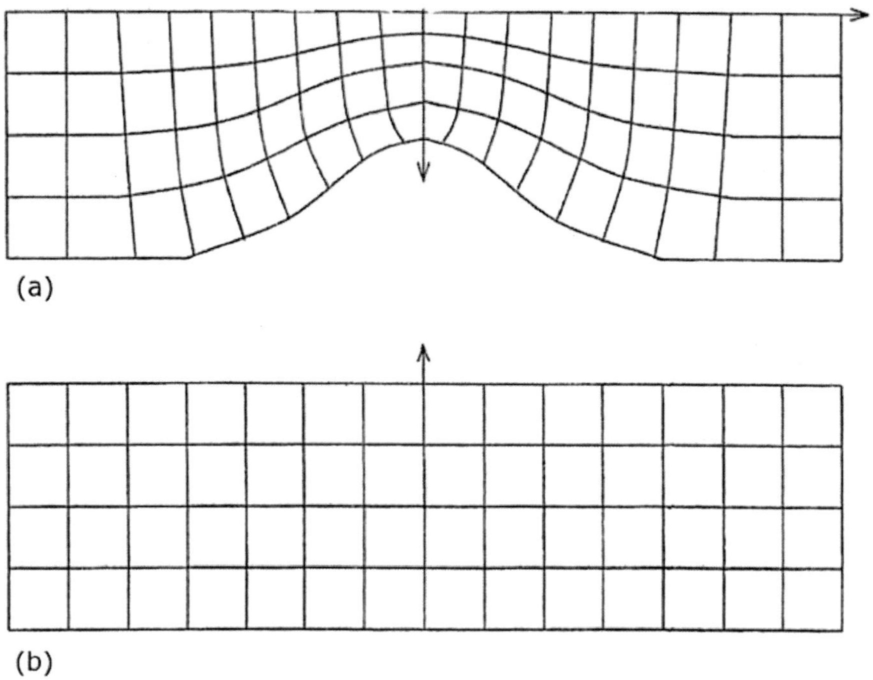

Fig. 3.3 Conformal transformation: **a** physical domain Ω; **b** mapped computational domain Ω^*

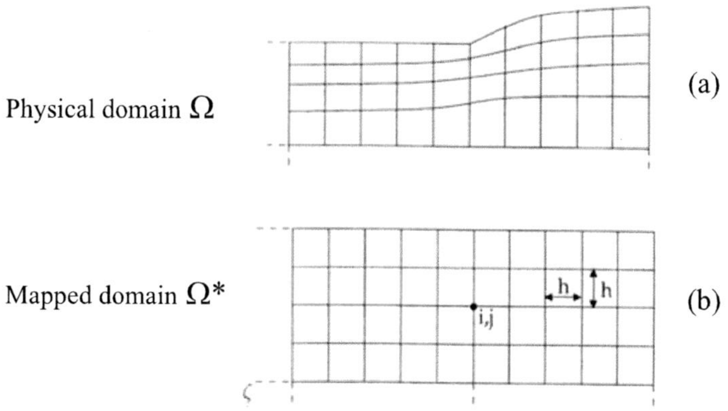

Fig. 3.4 Conformal transformation: **a** physical domain; **b** mapped computational domain

3.2.2 *General Curvilinear Coordinates*

For decades, domain transformations applied to numerical techniques have allowed automatic generation of curvilinear coordinate systems (q^i). The coordinate curves coincide with boundary of arbitrary shape, leading to define natural basis systems

$(\underline{e}_i)$ such that

$$\underline{e}_i = \frac{\partial M}{\partial q^i}. \tag{3.7}$$

These methods, increasingly developed, allow mesh generation for more various problems than those attainable by conformal mappings. For example, the use of curvilinear coordinates, initially developed by authors as Thompson et al. [3] for potential flows over airfoils, of Joukowski and Karman–Trefftz has led to numerical solutions quickly deemed in agreement with existing analytic solutions. This procedure, called *boundary-fitted coordinate schemes*, is mainly applied for Navier–Stokes equations.

It should be pointed out that the non-orthogonality of some boundary-fitted coordinate systems can affect the accuracy and the numerical stability [7]. The curvilinear coordinates are obtained for numerical solutions of an elliptic system of equations in the physical domain. Then, the solving of the governing equations is performed in a transformed domain rectangular in shape with discretization schemes defined for various cases as follows:

- non-stationary problems which can be associated to deformation of domains versus time, while the solving domain remains unchanged;
- flow problems involving unknown free surfaces.

In such conditions, two- and three-dimensional curvilinear systems have been proposed by different authors (e.g. Eiseman [10]) who developed a "multi-surface" method. Intermediate surfaces between the boundary surfaces have been introduced. Mathematical functions are provided in order to transform a physical domain into a rectangular one (e.g. Smith [11]).

Figure 3.5 presents flow domains involving a free surface, computed with boundary-fitted coordinate method (Haussling [9]), related to movement of waves encountering an obstacle.

These transformation methods generally concern the velocity–pressure $(\underline{V}, p)$ or stream function–vorticity (Ψ, ω) as primary unknowns, for a direct solving of the equations in the flow domain. The approaches are more or less complicated, in relation to the shape of the physical domain and the type of boundary conditions (Dirichlet, Neumann). Finite differences are mostly adopted in such cases.

Mesh generations in a physical domain Ω are often defined for finite element calculations. Elements of different types (triangular, quadrilateral) may be used. Their choice of shape does not necessarily imply definition of a simple global domain for solving the equations: as well-known, finite elements may be adapted to domains of complex shape. Numerous papers have been devoted to grid generation problems (e.g. [12–14]).

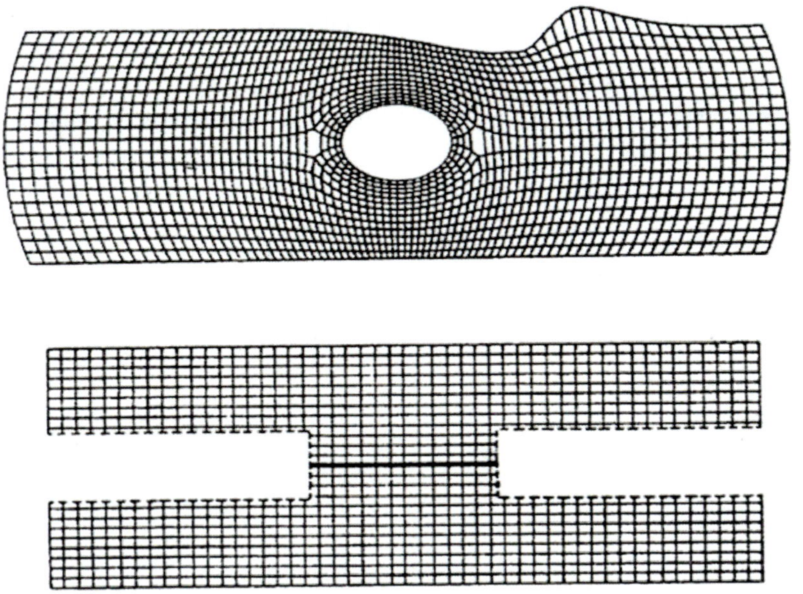
(a)
(b)

Fig. 3.5 Circular cylinder and translation: **a** physical domain Ω **b** transformed domain Ω^*

3.2.3 *Domain Transformations Based on Kinematic Concepts*

With reference to kinematic considerations related to streamlines, pathlines and stream functions, simple transformed domains for flow computations have been also studied in the literature. These approaches lead to express curvilinear coordinates for description of moving particles. Various methods, dating back several decades, have been proposed for perfect and viscous fluid flows. Examples of such studies are provided in Greywall works (1985, 1988) [15, 16] who defined a coordinate system (X, Y, Z) in two- and three-dimensional situations, for potential flows. This author has associated a system of variables (X, Y, Z) to Cartesian coordinates (x, y, z) using the following non-linear differential relations, in three-dimensional situations:

$$
\begin{aligned}
&\mathrm{E}_1\left(x, y, z, \frac{\partial \mathrm{X}}{\partial x}, \frac{\partial \mathrm{X}}{\partial y}, \frac{\partial \mathrm{X}}{\partial z}\right) = 0,\\
&\mathrm{E}_2\left(x, y, z, \frac{\partial \mathrm{X}}{\partial x}, \frac{\partial \mathrm{X}}{\partial y}, \frac{\partial \mathrm{X}}{\partial z}\right) = 0,\\
&\mathrm{E}_3\left(x, y, z, \frac{\partial \mathrm{X}}{\partial x}, \frac{\partial \mathrm{X}}{\partial y}, \frac{\partial \mathrm{X}}{\partial z}\right) = 0.
\end{aligned}
\tag{3.8}
$$

Mostly, the methods using streamlines to express coordinate systems resulting from the solving of differential systems are difficult to perform in non-Newtonian cases, owing to the complexity of the resulting equations. Such research should be directed towards simple approaches where variables as the stream function and the streamlines are concerned.

The explicit application of the stream function Ψ as a variable has been introduced by Duda and Vrentas [17]. The authors proposed a so-called *Protean coordinate system* with variables (X^1, X^2, X^3) for planar or axisymmetric two-dimensional strains. In their analysis, one coordinate surface is a stream surface and one coordinate is a streamline, leading to the cylindrical coordinates (r, θ, z) (Fig. 3.6):

$$\begin{aligned} X^1 &= \Psi, \\ X^2 &= \theta, \\ X^3 &= \xi. \end{aligned} \tag{3.9}$$

An interesting theoretical work of Protean coordinates has been proposed by Adachi [18, 19]. The method allows evaluation of kinematic tensors, particularly for the relative description of motion (codeformational case). Applications of these coordinate systems have been provided in numerical simulations involving constitutive equations, as those proposed by Papanastasiou et al. [20] with a viscoelastic

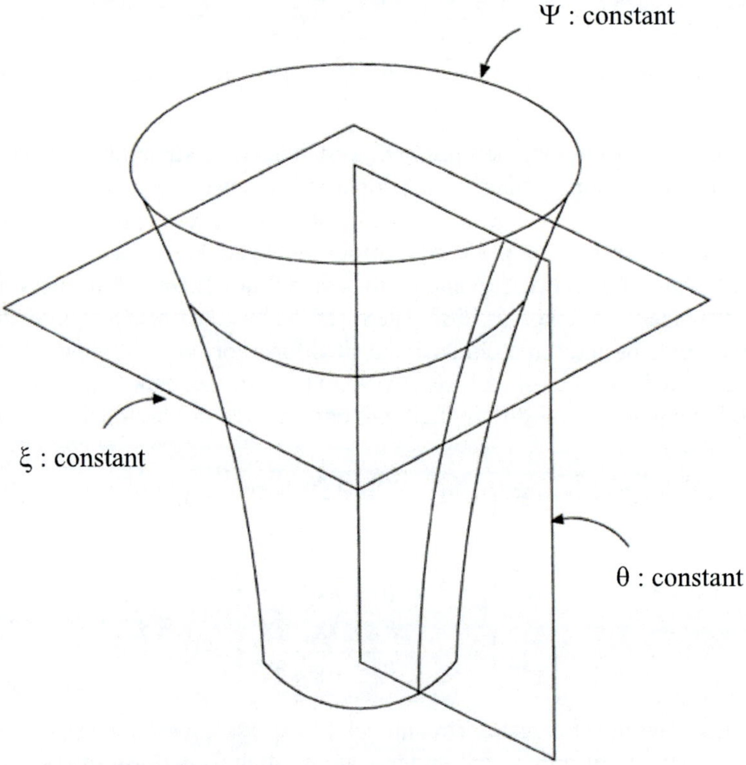

Fig. 3.6 Coordinate surfaces in a "Protean" coordinate system

model. Using Adachi's formalism, these authors have considered a memory-integral (K-BKZ) equation, in a two-dimensional extrusion problem.

Chung and co-authors [21, 22] have proposed an approach for calculation of stationary flows of non-Newtonian fluids based on the concept of streamline coordinate method. Axisymmetric confined and free-surface flows of Bingham fluids have been considered. The method does not allow computation of secondary flows but permits calculations with non-Newtonian fluids and possibilities to consider slip of the material along a duct wall.

3.3 Stream-Tube Method (STM) for Two-Dimensional Problems

The general elements and properties of the Stream-Tube Method (STM) in two-dimensional cases are now presented. The analysis is described here for incompressible deformable domains. Although the basic situations concern here open streamlines, closed streamlines can also be considered in particular cases. We firstly consider the method for two-dimensional problems related to planar or axisymmetric conditions.

STM is based on a non-conformal transformation of streamlines of a physical domain into a mapped domain where the streamlines are parallel straight lines. We start by presenting the elements related to deformations or flows involving open streamlines. The computations are performed in the mapped domain.

3.3.1 Transformation for Two-Dimensional Domains

For a two-dimensional study, we consider a bounded domain Ω where the following conditions are assumed:

(a) only open streamlines are encountered (no eddy regions),
(b) there exists a section z_I of the domain where the kinematics are known.

According to the first assumption, a transformed domain Ω^* of the physical domain Ω ($\Omega^* \rightarrow \Omega$) is considered such that the mapped streamlines are straight and parallel to a mean direction Oz of the moving particles (Fig. 3.7), by a non-conformal transformation T.

- The solving of the governing equations is carried out in Ω^*, used as computational domain.
- The equations are written with Cartesian variables (x, y, z) in the planar case and cylindrical coordinates (r, θ, z) for axisymmetric domains.

Thus, we may write the following one-to-one relationships by considering respective mapping functions k and f as follows:

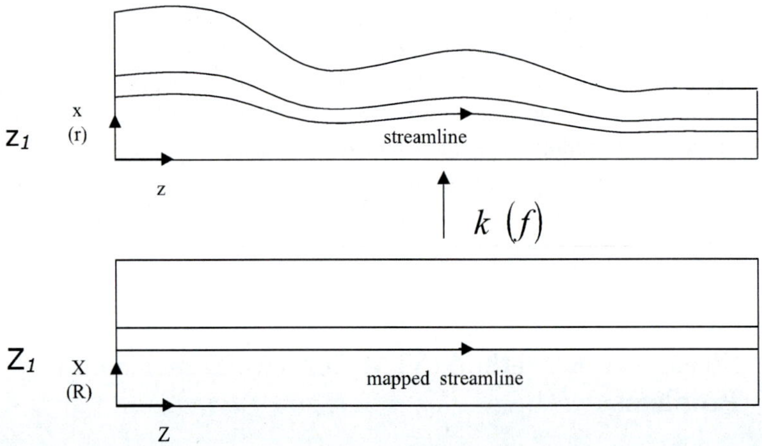

Fig. 3.7 Transformations in domain Ω^* related to original streamlines of the physical domain Ω. Upstream sections z_1 and Z_1 in the physical and mapped domains

$$\begin{aligned} x &= k(\mathrm{X}, \mathrm{Z}) \\ z &= Z \end{aligned} \quad \text{planar case,} \tag{3.10}$$

$$\begin{aligned} \mathrm{r} &= f(\mathrm{R}, \mathrm{Z}) \\ z &= Z \end{aligned} \quad \text{axisymmetric case.} \tag{3.11}$$

In these relations, we use Cartesian (x, y, z) or cylindrical (R, Θ, Z) coordinates, for the planar and axisymmetric domains, respectively.

The transformation function of Eqs. (3.10–3.11), expressed by k (planar case) or f (axisymmetric case), relates domains Ω^* and Ω. As shown in Fig. 3.7, the mapped domain is a rectangle where the streamlines are parallel and straight (no eddies). Domains Ω and Ω^* are limited by upstream and downstream sections of abscissae z_p and z_2, Z_p and Z_2, respectively.

In planar and axisymmetric cases, the quantities k'_X and f'_Rdenote the Jacobian Δ of the transformation $\mathcal{T} : \Omega^* \to \Omega$, defined by

$$\Delta = \left| \frac{\partial(x, y, z)}{\partial(X, Y, Z)} \right| \quad \text{or} \quad \Delta = \left| \frac{\partial(r, \theta, z)}{\partial(R, \Theta, Z)} \right|, \tag{3.12}$$

assumed to be non-zero. Comparisons of STM data with literature results (e.g. [23]) lead to specify that

- Flow conditions and geometrical configurations leading to divergence of the solving algorithms: these cases were found to be in agreement with appearance of secondary flow regions in the physical domain Ω.
- The restriction on problems involving only open streamlines has been overcome in several configurations from a STM analysis, investigated in next chapters. They concern domain decomposition and local transformation functions.

3.3.2 Basic Operators

3.3.2.1 Derivative Operators

Using Eqs. (3.10–3.11), the derivative operators related to the physical and mapped domains may be written as follows:

$$\begin{cases} \dfrac{\partial}{\partial x} = \dfrac{1}{k'_X} \cdot \dfrac{\partial}{\partial X} \\ \dfrac{\partial}{\partial z} = -\dfrac{k'_Z}{k'_X} \cdot \dfrac{\partial}{\partial X} + \dfrac{\partial}{\partial Z} \end{cases} \tag{3.13}$$

in the planar case and

$$\begin{cases} \dfrac{\partial}{\partial r} = \dfrac{1}{f'_R} \cdot \dfrac{\partial}{\partial R} \\ \dfrac{\partial}{\partial z} = -\dfrac{f'_Z}{f'_R} \cdot \dfrac{\partial}{\partial R} + \dfrac{\partial}{\partial Z} \end{cases} \tag{3.14}$$

for axisymmetric situations.

The stream function ψ related to planar or axisymmetric cases is constant on a streamline. We may write the following relations for the velocities:

$$\begin{cases} w = -\dfrac{\partial \Psi}{\partial \mathrm{x}} \\ u = \dfrac{\partial \Psi}{\partial \mathrm{z}} \end{cases} \tag{3.15}$$

in the planar case and

$$\begin{cases} u = \dfrac{1}{r}\dfrac{\partial \Psi}{\partial z} \\ w = -\dfrac{1}{r}\dfrac{\partial \Psi}{\partial r} \end{cases} \tag{3.16}$$

in axisymmetric conditions. When assuming a known velocity component w_p at the upstream section z_p, one may deduce the velocity components in the medium, in terms of w_p and the mapping functions k or f, respectively, for the planar and axisymmetric configurations:

$$\begin{cases} \mathrm{w} = \dfrac{\mathrm{w_p}}{k'_\mathrm{X}} \\ \mathrm{u} = k'_\mathrm{Z}.\mathrm{w} \end{cases} \tag{3.17}$$

and

$$\begin{cases} \mathrm{u} = \dfrac{\mathrm{w_p} R}{f \cdot f'_R} \quad (for \quad R \neq 0) \\ \mathrm{w} = \dfrac{\mathrm{w_p}}{(f'_\mathrm{R})^2} \quad (for \quad R \neq 0) \end{cases} \tag{3.18}$$

according to the relations:

$$u = k'_Z.w \text{ (planar case) } \quad \text{or} \quad u = f'_Z.w \text{ (axisymmetric case).} \tag{3.19}$$

When assuming the incompressibility property of the medium and using Eqs. (3.15) and (3.16), it may be easily shown that the mass conservation equation expressed by

$$\nabla.\underline{V} = 0 \tag{3.20}$$

is automatically verified by the formulation, for incompressible materials.

3.3.2.2 Time Evolution of Particles

It may be of interest to consider time evolution of particles in a deformable medium, notably for fluids with memory (Fig. 3.8).

It may be of interest to consider time evolution of particles in a deformable medium, notably for fluids with memory. In the STM approach, the time t related to a material point occupying the position M_0 at time t_0 depends only on the variable Z, in the mapped domain. Thus, we may write

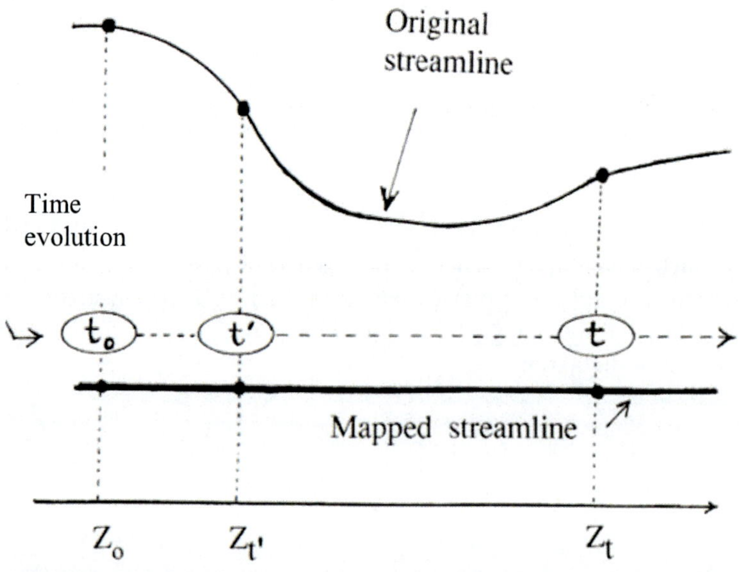

Fig. 3.8 Streamline of the two-dimensional physical domain and its corresponding mapped streamline for a two-dimensional problem. Reprinted by permission from Springer Nature Customer Service Centre GmbH from [32], Copyright 1993

$$t(Z) - t_0\,(Z_0) = \frac{1}{w_p(X)} \int\limits_{Z_0}^{Z} k'_X(X, Y, \xi)\, \mathrm{d}\xi \qquad (3.21)$$

in the planar case (mapping function k, Cartesian coordinates) and in axisymmetric conditions. The determination of time evolution is thus reduced to a one-dimensional problem.

3.3.3 *Natural and Reciprocal Bases with Curvilinear Coordinates*

3.3.3.1 The Planar Case

When using curvilinear coordinates *(X, Y, Z)*, the natural basis vectors $\underline{E}_X, \underline{E}_Y, \underline{E}_Z$ related to a point M of the medium are defined by the following relations:

$$\underline{E}_X = \frac{\partial \underline{M}}{\partial X}, \quad \underline{E}_Y = \frac{\partial \underline{M}}{\partial Y}, \quad \underline{E}_Z = \frac{\partial \underline{M}}{\partial Z}. \qquad (3.22)$$

With curvilinear coordinates (*X,Y,Z*), the natural basis vectors corresponding to a position M of the domain are expressed by

$$\underline{E}_{\mathrm{X}} = k'_{\mathrm{X}}\underline{\varepsilon}_{\mathrm{x}}, \quad \underline{E}_{\mathrm{Y}} = \underline{\varepsilon}_{\mathrm{y}}, \quad \underline{E}_{\mathrm{Z}} = k'_{\mathrm{Z}}\underline{\varepsilon}_{\mathrm{x}} + \underline{\varepsilon}_{\mathrm{z}}. \qquad (3.23)$$

According to the special case of such flow conditions in Cartesian coordinates, one may restrict the calculations and results to cases of two-dimensional vectors and matrices, using the Cartesian basic vectors ε_{x}, ε_{z} and the natural basis vectors $\underline{\mathrm{E}}_{\mathrm{X}}$ and $\underline{\mathrm{E}}_{\mathrm{Z}}$.

Thus, the transformation matrix $\mathbf{A}$ corresponding to Eq. (3.10) at time t is expressed as

$$\mathbf{A} = \begin{bmatrix} k'_X(t) & 0 \\ \\ k'_Z(t) & 1 \end{bmatrix}. \qquad (3.24)$$

To determine the vectors of the reciprocal basis $\underline{E}^i$, one uses the following relation:

$$\underline{E}^{\mathrm{i}} = \frac{\text{matrix of cofactors of } \left[A_{ij}\right]}{\left\| A_{ij} \right\|} \underline{E}_{\mathrm{j}}, \qquad (3.25)$$

where

- $\underline{E}_{\rm j}$ denotes the vectors of the natural basis;
- $\boldsymbol{A} = [A_{ij}]$, the transformed matrix between the natural basis and the Cartesian basis given by Eq. (3.24).

Thus, we get the following relations for the reciprocal basis vectors:

$$\underline{\rm E}_X = \frac{1}{{\rm k}'_{\rm X}}\varepsilon_{\rm X} - \frac{{\rm k}'_{\rm Z}}{{\rm k}'_{\rm X}}\varepsilon_{\rm Z}; \quad \underline{\rm E}_Z = \varepsilon_{\rm Z}. \tag{3.26}$$

The transfer matrix **B** from the reciprocal to the Cartesian basis is expressed as follows:

$$\mathbf{B} = \begin{bmatrix} \frac{1}{k'_X} & -\frac{k'_Z}{k'_X} \\ 0 & 1 \end{bmatrix}, \tag{3.27}$$

where $[B_{ij}]$ verifies the following relation:

$$\underline{\varepsilon}_i = B^j_{\ i}\underline{E}_j. \tag{3.28}$$

The metric tensors $\underline{\underline{\rm G}}_{\rm ij} = \underline{E}_i \bullet \underline{E}_j$ and $\underline{\underline{\rm G}}*^{ij} = \underline{E}^i \bullet \underline{E}^j$ associated to the natural and reciprocal basis $(\underline{\rm E}_{\rm i})$ and $(\underline{\rm E}^j)$, respectively, are written according to the following relations (where the symbol «•» denotes the scalar product):

$$\mathbf{G} = \begin{bmatrix} (k'_X)^2 & k'_X k'_Z \\ k'_X k'_Z & 1 + (k'_Z)^2 \end{bmatrix} \tag{3.29}$$

and

$$\mathbf{G}^* = \begin{bmatrix} \frac{1+(k'_Z)^2}{(k'_X)^2} & \frac{-k'_Z}{k'_X} \\ \frac{-k'_Z}{k'_R} & 1 \end{bmatrix}. \tag{3.30}$$

These expressions are related to the reference time *t*.

3.3.3.2 The Axisymmetric Case

When using the curvilinear coordinates (R, Θ, Z), the natural basis vectors $\underline{E}_R, \underline{E}_\Theta, \underline{E}_Z$ related to a point M of the medium, defined by the following relations:

$$\underline{E}_R = \frac{\partial \underline{M}}{\partial R}, \quad \underline{E}_\Theta = \frac{\partial \underline{M}}{\partial \Theta}, \quad \underline{E}_Z = \frac{\partial \underline{M}}{\partial Z}, \tag{3.31}$$

are expressed by

$$\underline{E}_R = f'_R \underline{e}_r, \quad \underline{E}_\Theta = f \underline{e}_\theta, \quad \underline{E}_Z = f'_Z \underline{e}_r + \underline{e}_z. \tag{3.32}$$

Similar to the planar case, the vectors of the reciprocal basis are obtained by the following relation:

$$\underline{E}^i = \frac{\text{matrix of cofactors of } [A_{ij}]}{\|A_{ij}\|} \underline{e}_j, \tag{3.33}$$

where

$\underline{e}_j$ denotes the vectors of the natural basis;

$[A_{ij}]$ is the transformed matrix between the natural basis and the cylindrical basis such that

$$\underline{E}_i = A_{ij} \underline{e}_j, \tag{3.34}$$

involving the 3 × 3 matrix:

$$\mathbf{A} = \begin{bmatrix} f'_R & 0 & 0 \\ O & f & 0 \\ f'_Z & 0 & 1 \end{bmatrix}. \tag{3.35}$$

Then, we get

$$\begin{aligned} \underline{E}^R &= \frac{1}{f'_R} \underline{e}_r - \frac{f'_Z}{f'_R} \underline{e}_z \\ \underline{E}^\Theta &= \frac{1}{f} \underline{e}_\theta \\ \underline{E}^Z &= \underline{e}_z. \end{aligned} \tag{3.36}$$

The metric tensors $\underline{\underline{G}}_{ij} = \underline{E}_i \bullet \underline{E}_j$ and $\underline{\underline{G}}^{*ij} = \underline{E}^i \bullet \underline{E}^j$ associated to the natural and reciprocal bases $(\underline{E}_i)$ and $(\underline{E}^j)$, where the symbol «•» denotes the scalar product, are given by the following 3 × 3 symmetric matrices:

$$\mathbf{G} = \begin{bmatrix} \left(f'_R\right)^2 & 0 & f'_R f'_Z \\ O & (f)^2 & 0 \\ f'_R f'_Z & 0 & 1+\left(f'_Z\right)^2 \end{bmatrix} \tag{3.37}$$

and

$$\mathbf{G}^* = \begin{bmatrix} \frac{1+\left(f'_Z\right)^2}{\left(f'_R\right)^2} & 0 & \frac{-f'_Z}{f'_R} \\ 0 & \frac{1}{(f)^2} & 0 \\ \frac{-f'_Z}{f'_R} & 0 & 1 \end{bmatrix}. \tag{3.38}$$

3.3.4 *Deformation Gradient Tensor*

The deformation gradient tensor sets up the correspondence between a deformed configuration at time τ and the present configuration at time t defined as reference.

3.3.4.1 The Planar Case

The deformation gradient tensor and its inverse are defined by the following relations:

$$\underline{\underline{\mathrm{F}}}_t(\tau) = \left[\frac{\partial \mathrm{x}_\tau^{\mathrm{i}}}{\partial \mathrm{x}_{\mathrm{t}}^{\mathrm{j}}}\right] \varepsilon_{\mathrm{i}}(\tau) \otimes \varepsilon_{\mathrm{j}}(\mathrm{t}) \tag{3.39}$$

and

$$\underline{\underline{\mathrm{F}}}_t^{-1}(\tau) = \left[\frac{\partial \mathrm{x}_{\mathrm{t}}^{\mathrm{i}}}{\partial \mathrm{x}_\tau^{\mathrm{j}}}\right] \varepsilon_{\mathrm{i}}(\mathrm{t}) \otimes \varepsilon_{\mathrm{j}}(\tau), \tag{3.40}$$

where τ denotes a time of the past history, t the present time and « ⊗» the tensorial product.

In Cartesian coordinates, the velocity gradient tensor $\underline{\underline{\mathrm{F}}}_t(\tau)$ and its inverse $\underline{\underline{\mathrm{F}}}_t^{-1}(\tau)$ are given by the following matrices:

$$\mathbf{F} = \begin{bmatrix} \frac{\partial X(\tau)}{\partial X(t)} & \frac{\partial X(\tau)}{\partial Z(t)} \\ \frac{\partial Z(\tau)}{\partial X(t)} & \frac{\partial Z(\tau)}{\partial Z(t)} \end{bmatrix} \tag{3.41}$$

and

$$\mathbf{F}^{-1}\begin{bmatrix} \frac{\partial X(t)}{\partial R(\tau)} & \frac{\partial X(t)}{\partial Z(\tau)} \\ \frac{\partial Z(t)}{\partial X(\tau)} & \frac{\partial Z(t)}{\partial Z(\tau)} \end{bmatrix}. \tag{3.42}$$

In the STM approach, one obtains the following expressions:

- for the matrix **F**:

$$\begin{aligned} &\frac{\partial X(\tau)}{\partial X(t)} = 1; \quad \frac{\partial X(\tau)}{\partial Z(t)} = 0 \\ &\frac{\partial Z(\tau)}{\partial X(t)} = -w(X(\tau), Z(\tau)) \int_{Z(\tau)}^{Z(t)} \frac{\partial w(X(\xi), \xi)}{\partial X(\xi)} \frac{d\xi}{[w(X(\xi), \xi)]^2} \\ &\frac{\partial Z(\tau)}{\partial Z(t)} = \frac{w(X(\tau), Z(\tau))}{w(X(t), Z(t))} \end{aligned} \tag{3.43}$$

- for the matrix $\mathbf{F}^{-1}$:

$$\begin{aligned} &\frac{\partial X(t)}{\partial X(\tau)} = 1; \quad \frac{\partial X(t)}{\partial Z(\tau)} = 0 \\ &\frac{\partial Z(t)}{\partial X(\tau)} = w(X(t), Z(t)) \int_{Z(\tau)}^{Z(t)} \frac{\partial w(X(\xi), \xi)}{\partial X(\xi)} \frac{d\xi}{[w(X(\xi), \xi)]^2} \\ &\frac{\partial Z(t)}{\partial Z(\tau)} = \frac{w(X(t), Z(t))}{w(X(\tau), Z(\tau))}, \end{aligned} \tag{3.44}$$

- where $w(X(t), Z(t)), w(X(\tau), Z(\tau))$ denote, respectively, the velocity in the main flow direction $\underline{E}_Z$ at the reference time t and at another time $\tau \leq t$.

3.3.4.2 The Axisymmetric Case

In the natural basis referred to variables $(R(t), \Theta(t), Z(t))$, the deformation gradient tensor and its inverse are expressed by the following matrix forms:

$$\mathbf{F} = \begin{bmatrix} \frac{\partial R(\tau)}{\partial R(t)} & \frac{\partial R(\tau)}{\partial \Theta(t)} & \frac{\partial R(\tau)}{\partial Z(t)} \\ \frac{\partial \Theta(\tau)}{\partial R(t)} & \frac{\partial \Theta(\tau)}{\partial \Theta(t)} & \frac{\partial \Theta(\tau)}{\partial Z(t)} \\ \frac{\partial Z(\tau)}{\partial R(t)} & \frac{\partial Z(\tau)}{\partial \Theta(t)} & \frac{\partial Z(\tau)}{\partial Z(t)} \end{bmatrix} \tag{3.45}$$

$$\mathbf{F}^{-1} = \begin{bmatrix} \frac{\partial R(t)}{\partial R(\tau)} & \frac{\partial R(t)}{\partial \Theta(\tau)} & \frac{\partial R(t)}{\partial Z(\tau)} \\ \frac{\partial \Theta(t)}{\partial R(\tau)} & \frac{\partial \Theta(t)}{\partial \Theta(t)} & \frac{\partial \Theta(t)}{\partial Z(t)} \\ \frac{\partial Z(t)}{\partial R(\tau)} & \frac{\partial Z(t)}{\partial \Theta(\tau)} & \frac{\partial Z(t)}{\partial Z(\tau)} \end{bmatrix}. \tag{3.46}$$

3.4 Velocity Gradient, Rate-of-Deformation and Vorticity Tensors in Two-Dimensional Cases

In the mapped domain of STM, the coordinates *R*, Θ and *X*, *Y* are independent of time.

3.4.1 The Planar Case

When considering a velocity vector $\underline{V}$ of components (*u*, *v*, *w*) in a Cartesian basis, the velocity gradient tensor $\underline{\nabla}\,\underline{\mathrm{V}}$ is expressed in this basis, by the following matrix, according to Eq. (1.54):

$$\nabla \mathbf{V} = \begin{bmatrix} \frac{\partial u}{\partial x} & \frac{\partial u}{\partial y} & \frac{\partial u}{\partial z} \\ \frac{\partial v}{\partial x} & \frac{\partial v}{\partial y} & \frac{\partial v}{\partial z} \\ \frac{\partial w}{\partial x} & \frac{\partial w}{\partial y} & \frac{\partial w}{\partial z} \end{bmatrix}. \tag{3.47}$$

In the planar case, the rate-of-deformation tensor $\underline{\underline{D}}$ is the symmetric part of the velocity gradient tensor and is given by following global matrix in the Cartesian basis (in three dimensions):

$$\mathbf{D} = \begin{bmatrix} \frac{\partial u}{\partial x} & 0 & (1/2)\left(\frac{\partial w}{\partial x} + \frac{\partial u}{\partial z}\right) \\ 0 & 0 & 0 \\ (1/2)\left(\frac{\partial w}{\partial x} + \frac{\partial u}{\partial z}\right) & 0 & \frac{\partial w}{\partial z} \end{bmatrix}. \tag{3.48}$$

The vorticity tensor $\underline{\underline{\Omega}}$ is expressed by the antisymmetric part of the velocity gradient matrix such that (in three dimensions):

$$\mathbf{R} = \begin{bmatrix} 0 & 0 & (1/2)\left(\frac{\partial u}{\partial z} - \frac{\partial w}{\partial x}\right) \\ 0 & 0 & 0 \\ -(1/2)\left(\frac{\partial u}{\partial z} - \frac{\partial w}{\partial x}\right) & 0 & 0 \end{bmatrix}. \tag{3.49}$$

To this matrix corresponds the rotation velocity vector $\underline{R}^V$ expressed by

$$\underline{R}^V = -(1/2)\left(\frac{\partial \mathrm{u}}{\partial \mathrm{z}} - \frac{\partial \mathrm{w}}{\partial \mathrm{x}}\right)\underline{\varepsilon}_2. \tag{3.50}$$

This relation indicates that the rotation velocity vector is normal to the (*z*,*x*) plane.

3.4.2 The Axisymmetric Case

In axisymmetric conditions, the basis $(\underline{e}_r, \underline{e}_\theta, \underline{e}_z)$ is associated to cylindrical coordinates (r, θ, z). With this basis, one may write the rate-of-deformation matrix **D** as follows:

$$\mathbf{D} = \begin{bmatrix} \frac{\partial u}{\partial r} & 0 & (1/2)\left(\frac{\partial w}{\partial r} + \frac{\partial u}{\partial z}\right) \\ 0 & \frac{u}{r} & 0 \\ (1/2)\left(\frac{\partial w}{\partial r} + \frac{\partial u}{\partial z}\right) & 0 & \frac{\partial w}{\partial z} \end{bmatrix}. \tag{3.51}$$

Here, the vorticity tensor $\underline{\underline{\Omega}}^c$ is expressed, in the basis $(\underline{e}_r, \underline{e}_\theta, \underline{e}_z)$, by the following matrix:

$$\boldsymbol{\Omega}^c = \begin{bmatrix} 0 & 0 & (1/2)\left(\frac{\partial u}{\partial z} - \frac{\partial w}{\partial r}\right) \\ 0 & 0 & 0 \\ -(1/2)\left(\frac{\partial u}{\partial z} - \frac{\partial w}{\partial r}\right) & 0 & 0 \end{bmatrix}. \tag{3.52}$$

The corresponding rotation velocity vector $\underline{R}^c$ is written as follows:

$$\underline{\mathrm{R}}^{\mathrm{c}} = -(1/2)\left(\frac{\partial \mathrm{u}}{\partial \mathrm{z}} - \frac{\partial \mathrm{w}}{\partial \mathrm{x}}\right)\underline{\varepsilon}_2. \tag{3.53}$$

According to this relation, the rotation velocity vector is normal to the (r,z) meridian plane.

3.4.3 Velocity Derivatives Versus the Mapping Functions

The following expressions, given by Eqs. (2.49)–(2.54) in terms of the mapping functions f or k and their derivatives, provide relations involving the upstream velocity component w_p:

- In the planar case,

$$\frac{\partial w}{\partial X} = \frac{w'_{pX}}{k'_X} - \frac{w_p \cdot k''_{XX}}{(k'_X)^2}, \quad \frac{\partial w}{\partial Z} = -\frac{w_p \cdot k''_{XZ}}{(k'_X)^2}$$

and

$$\frac{\partial u}{\partial X} = \frac{\partial w}{\partial X} \cdot k'_Z + w \cdot k''_{XZ}, \quad \frac{\partial u}{\partial Z} = \frac{\partial w}{\partial Z} \cdot k'_Z + w \cdot k''_{ZZ}. \tag{3.54}$$

- In the axisymmetric case,

$$\frac{\partial w}{\partial Z} = -\frac{w_p \cdot R \cdot (f'_R \cdot f'_Z + f \cdot f''_{RZ})}{(f \cdot f'_R)^2} \tag{3.55}$$

such that

$$\frac{\partial u}{\partial R} = \frac{\partial w}{\partial R} \cdot f'_Z + w \cdot f''_{RZ}, \tag{3.56}$$

$$\frac{\partial u}{\partial Z} = \frac{\partial w}{\partial Z} \cdot f'_Z + w \cdot f''_{Z^2}. \tag{3.57}$$

In axisymmetric conditions, the following relations should be considered on the centreline:

$$\frac{\partial w}{\partial R} = 0, \quad \frac{\partial w}{\partial Z} = -\frac{w_p \cdot f''_{RZ}}{(f'_R)^3}. \tag{3.58}$$

3.4.4 Momentum Conservation Equations in 2D Isothermal Cases

The dynamic conservation equations are now considered in planar and axisymmetric conditions. By using the derivative operators defined previously Eqs. (3.13)–(3.14), the momentum conservation equations can be written, with velocity components u and w, τ^{ij} the tensor components, ρ the mass density of the material and p the isotropic pressure.

With the mapping function k in the planar case, one writes

$$\begin{cases} \rho(k'_X \cdot \frac{\partial u}{\partial t} + u \cdot \frac{\partial u}{\partial X} + w \cdot k'_X \cdot \frac{\partial u}{\partial Z} - w \cdot k'_Z \cdot \frac{\partial u}{\partial X}) = \frac{\partial \tau 11}{\partial X} + k'_X \cdot \frac{\partial \tau 13}{\partial Z} - k'_Z \cdot \frac{\partial \tau 13}{\partial X} - \frac{\partial P}{\partial X} \\ \rho(k'_X \cdot \frac{\partial w}{\partial t} + u \cdot \frac{\partial w}{\partial X} + w \cdot k'_X \cdot \frac{\partial w}{\partial Z} - w \cdot k'_Z \cdot \frac{\partial w}{\partial X}) = \frac{\partial \tau^{13}}{\partial X} + k'_X \cdot \frac{\partial \tau 33}{\partial Z} - k'_Z \cdot \frac{\partial \tau 33}{\partial X} - k'_X \cdot \frac{\partial p}{\partial Z} + k'_Z \cdot \frac{\partial p}{\partial X} \end{cases}, \tag{3.59}$$

where the indices 1 and 3 of the stress tensor correspond to variables x and z, respectively.

In axisymmetric conditions, the following equations are obtained:

$$\begin{cases} \rho\left(f'_R \cdot \dfrac{\partial u}{\partial t} + u \cdot \dfrac{\partial u}{\partial R} + w \cdot f'_R \cdot \dfrac{\partial u}{\partial Z} - w \cdot f'_Z \cdot \dfrac{\partial u}{\partial R}\right) \\ \quad = \dfrac{\partial T^{11}}{\partial R} + f'_R \cdot \dfrac{\partial T^{13}}{\partial Z} - f'_Z \cdot \dfrac{\partial T^{13}}{\partial R} + \dfrac{f'_R}{f} \cdot T_{11} - \dfrac{\partial p}{\partial R} \\ \rho\left(f'_R \cdot \dfrac{\partial w}{\partial t} + u \cdot \dfrac{\partial w}{\partial R} + w \cdot f'_R \cdot \dfrac{\partial w}{\partial Z} - w \cdot f'_Z \cdot \dfrac{\partial w}{\partial R}\right) \\ \quad = \dfrac{\partial T^{13}}{\partial R} + f'_R \cdot \dfrac{\partial T^{33}}{\partial Z} - f'_Z \cdot \dfrac{\partial T^{33}}{\partial R} + \dfrac{f'_R}{f} \cdot T^{13} - f'_R \cdot \dfrac{\partial p}{\partial Z} + f'_Z \cdot \dfrac{\partial p}{\partial R}. \end{cases} \tag{3.60}$$

Here, the indices 1 and 3 are related to R and z, respectively.

The mapping function k (or f) and the pressure p are the two primary unknowns of the problem. For an incompressible material, the mass conservation is automatically verified and only the equilibrium equations should be written in the isothermal case, using a summarized form.

Find (k or f, p) such that

$$m = 1, 2 \quad \mathrm{E_m}(k, p) = 0 \quad \text{or} \quad E_m(f, p) = 0. \tag{3.61}$$

In Eq. (3.61), the stress components are expressed in terms of the mapping function and its partial derivatives. The writing of these relations corresponds to differential equations with the primary variables (k,p) (or f,p) called "primary formulation". Otherwise, the stress components provide additional equations in terms of the mapping function corresponding to a "mixed formulation" as

$$\mathrm{T^{ij}} = \Gamma^{\mathrm{ij}}(k, k'_{\mathrm{X}}, k'_{\mathrm{Z}}, k''_{\mathrm{X}^2}, k''_{\mathrm{Z}^2}, k''_{\mathrm{XZ}}, \ominus) = 0 \tag{3.62}$$

or

$$T^{ij} = \Gamma^{ij}(f, f'_R, f_Z, f''_{\mathrm{R}^2}, f'_{\mathrm{Z}^2}, f'_{\mathrm{RZ}}, \ominus) = 0 \tag{3.63}$$

in planar and axisymmetric cases, respectively.

Equations (3.61) to (3.63) are non-linear in terms of mapping functions and should be solved numerically, even in the Newtonian case or for another linear constitutive model. However, it should be pointed out that

- Solving the equations in a simple computational geometry Ω^* leads to consider simple discretization schemes for the unknowns.
- Different numerical experiments with the STM approach do not reveal significant problems arising in such non-linear cases, notably when a complex constitutive model is retained.
- When the physical domain Ω is assumed to be simply or doubly connected (Fig. 3.9) with regard to the streamlines, STM still allows possibility to compute locally the flow by considering only the so-called "peripheral stream tube" [23]. From this property, it is then possible to determine successively stream tubes,

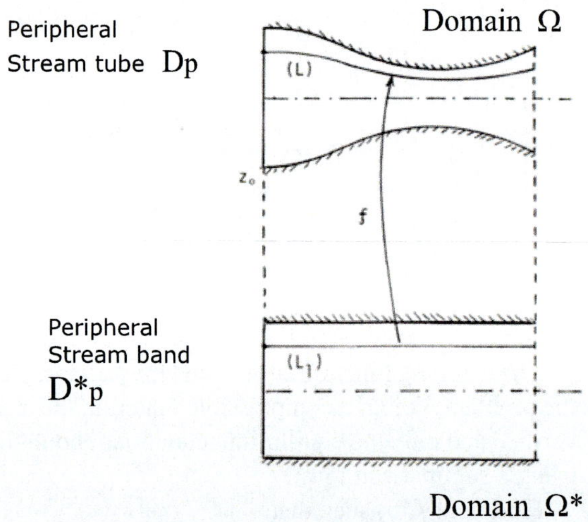

Fig. 3.9 Simply connected domain: peripheral stream tube D_p in the physical domain Ω and its transformed stream band D_p^* in domain Ω

streamlines and flow characteristics from the boundary zone to the centreline of the domain. The case of a two-dimensional doubly connected flow domain is investigated in Subsect. 3.3.4.2.

3.4.5 *Specific Features in Stream-Tube Method*

Solving the governing equations generally requires the writing of boundary conditions. In this section, some particular properties of the STM formulation are considered, leading to write additional boundary conditions equations. The assumption of a fluid of constant density is made, without loss of generality for the applications.

3.4.5.1 Momentum Conservation for Fluid Dynamics Problems

For a material of constant density m_0, the momentum conservation Eq. (2.45) related to a volume ω limited by a surface $\partial\omega$, under a body force vector $\underline{\mathrm{B}}$, may be written in the following form:

$$\int\limits_{\omega} m_0 \frac{D}{Dt}\underline{V}\,\mathrm{d}v = \int\limits_{\partial\omega} \underline{\underline{\sigma}}\,\underline{n}\,\mathrm{d}a + \int\limits_{\omega} m_0 \underline{\mathrm{B}}\,\mathrm{d}v, \tag{3.64}$$

where $\underline{n}$ denotes the normal unit vector to the limiting surface $\partial\omega$. When considering a volumetric vector $\underline{\mathrm{B}}$, the coefficient m_0 can be removed from the equation.

In steady-state conditions, Eq. (3.64) becomes

$$\int_{\omega} \mathrm{m}_0 \underline{V} . \underline{\underline{\nabla}}\, \underline{V} \, \mathrm{d}\nu = \int_{\partial\omega} \underline{\underline{\sigma}}\, \underline{n} \, \mathrm{d}a + \int_{\omega} \mathrm{m}_0 \, \underline{\mathrm{B}} \, \mathrm{d}\nu. \tag{3.65}$$

This momentum conservation equation is particularly well adapted to STM analysis. In this method, where stream tubes are considered, the control surfaces are thus defined from the limiting surfaces of these volumes or by their sections. Using the stream-tube method, the solving technique will concern development of such approaches related to the integral Eqs. (3.65) and (3.66), for geometries encountered in practical problems.

3.4.5.2 Peripheral Stream Tube: Equations for Flows with Open Streamlines—Simply Connected Domain

It was previously underlined that the mass conservation equation is automatically verified for incompressible materials in the STM approach. In isothermal cases, only the dynamic equations should be written.

The solution can be computed by considering successively the sub-domains of the total domain. A sub-domain may be specifically studied provided the action of the complementary domain is taken into account. Accordingly, the peripheral stream tube $\mathrm{D_p}$, involving the wall of a domain Ω, which requires specifications previously pointed out, can be the only domain to be computed with the dynamic equations. The action of the complementary domain (Fig. 3.9) should be considered. The sub-domain $\mathrm{D_{p-1}}$, adjacent to the stream tube $\mathrm{D_p}$, may then be only studied in the computations by involving the results from the peripheral tube $\mathrm{D_p}$ (Fig. 3.9), and so on for the other sub-domains of the total physical and mapped domains.

Concerning the peripheral stream tube, the dynamic equations are written in the computational stream band D_p^* of Ω^*. The (inside) unknown boundary is then determined by writing the action of the inside complementary domain according to the following integral equation:

$$\left[\int_{\partial \mathrm{D}} \underline{\underline{\tau}} \bullet \underline{\mathrm{n}} \, \mathrm{ds} \right] \bullet \underline{\mathrm{E}}_{\mathrm{z}} = 0, \tag{3.66}$$

where $\underline{n}$ denotes the outer normal unit vector to the surface of the peripheral stream tube. In the planar or axisymmetric case, the symmetry of the domain leads to write Eq. (3.66) for the Z-axis (Fig. 3.10a).

The (3.6) scalar equation, expressed in terms of the mapping function and variables (X,Z) or (R,Z) of the transformed domain Ω^*, involves non-linearly the unknowns related to the peripheral stream tube $\mathrm{D_p}$. It is a non-linear boundary condition equation. Thus, concerning the peripheral stream tube, the system to be solved is formed by

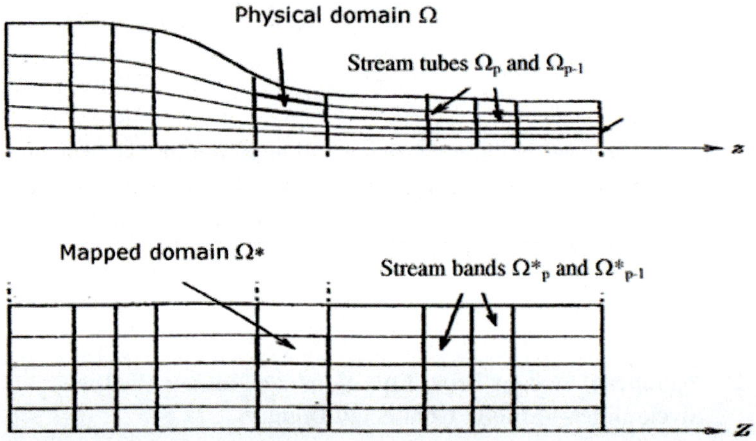

Fig. 3.10 Peripheral stream tube, stream tube D_{p-1} in the physical domain Ω and their transformed stream bands D^*_{p-1} and D^*_{p-1} in domain Ω*

(a) Equations (3.59) and (3.61) for the planar case;
(b) Equations (3.60) and (3.62) for the axisymmetric situation.

Systems of equations to be solved are of two types:

- an over-determined system of equations;
- a close system of Eqs. (3.59)–(3.61) or (3.60)–(3.62) associated to the constraint (3.67). From a theoretical viewpoint, such situation may be considered as an optimization problem under constraint.

Here $\underline{n}$ denotes the outer normal unit vector to the surface of the peripheral stream tube. In the planar or axisymmetric case, the symmetry of the domain leads to only write Eq. (3.66) for the Z-axis (Fig. 3.10a).

The (3.66) scalar equation, expressed in terms of the mapping function and variables (*X*,*Z*) or (*R*,*Z*) of the transformed domain Ω*, involves non-linearly the unknowns related to the peripheral stream-tube D_p. It is a non-linear boundary condition equation. Thus, concerning the peripheral stream tube, the system to be solved is formed by

(d) Equations (3.59) and (3.61) for the planar case;
(e) Equations (3.60) and (3.62) for the axisymmetric situation.

Systems of equations to be solved are of two types:

- an over-determined system of equations;
- a close system of the Eqs. (3.59)–(3.61) or (3.60)–(3.62) associated to the constraint (3.66). From a theoretical viewpoint, such situation may be considered as an optimization problem under constraints.

The approach of flow computations on successive stream tubes cannot be developed for doubly connected geometries as the annular two-dimensional domain shown

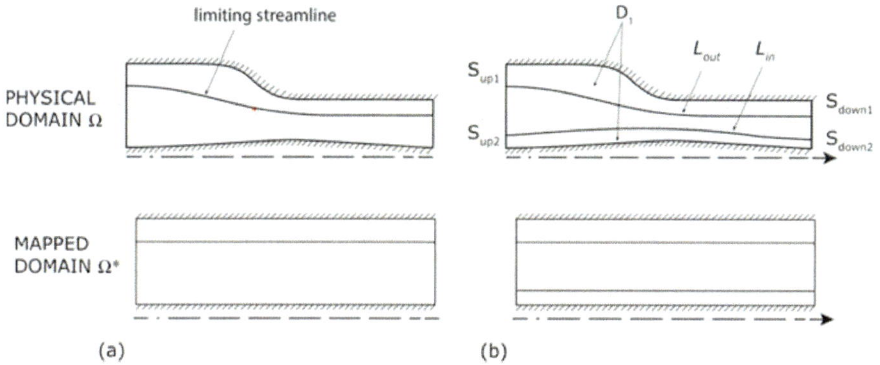

Fig. 3.11 **a** Doubly connected domain Ω; **b** Sub-domain D_1 of Ω limited by inner and outer streamlines L_{in} and L_{out}

in Fig. 3.11, for planar and axisymmetric cases. However, such situation may be considered as depicted in the following sub-section, in axisymmetric conditions.

3.4.5.3 Considerations on an Axisymmetric Peripheral Stream Tube, for a Doubly Connected Domain Ω Limited by Open Streamlines

Figure 3.11a concerns a doubly connected axisymmetric flow domain Ω. As already pointed out, the approach proposed in the previous sub-Sect. 3.3.4.1 cannot be retained for computing in the outer region limited by the wall and a close streamline *L*, owing to the inner limiting surface. Such inner boundary does not allow to consider only equations related to the outer surface. The problem can be solved as follows.

The doubly connected sub-domain D_1 of Ω is limited inward and outward by streamlines L_{in} and L_{out}, respectively. Given a point O of the physical region, the action of the complementary domain of D_1 is expressed by the following torsor $[\underline{R}, \underline{M}]$:

$$\mathrm{R} = \left[\int\limits_{\partial D_1} \underline{\underline{\sigma}} \bullet \underline{n}\,\mathrm{ds}\right] = \left[\int\limits_{\partial D_1} \left(-p\,\underline{\underline{I}} + \underline{\underline{\tau}}\right) \bullet \underline{n}\,\mathrm{ds}\right] = 0 \tag{3.67}$$

and

$$\underline{\mathrm{M}}(\mathrm{M}) = \left[\int\limits_{\partial D_1} \underline{\mathrm{OM}} \times \underline{\underline{\sigma}} \bullet \underline{n}(\mathrm{M})\mathrm{ds}\right] = 0, \tag{3.68}$$

where $\partial \mathsf{D}$ denotes the boundary of the complementary domain of the stream tube D retained for the computations and $\underline{n}$ the unit normal vector to the surface $\partial \mathsf{D}$.

According to the axisymmetry of domain D, Eqs. (3.67)–(3.68) are reduced to the following relation:

$$\left[\int_{\partial D_1} \left(-p\,\underline{\underline{I}} + \underline{\underline{\tau}}\right) \cdot \underline{n}\, ds\right] . \underline{\varepsilon}_z = 0. \tag{3.69}$$

In this equation, $\underline{\varepsilon}_z$ stands for the unit vector in the z-direction.

As seen in Fig. 3.11b, two stream tubes are considered. The outer one is limited transversely by the planar upstream and downstream surfaces S_{up1} and S_{dow1}, by the lateral surfaces generated by the outer wall line and the downstream surfaces S_{up2} and S_{dow2} and by the surfaces generated by the inner wall line and the streamline L_{in}. The complementary domain of both stream tubes concerns the respective complementary upstream and downstream surfaces of $S_{up1} \cup S_{up2}$ and $S_{dow1} \cup S_{dow2}$ in the one hand, and by the surfaces generated by the streamlines L_{in} and L_{out}.

The normal unit vectors at the surface of the complementary domain of D_1 are defined as follows:

- $\underline{n}_{up}$, perpendicular to the upstream section;
- $\underline{n}_{dow}$, perpendicular to the downstream section;
- $\underline{n}_{out}$, perpendicular to the lateral outer surface of the complementary domain;
- $\underline{n}_{in}$, perpendicular to the lateral inner surface of the complementary domain.

At the upstream and downstream sections of the complementary domain, the normal unit vectors are parallel to the z-axis and may be expressed by the following components:

$$n_{up} = (0, 0, -1), \quad n_{dow} = (0, 0, 1). \tag{3.70}$$

More complex expressions are obtained for the unit vectors normal to the non-planar lateral surfaces. Thus, the outer and inner unit normal vectors are given by the following components:

$$n_{out} = \left(\frac{1}{\sqrt{1+\left(f'_Z\right)^2}}, 0, \frac{-f'_Z}{\sqrt{1+\left(f'_Z\right)^2}}\right), \tag{3.71}$$

$$n_{in} = \left(\frac{-1}{\sqrt{1+\left(f'_Z\right)^2}}, 0, \frac{f'_Z}{\sqrt{1+\left(f'_Z\right)^2}}\right).$$

The expression of a surface element ds in curvilinear coordinates is given by

$$ds = \sqrt{1 + \left(f_Z'\right)^2}.f.dZ.d\theta. \tag{3.72}$$

The dynamic equations (3.60) and Eq. (3.69) are then used to compute the unknown mapping function f and the pressure p. Computations referred to sub-domains in STM may be considered with an adaptive solution procedure (e.g. [24]), starting from results on the peripheral stream tube used as boundary conditions.

3.4.5.4 Peripheral Stream Tube: Confined Flows with Recirculations in a Contraction Geometry

1 - *The flow domain*

Figure 3.12a presents an example of abrupt axisymmetric contraction domain Ω of radius r_0 (in the half-plane) where recirculations are encountered. The mapped computational domain Ω^* is shown in Fig. 3.12b.

The sections z_0 and z_1 (Z_0 and Z_1 in the mapped domain Ω^*) are assumed to limit upstream and downstream Poiseuille flows. One assumes that vortices only exist close to the wall: consistency of this assumption is provided in numerous experimental and numerical studies (e.g. Crochet et al. [25], Owens and Phillips [26]) (Fig. 3.12a).

Then, one may write, for a transformed domain Ω^* limited by abscissa r_b such that $r_b < r_0$, the upstream radius of the physical domain Ω, the property:

$$\text{for} \quad M \in \Omega^* : f_R' \neq 0. \tag{3.73}$$

To compute the unknowns, we need to specify boundary condition relations, together with the conservation equations.

As mentioned earlier, we may compute the flow by considering successive stream tubes in Ω (or stream bands in Ω^*). In the following, we shall only restrict ourselves to the *peripheral stream tube* (B) shown in Fig. 3.12a, without loss of generality for computation of unknowns in the other stream bands.

One now considers the boundary limiting surfaces $\partial\Omega$ of the whole physical domain Ω, ∂S and ∂C of sub-domains (S) and (C), which denote the outer and inner complementary domains of the so-called peripheral stream-tube B (to be considered), respectively (Fig. 3.12a). The boundaries $\partial\Omega$, ∂S and ∂C may be decomposed as follows:

$$\begin{aligned} &\partial\Omega = S_0 \cup S_L \cup S_1, \; \partial S = S_0^b \cup S_L \cup S_1^b \cup S_L^*, \\ &\partial C = S_0^c \cup S_L^c \cup L_c \cup S_1^b, \end{aligned} \tag{3.74}$$

In the mapped domain D^*, the peripheral stream band B_p corresponds to the rectangular sub-domain limited by the lines $Z_0 = z_0$, $Z_1 = z_1$ (perpendicular to the z-axis), and the mapped streamlines of origins $R_b = r_b$ and $R^* = r^*$. In such problems,

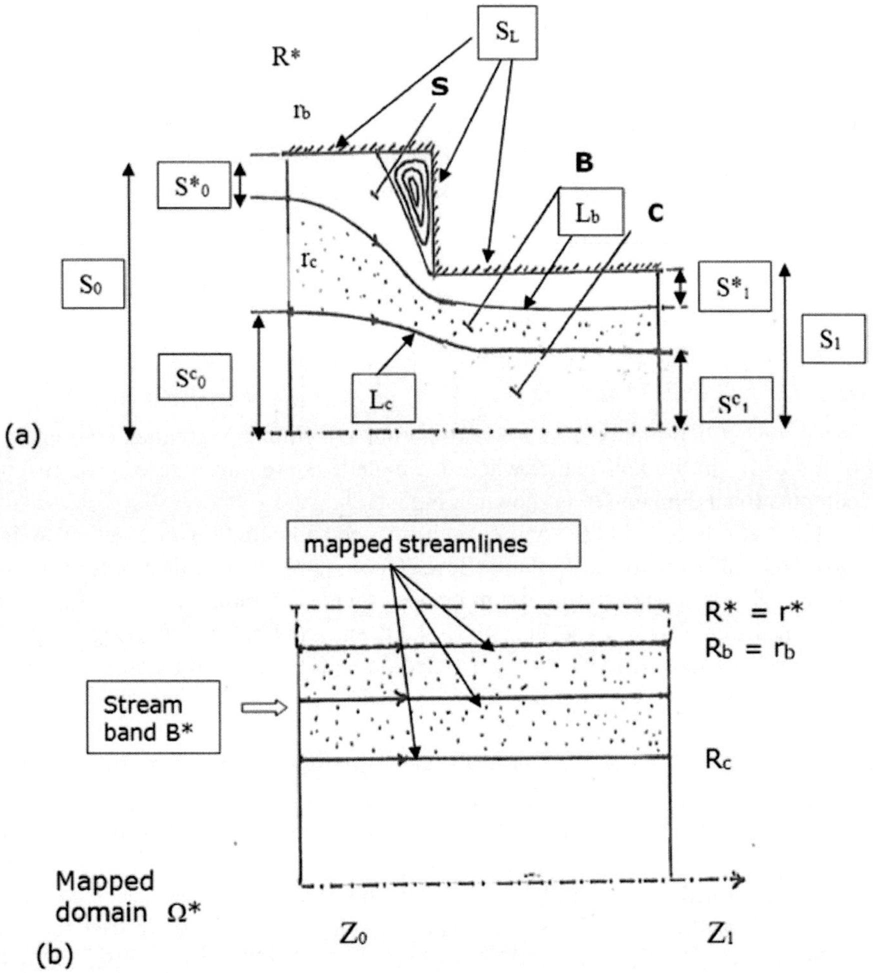

Fig. 3.12 **a** Sub-domains (S), (B) and (C) in domain Ω **b** Domain Ω*

we adopt a mixed formulation where the unknowns are f, p and the components of the stress tensor $\underline{\underline{\tau}}$ related to the constitutive equation.

The boundary condition equations to be involved for domain B are depicted as follows:

(1) Simple equations related to the knowledge of the transformation f at sections Z_0 and Z_1 as well as at the boundary S_L of the total domain D, the pressure p at the upstream section Z_0 and the stress components of the rheological tensor $\underline{\underline{\tau}}$ at sections Z_0 and Z_1. The total stress tensor $\underline{\underline{\sigma}}$ involves the pressure p and the rheological stress tensor according to the following equation:

$$\underline{\underline{\sigma}} = -\mathrm{p}\,\underline{\underline{I}} + \underline{\underline{\tau}}. \tag{3.75}$$

(2) Integral boundary condition equations for the unknown streamlines L^b and L^c.

The case of the equation corresponding to the inner streamline L^c, already considered in previous papers [27–29], is recalled here. The relevant equation represents the action of the complementary domain (C) $^{\cup}$ (S) on the stream tube (B). It is now wanted to write a boundary condition equation for the outer streamline L* corresponding to the action of the complementary domain (S). To this end, one should successively compute the integral momentum equation for domain D (which involves line S_L) and S (involving the line L*).

- Considering the domain D, the integral form of the momentum equation law leads to the following relation:

$$\int_{\partial D} \underline{\underline{\sigma}}(f,p) \bullet \underline{n}(f)\,ds = \int_{S_0} \underline{\underline{\sigma}}(f,p) \bullet \underline{n}(f)\,ds + \int_{S_L} \underline{\underline{\sigma}}(f,p) \bullet \underline{n}(f)\,ds + \int_{S_1} \underline{\underline{\sigma}}(f,p) \bullet \underline{n}(f)\,ds, \tag{3.76}$$

- where $\underline{n}$ denotes the outer normal unit vector to the surface under consideration, in Cartesian coordinates. Owing to the symmetry of the flow, only a scalar equation derived from Eq. (3.76) has to be satisfied, we can write

$$\left[\int_{S_0} \underline{\underline{\sigma}}(f,p) \cdot \underline{n}(f)\,ds + \int_{S_L} \underline{\underline{\sigma}}(f,p) \cdot \underline{n}(f)\,ds + \int_{S_1} \underline{\underline{\sigma}}(f,p) \cdot \underline{n}(f)\,ds\right] \cdot \underline{e}'_z = 0, \tag{3.77}$$

- where $\underline{e}'_z$ denotes the unit vector along the z-axis.
- The terms at the upstream section being known, as also the components of the tensor $\underline{\underline{\tau}}$ at the downstream section z_1, Eq. (3.77) may be written as follows:

$$\left[\int_{S_L} \underline{\underline{\sigma}}(f,p) \bullet \underline{n}(f)\,ds\right] . \underline{e}'_z = G(p_1), \tag{3.78}$$

- where p_1 denotes the unknown pressure value at a point M_1 of the downstream section Z_1. In a similar way, one also uses the boundary condition equation for the line L_b:

$$\left[\int_{S_{L_b}} \underline{\underline{\sigma}}(f,p) \bullet \underline{n}(f)\,ds\right] . \underline{e}'_z = G^*(p_1), \tag{3.79}$$

- which expresses that the left-hand side of this equation may be considered as a function of the unknown downstream pressure value p_1. Determination of function G* is possible by writing the integral form of the momentum equation applied to the closed surface ∂S.
- Therefore, taking into account Eq. (3.78), it can be written as

$$\left[\int_{S_{L_b}} \underline{\underline{\sigma}}(f,p) \bullet \underline{n}(f)\,ds\right].\underline{e}'_z = G^*(p_1)$$

$$= -\left[\int_{S_0^b} \underline{\underline{\sigma}}(f,p) \bullet \underline{n}(f)\,ds + \int_{S_1^b} \underline{\underline{\sigma}}(f,p) \bullet \underline{n}(f)\,ds\right].\underline{e}'_z - G(p_1). \quad (3.80)$$

- According to these results, a similar numerical procedure to that adopted for non-circulatory flows can be proposed for determining the unknowns of the problem presented in Fig. 3.12 (contraction flows with eddies).

3.5 Stream-Tube Method and Constitutive Equations

Different types of constitutive equations considered in scientific and engineering problems can be adopted in STM situations, notably viscoelastic models of the integral type. In this section, besides Newtonian and inelastic constitutive equations, one reconsiders some theoretical models requiring particular approaches for the numerical computations, notably those requiring particle-tracking approaches. For fluid materials, the models are concerned by two types of kinematic tensors:

- the rate-of-deformation tensor, notably for many inelastic constitutive equations as those for the purely viscous fluids and materials obeying viscoelastic corotational equations;
- strain tensors as, for example, the Cauchy tensor, for viscoelastic codeformational models.

3.5.1 *Newtonian and Inelastic Rheological Models*

The Newtonian fluid is defined by Eq. (3.50): $\underline{\underline{\tau}} = 2\eta\underline{\underline{D}}$, where $\underline{\underline{D}}$ denotes the rate-of-deformation tensor. Purely viscous shear-thinning fluids are often concerned by the constitutive Eq. (2.54): $\underline{\underline{\tau}} = 2\eta\left(\underline{\underline{D}}\right)\underline{\underline{D}}$. Expressions of the matrix of $\underline{\underline{D}}$ are given by Eqs. (3.49) and (3.52) in planar and axisymmetric cases. The elements of the corresponding matrices are then expressed in terms of the mapping functions

k or f and their derivatives matrices. The dynamic equations (3.60) or (3.61) are written together with the constitutive equations, leading to solve a non-linear system involving the mapping function and its derivatives.

3.5.2 *Differential Models*

The writing of such equations with STM does not imply particular difficulties, as considered in a two-dimensional swell problem with the Oldroyd-B fluid (e.g. [30], swell problem). The differential operators of the constitutive are treated in the same way than the approaches provided in the literature (upwind scheme [25, 26]), where the unknowns are involved.

3.5.3 *Memory-Integral Models*

Such constitutive equations should be handled with care for both corotational and codeformational types. Formulation of these models requires to consider the material points on their trajectories, in stationary and non-stationary conditions (e.g. Bird et al., [31]). Two types of such models, based on rate-of-strain and deformation tensors, are now considered.

3.5.3.1 Corotational Models: The Corotational Frame

Such equations, where the stress tensor is expressed in terms of the rate-of-deformation tensor $\underline{\underline{D}}$, require determination of the corotational frame $(\mathbf{e}^*(t))$ at every time. The vorticity vector $\underline{R}$ corresponds to the instantaneous rotational velocity of the frame $(\mathbf{e}^*(t))$, with respect to the fixed reference frame $(\boldsymbol{\varepsilon}_i)$. It may be expressed as follows:

$$\underline{R}((\mathbf{e}^*(t))/(\boldsymbol{\varepsilon})) = \omega_1\underline{\varepsilon}_1 + \omega_2\underline{\varepsilon}_2 + \omega_3\underline{\varepsilon}_3. \tag{3.81}$$

In the planar and axisymmetric situations, the vorticity vector $\underline{R}$ is given by the respective Eqs. (3.48) and (3.51), according to the following relations:

$$\underline{R} = -(1/2)\left(\frac{\partial \mathrm{u}}{\partial \mathrm{z}} - \frac{\partial \mathrm{w}}{\partial \mathrm{x}}\right)\underline{\varepsilon}_2 \quad \text{(Cartesian coordinates)},$$
$$\underline{R} = -(1/2)\left(\frac{\partial \mathrm{u}}{\partial \mathrm{z}} - \frac{\partial \mathrm{w}}{\partial \mathrm{x}}\right)\underline{\mathrm{e}}_2 \quad \text{(cylindrical coordinates)}. \tag{3.82}$$

In both cases, this vector is normal to the reference deformation planes, (z,x)- and (z,r)-meridian plane, according to the basis where the kinematics are written.

The components of the strain-rate tensor are given in terms of the mapping functions f or k, considered in the two-dimensional case The calculation of the stresses from a corotational constitutive equation of the type of Eq. (2.82) or (2.83) requires the stress tensor to be expressed in the corotating frame $(\mathbf{e}^*(t))$ before being calculated again in the reference frame (Cartesian variables or cylindrical coordinates).

We may write

$$\text{Rotation}\left[(\mathbf{e}^*(t))/(\mathbf{e})\right] = \left[(\mathbf{e}^*(t))/(\boldsymbol{\varepsilon})\right] + \text{Rotation}[(\boldsymbol{\varepsilon})/(\mathbf{e})]. \tag{3.83}$$

Since the rotation of the Cartesian frame $(\boldsymbol{\varepsilon})$ versus the cylindrical frame $(\mathbf{e})$ is zero (no rotation), we get, from Eq. (3.81),

$$\text{Rotation}\left[(\mathbf{e}^*(t))/(\mathbf{e})\right] = -(1/2)\left(\frac{\partial u}{\partial z} - \frac{\partial w}{\partial x}\right)\underline{e}_2. \tag{3.84}$$

Putting

$$\eta = -(1/2)\left(\frac{\partial u}{\partial z} - \frac{\partial w}{\partial x}\right), \tag{3.85}$$

we can, according to the rotation, define at every time τ the corotational frame $(\mathbf{e}^*(\tau))$ from a rotation tensor $\underline{\underline{R}}(\tau)$. Its corresponding matrix is expressed in the cylindrical basis $(\mathbf{e})$ as follows:

$$\mathbf{R}(\tau) = \begin{bmatrix} \cos\Theta(\tau) & 0 & -\sin\Theta(\tau) \\ 0 & 0 & 0 \\ \sin\Theta(\tau) & 0 & \cos\Theta(\tau) \end{bmatrix}, \tag{3.86}$$

with

$$\Theta(\tau) = (1/2)\int_{t_0}^{\tau} \eta(\xi)\mathrm{d}\xi = -(1/2)\int_{t_0}^{\tau} \left(\frac{\partial v}{\partial r} - \frac{\partial u}{\partial z}\right)(\xi)\,\mathrm{d}\xi. \tag{3.87}$$

Then we get

$$\mathbf{e}^*(\tau) = \mathrm{R}(\tau)\,\mathbf{e}. \tag{3.88}$$

If $\mathbf{D}^*$ denotes the matrix of the rate-of-deformation tensor $\underline{\underline{D}}$ in the corotational basis $(\mathbf{e}^*)$ at time τ, we may write

$$\mathrm{D}^{ij}(\tau) = \left(\mathrm{R}(\tau)\,\mathrm{D}(\tau)\,\mathrm{R}^{-1}(\tau)\right)^{ij}. \tag{3.89}$$

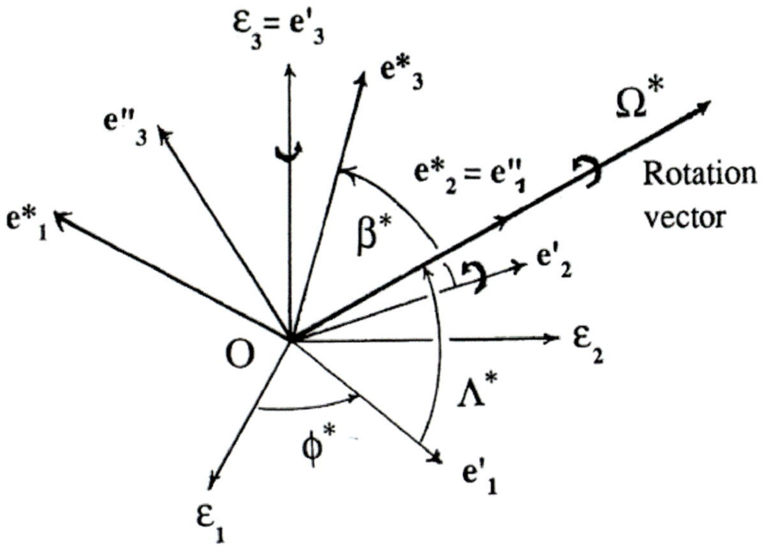

Fig. 3.13 Cartesian frame ($\boldsymbol{\varepsilon}_i$), cylindrical orthonormal reference frame ($\mathbf{e}_i$) and corotating frame ($\mathbf{e}^*_i$) for an axisymmetric flow field. Reprinted by permission from Springer Nature Customer Service Centre GmbH from [32], Copyright 1993

Using Eqs. (3.49), (3.86) and (3.89), we obtain the components of tensor $\underline{\underline{D}}$ in the corotational frame ($e^*(\tau)$) as (Fig. 3.13)

$$
\begin{aligned}
D^{*11} &= \frac{\partial u}{\partial r}\cos^2\Theta(\tau) - (1/2)\left(\frac{\partial w}{\partial r} + \frac{\partial u}{\partial z}\right)\sin 2\Theta(\tau) + \frac{\partial w}{\partial z}\sin^2 2\Theta(\tau)\\
D^{*22} &= \frac{u}{r}\\
D^{*33} &= \frac{\partial u}{\partial r}\sin^2\Theta(\tau) + (1/2)\left(\frac{\partial w}{\partial r} + \frac{\partial u}{\partial z}\right)\sin 2\Theta(\tau) + \frac{\partial w}{\partial z}\cos^2 2\Theta(\tau)\\
D^{*13} &= D^{31} = (1/2)\left(\frac{\partial w}{\partial r} + \frac{\partial u}{\partial z}\right)\sin 2\Theta(\tau) + (1/2)\left(\frac{\partial w}{\partial r} + \frac{\partial u}{\partial z}\right)\cos 2\Theta(\tau)\,\frac{\partial u}{\partial r}\cos 2\Theta(\tau)\\
D^{*23} &= D^{*31} = D^{*12} = D^{*21} = 0.
\end{aligned}
\tag{3.90}
$$

Constitutive corotational equations are generally considered as integral forms involving the rate-of-deformation tensor $\underline{\underline{D}}$ ([Bird et al. [31]). Starting from the rheological law initially expressed in the corotational frame ($\mathbf{e}^*(\tau)$), in the two-dimensional case, the tensor components are then evaluated in a fixed convenient reference frame with generally Cartesian, cylindrical or spherical coordinates. Denoting by $\underline{\underline{S}}$ the stress tensor to be referred in a frame noted (**e'**), with the same notation similar to that used previously, the stress components are obtained with an equation similar to (3.81). Since such rheological models are expressed versus time, the evolution versus t is obtained by the particle-tracking method related to the STM approach.

3.5.3.2 Codeformational Models: An Example with Cylindrical Coordinates

To describe memory-integral equations involving the strain history, the Cauchy and Finger tensors $\underline{\underline{C}}_t$ and $\underline{\underline{C}}_t^{-1}$ (Eqs. (2.19) and (2.20)) are used in the Cartesian basis ($\boldsymbol{\varepsilon}_i$) or another reference frame, without additional operations as those required for corotational models. We consider an example for codeformational constitutive equations, in an axisymmetric problem, where the basic kinematic quantities are the displacement functions. As pointed out previously, the respective Cauchy and Finger objective strain tensors are related to the deformation gradient tensor $\underline{\underline{F}}$ by the equations

$$\underline{\underline{C}}_t(t') = \underline{\underline{F}}_t^T(t').\underline{\underline{F}}_t(t'), \quad \underline{\underline{B}} = \underline{\underline{C}}_t^{-1}(t') = \left[\underline{\underline{F}}_t^{\mathrm{T}}(t').\underline{\underline{F}}_t^{\mathrm{T}}(t')\right]^{-1}, \tag{3.91}$$

with

$$[F_t(\tau)]^{mj} = \frac{\partial x_\tau^m}{\partial x_{t'}^{\mathrm{j}}}. \tag{3.92}$$

In STM, the evaluation of the kinematic quantities is performed, as specified previously, in the computational mapped domain Ω^* where the mapped pathlines are rectilinear. Particularly, the time evolution for the particle M may be expressed as follows by an equation of the type of (3.21):

$$\tau - \mathrm{t}_0(\mathrm{Z}_0) = -\frac{1}{\Psi^*(R)}\int_{\mathrm{Z}_0}^{\mathrm{Z}_\tau} f(R,\zeta)\, f_R'(R,\zeta)\, d\zeta. \tag{3.93}$$

The sections Z_0 and Z_t correspond to positions P_0 and P_τ, respectively. The reference time t_0 will be related to the position of the particle P at the limiting upstream section $z_0 = Z_0$.

We start from Adachi's work [18, 19]. We use the vector positions $\underline{X}_t$ and $\underline{X}_\tau$ and write again Eqs. (3.32) and (3.37), corresponding to the natural and reciprocal basis vectors for the cylindrical variables (R,Θ,Z):

$$\underline{E}_R = f_{\mathrm{R}}'\underline{\mathrm{e}}_{\mathrm{r}}, \quad \underline{E}_\Theta = f\,\underline{e}_\theta, \quad \underline{E}_Z = f_Z'\underline{\mathrm{e}}_{\mathrm{r}} + \underline{\mathrm{e}}_{\mathrm{z}} \tag{3.94}$$

$$\underline{E}^R = \frac{1}{f_R'}\underline{e}_r - \frac{f_Z'}{f_R'}\underline{\mathrm{e}}_{\mathrm{z}}, \quad \underline{E}^\theta = \frac{1}{f}\underline{e}_\theta, \quad \underline{E}^Z = \underline{e}_z. \tag{3.95}$$

This approach leads to express the deformation gradient tensor in the frame corresponding to (R,Θ,Z) on the matrix form:

$$\mathbf{F}_t(\tau) = \begin{bmatrix} 1 & 0\,1 & \\ 0 & 1\,0 & \\ \frac{\partial Z_\tau}{\partial R_t} & 0 & \frac{\partial Z_\tau}{\partial Z_t} \end{bmatrix} \tag{3.96}$$

with

$$\frac{\partial Z_\tau}{\partial Z_t} = \frac{w(R, Z_\tau)}{w(R, Z_\tau)}, \quad \frac{\partial Z_\tau}{\partial R_t} = \frac{w(R, Z_t)}{w(R, Z_\tau)} \int\limits_{Z_t}^{Z_\tau} \frac{\partial w(R, Z_\xi)}{\partial R} d\xi. \tag{3.97}$$

From the definition of the mapping function f, the Cauchy and Finger tensor components can be expressed in terms of f and its derivatives. The deformation gradient tensor $\underline{\underline{F}}_t(\tau)$ may be computed by using the natural and reciprocal bases related to Eqs. (3.94) and (3.95). In cylindrical coordinates, the matrix components may be written [32] as

$$F^{11} = \frac{f_R'(\mathrm{R}, \mathrm{Z}_\tau)}{f_R'(\mathrm{R}, \mathrm{Z}_t)} + \frac{f_Z'(\mathrm{R}, \mathrm{Z}_\tau)}{f_R'(\mathrm{R}, \mathrm{Z}_t)} \frac{\partial z_\tau}{\partial \mathrm{R}}, \tag{3.98}$$

$$F^{13} = -\frac{f_R'(\mathrm{R}, \mathrm{Z}_\tau) f_Z'(\mathrm{R}, \mathrm{Z}_t)}{f_R'(\mathrm{R}, \mathrm{Z}_t)} - \frac{f_Z'(\mathrm{R}, \mathrm{Z}_\tau) f_Z'(\mathrm{R}, \mathrm{Z}_t)}{f_R'(\mathrm{R}, \mathrm{Z}_t)} + f_Z'(\mathrm{R}, \mathrm{Z}_t) \frac{\partial \mathrm{Z}_\tau}{\partial \mathrm{Z}_t}, \tag{3.99}$$

$$F^{22} = -\frac{f_R'(R, Z_\tau)}{f_R'(R, Z_t)}, \tag{3.100}$$

$$\begin{aligned} \mathrm{F}^{31} &= \left[\frac{\Psi^*(\mathrm{R})}{f_R'(\mathrm{R}, \mathrm{Z}_t)} \frac{d\Psi^*(\mathrm{R})}{d\mathrm{R}} \frac{f(\mathrm{R}, \mathrm{Z}_t)}{f_R'(\mathrm{R}, \mathrm{Z}_\tau)} \right] \times \int_{\mathrm{Z}_t}^{Z_\tau} \frac{f_Z'(\mathrm{R}, \zeta)}{f_R'(\mathrm{R}, \zeta)} d\zeta \\ &- \frac{1}{f(\mathrm{R}, \mathrm{Z}_t) f_R'(\mathrm{R}, \mathrm{Z}_\tau)} \int_{Z_t}^{Z_\tau} \left[f_R'^2 + f f_{R^2}'' \right] d\zeta, \end{aligned} \tag{3.101}$$

$$F^{33} = -\frac{f_Z'(R, Z_t)}{f_R'(R, Z_t)} \left[\Psi^*(R) \frac{d\Psi^*(R)}{dR} \right] \frac{1}{f'(R, Z_t) f_R'(R, Z_\tau)} \times \int_{Z_\tau}^{Z_t} \frac{f_Z'(R, \zeta)}{f_R'(R, \zeta)} d\zeta \tag{3.102}$$

$$- \frac{1}{f(R, Z_t)\, f'(R, Z_\tau)} \times \int_{Z_t}^{Z_\tau} \left[(f_R')^2 + f\, f_{R^2}'' \right] d\zeta + \frac{f(R, Z_t)\, f'(R, Z_t)}{f(R, Z_\tau)\, f'(R, Z_\tau)},$$

$$F^{12} = F^{21} = F^{23} = F^{32} = 0. \tag{3.103}$$

It should be pointed out that in STM, Eqs. (3.97)–(3.102) related to the mapped rectilinear streamlines L involve the property:

$$\forall X(R, \Theta, Z)_{(\tau)} \in L : R = R_0. \tag{3.104}$$

The Cauchy and Finger tensor components in the cylindrical basis ($\boldsymbol{e}_i$) may be evaluated in terms of the mapping functions and its derivatives, using Eqs. (3.91) and (3.98)–(3.103).

3.6 Concluding Remarks

- Stream-Tube Method (STM) is a geometric approach based particularly on kinematic concepts.
- In two-dimensional flows without eddies, an unknown mapping function should be determined in a rectangular domain where streamlines are parallel and straight.
- Mass conservation is automatically verified from the formulation. Only the dynamic equations, non-linear, should be written.
- Computations can be performed only for the so-called "peripheral stream tube", adjacent to the boundary of the physical domain.

The following remarks are of interest with regard to the use of constitutive corotational and codeformational memory-integral equations.

- Both corotational and codeformational models may involve parameter values fitting satisfactorily with data of materials as polymers.
- Though memory-integral equations lead to more satisfactory predictions for industrial applications, more complex expressions are considered for numerical integrations.

References

1. Zienkiewicz OC (1977) The finite element method. McGraw-Hill
2. Gourdin A, Boumahrat M (1989) Méthodes Numériques appliquées, Technique et Documentation. Editions Lavoisier, Paris
3. Thompson JF, Thames FC, Mastin CW (1974) Automatic numerical generation of body-fitted curvilinear coordinate systems for field containing any number of arbitrary two-dimensional bodies. J Comp Phys 15, 299–319
4. Eiseman PR (1979) A multi-surface method of coordinate generation. J Comput Phys 33:118–150
5. Lee JS, Fung YC (1970) Flow in locally constricted tubes at low Reynolds numbers. J Appl Mech 9–16
6. Ryan ME, Dutta A (1981) A finite-difference simulation of extrudate swell. In: Proceedings of the 2nd world congress of chemical engineering, vol 6. Montreal, pp 277–281
7. Nickell RE, Tanner RI, Caswell B (1974) The solution of viscous incompressible jet in free-surface flows using finite-element methods. J Fluid Mech 65, 169–206
8. Huang W, Russel RD (2010) Adaptive moving mesh methods. Springer, Berlin
9. Haussling HJ, Coleman RM (1981) A method for generation of orthogonal and nearly elongational boundary-fitted coordinate systems. J Comput Phys. 43:373–381
10. Smith RE (1982) Algebraic grid generation. Appl Math Comput 10(11):137–170

11. Haussling HJ (1982) Solution of nonlinear water problems using boundary-fitted coordinate systems. Appl Math Comp 10/11, Nashville, Tenn, 385–407
12. Chinnaswamy C, Amadei B, Illangasek TH (1991) A new method for finite-element transitional mesh generation. Int J Numer Methods Eng 31:1253–1270
13. Lo SH, Lee CK (1992) On using mesh of mixed element types in adaptive finite element analysis. Finite Elem Anal Des 11:307–336
14. Lau TS, Lo SH, Lee CK (1997) Generation of quadrilateral mesh over analytical curved surfaces. Finite Elem Anal Des 27:251–272
15. Greywall MS (1985) Streamwise computation of two-dimensional incompressible potential flows. J Comput Phys 59:224–231
16. Greywall MS (1988) Streamwise computation of three-dimensional incompressible potential flows. J Comput Phys 78:178–193
17. Duda JL, Vrentas JS (1973) Entrance flows of Non-Newtonian fluids. Trans Soc Rheol 17:89–108
18. Adachi K (1983) Calculation of strain histories with Protean coordinates systems. Rheol Acta 22:326–335
19. Adachi K (1986) A note on the calculation of strain histories in orthogonal streamline coordinate systems. Rheol Acta 25:555–563
20. Papanastasiou AC, Scriven LE, Macosco CW (1987) A finite element method for liquid with memory. J Non-Newton Fluid Mech 22:271–288
21. Chung SG, Lu SC-Y, Richmond O (1989) Explicit streamline method for steady flows of non-Newtonian matter. Phys Rev A 39(5):2728–2730
22. Chung SG, Kuwahara K (1993) Explicit streamline method for steady flows of Non-Newtonian matter, history dependence and free surfaces. J Comput Phys 104, 444–450
23. Clermont JR, de la Lande ME, Pham Dinh T, Yassine A (1991) Analysis of plane and axisymmetric flows of incompressible fluids with the stream tube method: numerical simulation by trust region optimization algorithm. Int J Num Meth Fluids 13, 371–399
24. Normandin M, Clermont JR, Mahmoud A (2000) Extrusion de fluide newtonien en sortie de filières annulaires axisymétriques. Simulation numérique par la Méthode des Tubes de Courant. Les Cahiers de Rhéologie, 16(4):1–15
25. Crochet MJ, Davies AR, Walters K (1984) Numerical simulation of non-newtonian flow. Elsevier, Amsterdam
26. Owens RG, Phillips TN (2002) Computational rheology. Imperial College Press
27. Clermont JR, de la Lande ME (1991) On the use of the stream-tube method in relation to apparition of secondary flows in plane and axisymmetric geometries. Mech Res Comm 18(5):303–310
28. Clermont JR, de la Lande ME (1993) Calculation of main flows of a memory-integral fluid in an axisymmetric contraction at high Weissenberg numbers. J Non-Newt Fluids Mech 46:89–110
29. Bereaux Y, Clermont JR (1995) Numerical simulation of complex flows of Non-Newtonian fluids using the stream-tube method and memory-integral constitutive equations. Int J Numer Methods Fluids 21, 371–389
30. Clermont JR, Normandin M (1993) Numerical simulation of extrudate swell for Oldroyd-B fluids using the stream-tube analysis and a streamline approximation. J Non-Newt Fluids Mech 50:193–215
31. Bird RB, Armstrong RC, Hassager O (1987) Dynamics of polymer liquids. Wiley
32. Clermont JR (1993) Calculation of kinematic histories in two- and three-dimensional flows using streamline coordinate functions. Rheol Acta 32, 82–93. [Springer]

Chapter 4
Stream-Tube Method in Two-Dimensional Problems

4.1 Introduction

From basic STM equations and applications on two-dimensional cases, the present chapter concerns numerical studies in planar and axisymmetric flows. As depicted in Chap. 3 for flows involving only open streamlines, their mapped lines are parallel to the centreline in axisymmetric domains, or to the symmetry axis in planar configurations. Different formulae related to the kinematic and dynamic equations may be written for computing the velocities and stresses, under boundary conditions. According to the simplicity of the mapped domain adopted for solving the equations, various discretization techniques may be retained for discretizing the governing equations and unknowns. The governing equations can be resolved by adopting, as primary unknowns, the mapping functions and the pressure involved in the dynamic equations. The stream function is a constant along the rectilinear mapped streamlines. However, non-linearities arising even in the Newtonian case can lead to mixed formulations. In this chapter, different types of discretization are described for approximating the equations and unknowns. In mapped domain Ω^*, simple meshes and elements can be defined on these rectilinear lines, for determining flow characteristics in the physical domain.

Selected examples are close to situations and phenomena which can be encountered in polymer processing and fluid industry.

4.2 Formulations: Boundary Conditions

According to the non-linearity of the governing equations in primary or mixed formulations, the dynamic equations of variables (X, Z) or (r, Z) in the transformed domain are required to be solved by robust and efficient algorithms. The Newton–Raphson method may be adopted in some cases, but optimization codes are generally retained for the computations. The dynamic equations and simple expressions at the

J.-R. Clermont and A. Ammar, *Stream-Tube Method*,
https://doi.org/10.1007/978-3-030-65470-2_4

boundaries are sometimes completed by non-linear integral equations at the limiting surfaces, as shown in Chap. 3. This chapter also provides detailed examples on STM flows, under different boundary conditions, for linear (Newtonian), differential and integral constitutive equations. Several approaches concern finite differences and finite elements.

4.2.1 Primary and Mixed Formulations

Incompressible materials are considered in the studies. The momentum and constitutive equations, defined by the choice of a rheological model, are discretized with a grid on the mapped parallel streamlines. In two-dimensional flows, the following formulations may be concerned for solving the problem:

(i) The primary formulation (mapping function f or k, pressure p): in the equations, the stress components are written in terms of the transformation function and its derivatives. This formulation leads to a system of partial differential equations or integro-differential equations.
(ii) The mixed formulation of unknowns: f or k, pressure p, τ^{ij} components of the stress tensor): (k, p, τ^{11}, τ^{13}) or (f, p, τ^{11}, τ^{13}) with (k, p) and (f, p) in the planar ($x = 1, y = 2, z = 3$) and axisymmetric cases ($r = 1, \theta = 2, z = 3$), respectively. Additional equations are written, when expliciting the stress components. This formulation leads to a greater number of discretized equations, but reduces the degree of the derivatives.

Owing to the non-linearity of the governing equations, solving the equations should be carefully considered. One may also keep in mind that small changes of a streamline may lead to significant variations in the stresses. In an iterative numerical process, this requires, as far as possible, to satisfy the following requirements:

(i) define an initial estimate close to the solution to avoid divergence of the numerical procedure;
(ii) adopt a robust and efficient algorithm to ensure convergence of the numerical process even for a given initial estimate rather far from the solution.

Finite differences or finite elements may be adopted in the calculations, using a weak formulation when necessary. Though the non-linearity of the governing equations has been pointed out, one may underline the following features of the applications:

- When solving the equations in the rectangular mapped domain Ω^*, the discretization of the continuous problem is simplified when compared to that resulting from

a classic velocity–pressure formulation. As well known, such approach leads to computation in the physical domain Ω, where complex boundaries may be encountered.
- In the context of the stream-tube method, numerical experiments made with suitable solving algorithms show that the non-linear problem with a complex constitutive equation has scarcely led to additional difficulties. Such cases are encountered with some memory-integral constitutive equations, which has led to adopt specific elements when discretizing the equations.

4.2.2 Boundary Condition Equations

Specific features of stream-tube method are now considered. The fluid is assumed to be of constant density, without loss of generality for the applications. As seen in Chap. 3, sub-domains of the total flow domain may be computed separately, from the boundary (wall, moving free surface). Specific conditions require a simply connected global domain for such possibilities. As pointed out previously, in some cases, STM flow computations can be performed in geometries where eddies are close to the wall, as for the flow in an abrupt contraction or in a duct with an expansion region (Fig. 4.1). Different boundary condition equations will be expressed in examples proposed in this chapter.

4.2.2.1 Classic Boundary Condition Equations

Owing to the automatic verification of mass conservation for incompressible materials, the primary unknowns (k, p) in planar cases or (f, p) in axisymmetric problems are considered. When compared to classical methods, the velocity boundary conditions become those related to the mapping function k (or f). Requirements for the pressure remain the same than those referred to classic velocity–pressure formulations. Depending on the problem investigated, natural boundary condition equations may also be written.

4.2.2.2 Specific Boundary Condition Equations

In Chap. 3, possibilities of computing flows in sub-domains have been shown for simply connected two-dimensional geometries. Thus, integral forms can be written, leading to non-linear boundary condition equations. From a practical viewpoint, it has been shown that numerical tests on several situations have led to consistent results. Such techniques are considered in some examples provided in this chapter.

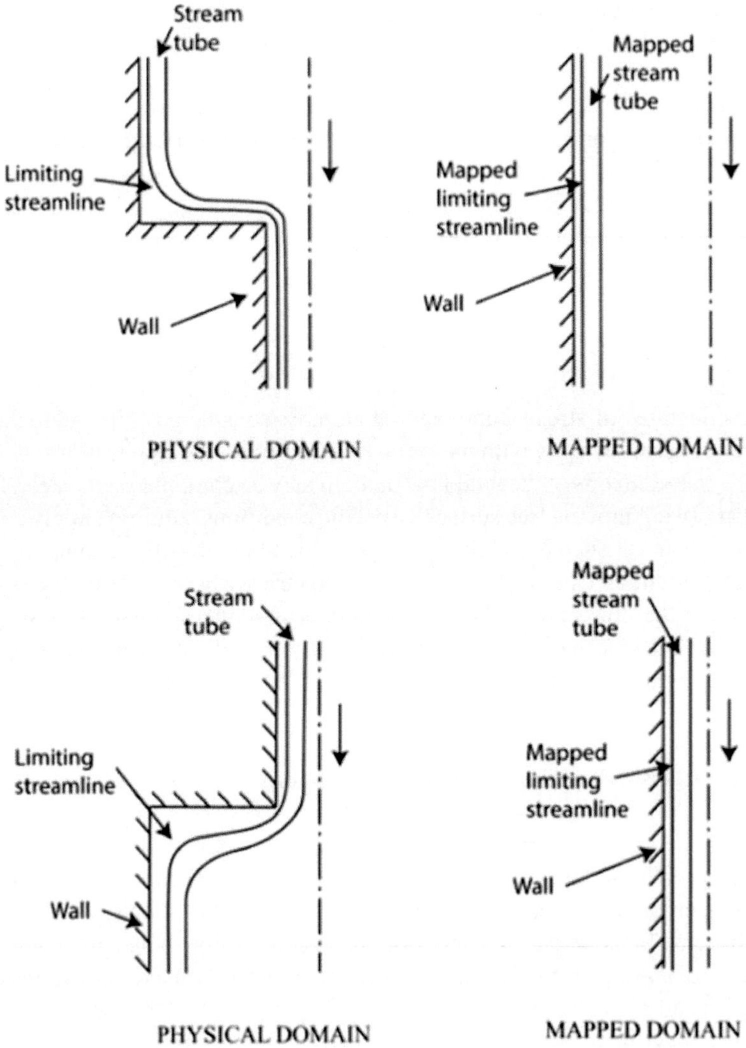

Fig. 4.1 Two-dimensional flow domains with a stream tube close to the boundary: **a** abrupt contraction; **b** abrupt expansion

4.3 Discretization

4.3.1 Approximating the Unknowns

The streamlines to be approximated, from variables adopted in the mapped computational domain Ω^* of parallel lines, may be considered with different choices for discretizing the unknowns. In some cases, approximations can be related to a guess of the physical shape of streamlines related to the physical flow configuration. Generally, in STM applications, finite-difference discretizations may be carried up. Finite elements for the equations and unknowns (*k, p*) or (*f, p*) are also defined for mixed formulations.

(i) Analytical approximating expressions
Approximating relations may be based on a guess of the physical shape of the streamlines related to the flow configuration expected from possible experimental data. Such situations will be considered further in swelling flow problems, starting from a guess of the streamlines expressed by analytical equations, thus involving a limited number of unknowns. The governing equations, still non-linear, can be solved by considering appropriate methods as those related to least squares techniques.

(ii) Finite difference interpolations
The STM governing Eqs. (3.61) and (3.62) in the respective planar and axisymmetric situations involve first- and second-order partial derivatives of the mapping functions or k or f as also first-order derivatives of the pressure. According to existence of parallel streamlines in domain Ω^*, simple derivative schemes for functions k or f, limiting possible errors in comparison with non-rectangular meshes, may be defined for the grid.

(iii) Finite elements
Finite element discretizations can be defined on the rectangular grid built in the mapped domain. According to such geometrical conditions, Hermite elements, allowing approximations of the mapped functions k or f, the pressure p and derivatives of the unknowns, can be adopted for the calculations. A weak finite element formulation may be retained for solving the governing equations.

4.3.2 Finite Differences

In computational analysis, the finite-difference method is a standard technique for providing approximate solutions in problems concerned by partial derivative equations. A numerical scheme is defined such that the values of the unknown functions are linked for points of the computational domain, assumed to be close for ensuring efficient determination of solutions. Convergence of the numerical scheme can be obtained when the distance of the points of discretization is small enough.

Figure 4.2 presents an example of mesh adopted for the transition from adherence-to-slip (called "stick-slip" in the literature) problem by finite-difference calculations [1]. In this case, the material adheres to the wall up to the section $z = 0$ and slips beyond this value. Different numerical tests have led to retain a regular mesh in the X (or R)-direction while the grid is refined versus Z-direction, in relation to velocity changes in the vicinity of $z = 0$. Generally, grid modifications should be made according to variations in shape of the flow domain that can lead to significant gradients of physical variables.

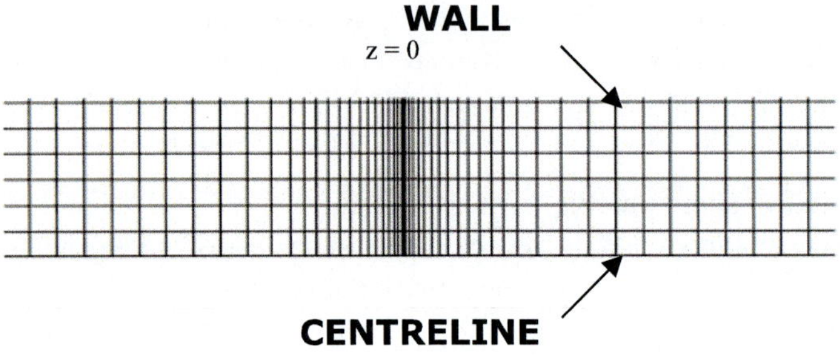

Fig. 4.2 Grid in a mapped domain Ω^* for finite-difference computations

The non-regular grid considered in the Z-direction allows to ensure more accurate approximations of the unknown variables in the vicinity of adherence-to-slip transition section ($z = 0$). Different approximating equations concerning the derivatives have been proposed in the literature for the equations and unknowns (e.g. Smith [2]).

4.3.3 Mesh Elements

In this section, discretizing elements used in the computations are presented for various domains and flow conditions.

4.3.3.1 Trapezoidal Element for a Primary Formulation

Elements of type 1, defined in a stream band Ω_i^* to discretize the mapping function f and the pressure p in a stream band, are considered for an axisymmetric domain (Fig. 4.3), in a primary formulation. It should be pointed out that the approximate expressions defined for these elements do not correspond to those related to a finite element method, according to [3].

In a stream band partially shown in Fig. 4.3, upstream and downstream sections $Z_0 = z_0$ and $Z_1 = z_1$ are defined.

Without loss of generality for the planar case, one considers here an axisymmetric geometry. For each element (i) of the stream band, the unknowns at a point M are

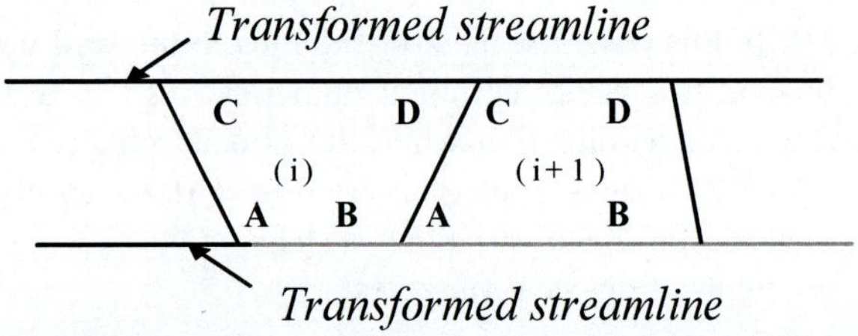

Fig. 4.3 Elements in a stream band Ω_i^* of a mapped domain Ω^* for the transformation function and the pressure

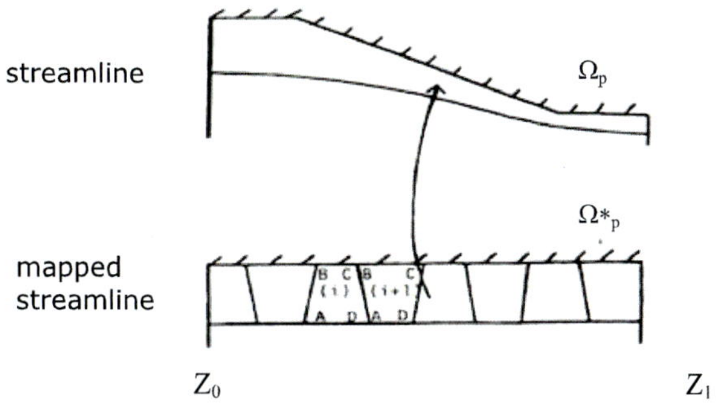

Fig. 4.4 Elements in a peripheral stream band of the mapped computational domain Ω*

assumed to be expressed in terms of second-order polynomials of R and Z as

$$\begin{aligned} f_M =& f_{B^i} + (R_M - R_{B_i})x_1^i + (Z_M - Z_{B_i})x_2^i + (R_M - R_{B_i})^2 x_3^i \\ &+ (Z_M - Z_{B_i})^2 x_4^i + (R_M - R_{B_i})(Z_M - Z_{B_i})x_5^i \end{aligned} \tag{4.1}$$

$$\begin{aligned} p_M =& p_{B^i} + (R_M - R_{B_i})x_6^i + (Z_M - Z_{B_i})x_7^i + (R_M - R_{B_i})^2 x_8^i \\ &+ (Z_M - Z_{B_i})^2 x_9^i + (R_M - R_{B_i})(Z_M - Z_{B_i})x_{10}^i, \end{aligned} \tag{4.2}$$

where (R_M, Z_M) and (R_{B_i}, Z_{B_i}) are the respective coordinates of points M and B_i in the element (i). The superscript i is related to the element (i), for which 10 unknowns are involved. When governing equations and boundary conditions are defined in the problem under consideration, additional compatibility equations are written, as, for example, for data in points $B_i \cong A_{i+1}$ (Fig. 4.3). Figure 4.4 presents, in the physical and mapped domains, the peripheral stream tube Ω_p and its transformed stream band Ω_{*p}, in order to determine numerically the inside boundary streamline, by solving the governing equations in the peripheral sub-domain and related boundary conditions and using Eqs. (4.1)–(4.2).

4.3.3.2 Rectangular Elements for Mixed Formulations

Discretizations defined in mixed formulations concern rectangular sub-domains where Hermite elements or others of the same type are built on the stream bands of the mapped domain. As in the previous case, these elements are not considered in a weak formulation required in finite element techniques.

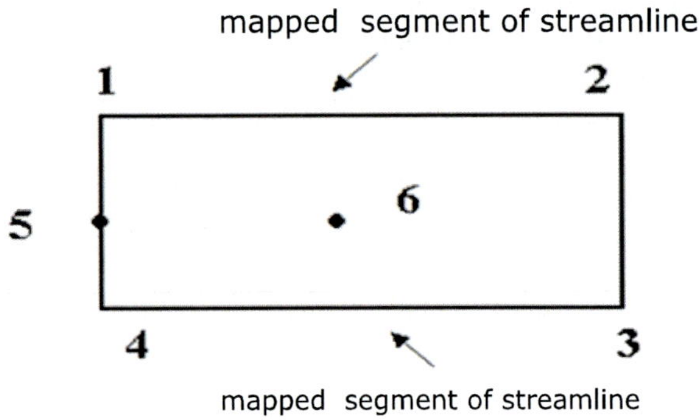

Fig. 4.5 Local element in a stream band of the mapped domain Ω[*]

A Specific Rectangular Six-Point Element in a Stream Band

A local element and its node positions are presented in Fig. 4.5 for a stream band proposed, after some numerical tests, by Clermont and de la Lande [2], for viscoelastic flows in an axisymmetric 4:1 contraction. For this rectangular element, the mapping function f, assumed to be quadratic, is expressed by the following equation:

$$f^{(e)}(x, y) = \sum_{i=1}^{i=6} \Phi_i(x, y) f_i^{(e)}, \tag{4.3}$$

where (e) denotes a local element of the stream band and Φ_i are the six basis functions related to nodal values f_i associated to local coordinates (x, y), for points (1, 2, 3, 4, 5, 6), shown in Fig. 4.5.

Using the following equations:

$$\Delta R = R_{(1)}^{(e)} - R_{(4)}^{(e)}, \quad \Delta Z = Z_{(2)}^{(e)} - Z_{(1)}^{(e)}, \tag{4.4}$$

the basis functions may be written as

$$\Phi_1(x, y) = 1 + (3x/\Delta R) + \left[2x^2/(\Delta R)^2\right] - \left[y^2/(\Delta Z)^2\right] - xy/(\Delta R \cdot \Delta Z), \tag{4.5}$$

$$\Phi_2(x, y) = \left[y^2/(\Delta Z)^2\right] + [xy/(\Delta R \cdot \Delta Z)], \tag{4.6}$$

$$\Phi_3(x, y) = (-y/\Delta Z) + \left[y^2/(\Delta Z)^2\right] - [xy/(\Delta R \cdot \Delta Z)], \tag{4.7}$$

$$\Phi_4(x, y) = (x/\Delta R) + (y/\Delta Z) + \left[2x^2/(\Delta R)^2\right] - \left[y^2/(\Delta Z)^2\right] + xy/(\Delta R \cdot \Delta Z), \tag{4.8}$$

---

[*]Reprinted from reference [3]. Copyright 1993, with permission from Elsevier

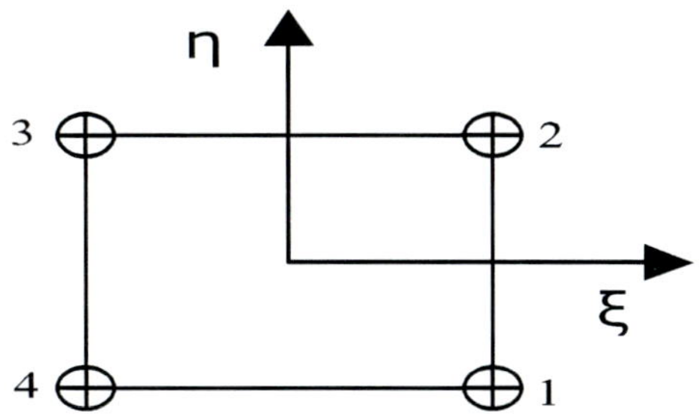

Fig. 4.6 Local Hermite element (**He$_1$**) for the mapping function f

$$\Phi_5(x, y) = 4\Big[(-x/\Delta R) - (y/(\Delta Z)) - \big[x^{2:}/(\Delta R)^2\big] + \Big[y^{2:/(\Delta Z)^2}\Big]\Big], \quad (4.9)$$

$$\Phi_6(x, y) = 4\big[\big(y/(\Delta Z) - \big(y^2/(\Delta Z)^2\big)\big)\big]. \quad (4.10)$$

The first and second derivatives of the mapping function f can be readily obtained from derivatives of the basic functions Φ_i, taking into account the following relations:

$$f'_R = f'_X X'_R = -f'_x; \quad f'_Z = f'_y y'_Z = f'_y. \quad (4.11)$$

The Classic Hermite Element (He$_1$)

For STM computations, the classic four-point Hermite element (e_H) [4], in a two-dimensional space with four nodes of interpolation, has been also adopted (Fig. 4.6) to approximate the unknowns, relating to the rectangular shape of the stream bands. The three nodal variables are (f,f'_R,f'_Z), with a total of 12 degrees of freedom. For example, in the axisymmetric case, a function f of two variables is expressed, at a point M of the element, by the following relation:

$$M \in (e_H): \quad f_M = \sum_{m=1}^{m=6} \Phi_m(x, y) \cdot f_m^e \quad (4.12)$$

(f,f'_R,f'_Z) at points 1, 2, 3, 4.

Modified Hermite Element (He$_2$)

In order to avoid implication of recirculating flow regions when computing the main flow in the vicinity of the corner, where the Jacobian of the mapping function vanishes, a Hermite element [4] has been built in the stream band adjacent to the wall, as shown in Fig. 4.7. Thus, the transformation function f is expressed in terms of the local variables (ξ, η) by the following relations:

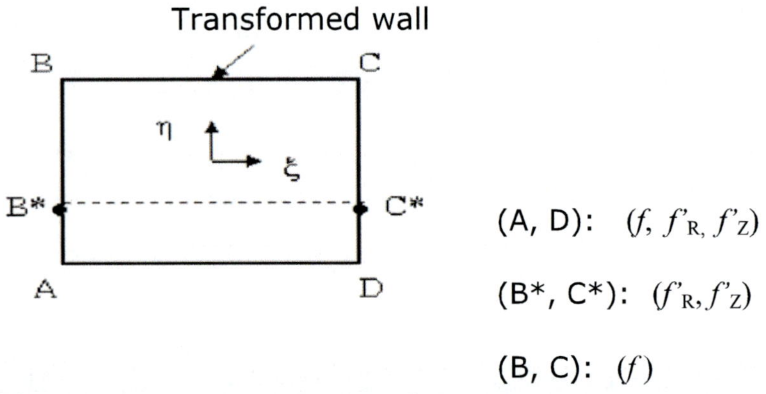

Fig. 4.7 Modified Hermite element in the (peripheral) stream tube adjacent to the wall of a mapped domain Ω*, with correspondence of the function f and its derivatives at the mesh points. © 1995 by John Wiley & Sons, Ltd: reproduced from Ref. [5] by permission of John Wiley & Sons, Ltd

Table 4.1 Correspondence between geometric nodes, nodal values (f, $\partial f/\partial \xi$ and $\partial f/\partial \eta$) and basis functions (H_1 to H_{12}) for the modified Hermite element

Geometric nodes	f	$\partial f/\partial \xi$	$\partial f/\partial \eta$
A	H_1	H_2	H_3
D	H_4	H_5	H_6
C	H_7		
B	H_8		
B*		H_9	H_{10}
C*		H_{11}	H_{12}

$$f_{(e)}(\xi,\eta) = \sum_{j=1}^{j=12} H_j(\xi,\eta) f_*^{(e)} j, \tag{4.13}$$

where $H_j(\xi,\eta)$ ($j = 1,\ldots, 12$) denote basis functions defined in six points (A, B, C, D, B*, C*). In Eq. (4.13), the quantities $\boldsymbol{f}^*_j$ denote the values of the function f or those of its derivatives $\boldsymbol{f}^*_R$ and $\boldsymbol{f}^*_Z$, in the axisymmetric case. The first spatial derivatives of f are specified at nodes B* and C*. The triplets (f,f'_R,f'_Z) are considered at points A and D.

The correspondence between the basis functions, the quantities $\boldsymbol{f}^*_j$ and the nodes of an element of type (**He$_2$**) is presented in Table 4.1 [5].

For the element (**He$_2$**), the basic functions may be written as follows:

$$\begin{aligned} H_1(\xi,\eta) = {} & (\eta^2-1)\left[2\eta - 3(\eta^2-1)\right]/)\left[8(3\alpha^*-1)\right] - (\eta-1)/4 \\ & + \xi(\xi^2-1)]\left\{1 - (\alpha^*-1)^2(\eta+1)/\left[4(3\alpha^*-1)\right]\right\}/4 \\ & - (\xi/2)\left\{1 + (\eta+1)^2(2\eta+1)/\left[4(3\alpha^*-1)\right]\right\}, \end{aligned} \tag{4.14}$$

$$H_2(\xi,\eta) = (\xi^2-1)(\eta-\alpha^*)(1-\xi)/[4(\alpha^*+1)], \tag{4.15}$$

$$\begin{aligned} H_3(\xi,\eta) = {} & 2(\eta^2-1)[3\alpha^{*2}-2\alpha^*\eta-1)]/[4(3\alpha^*-1)(\alpha^*+1)] \\ & -(\alpha^*-1)^2(\xi^2-1)\xi(\eta+1)/[8(3\alpha^*-1)], \end{aligned} \tag{4.16}$$

$$\begin{aligned} H_4(\xi,\eta) = {} & (\eta^2-1)[2\eta-3(\eta^2-1)]/[8(3\alpha^*-1)]-(\eta-1)/4 \\ & +(\eta\xi/2)[1+(\eta+1)^2(2\eta\xi-3\alpha^*-1)]/[4(3\alpha^*-1)], \end{aligned} \tag{4.17}$$

$$H_5(\xi,\eta) = -(\xi\eta^2-1)(\eta-\alpha^*)/[4(\alpha^*+1)], \tag{4.18}$$

$$\begin{aligned} H_6(\xi,\eta) = {} & -(\eta^2-1)[3\alpha^{*2}-2\alpha^*(\eta-1)]/[4(3\alpha^*-1)(\alpha^*+1)] \\ & +(\alpha^*-1)^2(\xi^2-1)\xi(\eta+1)/[8(\alpha^*-1)], \end{aligned} \tag{4.19}$$

$$\begin{aligned} H_7(\xi,\eta) = {} & -(\eta^2-1)[2\eta-3(3\alpha^*-1)]/[8(3\alpha^*-1)] \\ & +(\eta+1)]/4-(\alpha^*+1)^2\xi(\xi^2-1)(\eta+1)/[16(3\alpha^*-1)] \\ & -\xi(\eta+1)^2(2\eta-3\alpha^*-1)/[8(3\alpha^*-1)], \end{aligned} \tag{4.20}$$

$$\begin{aligned} H_8(\xi,\eta) = {} & -(\eta^2-1)[2\eta-3(3\alpha^*-1)]/[8(3\alpha^*-1)] \\ & +(\eta+1)/4+(\alpha^*+1)^2\xi(\xi^2-1)(\eta+1)/[16(3\alpha^*-1)] \\ & +\xi(\eta+1)^2(2\eta-3\alpha^*-1)/[8(3\alpha^*-1)], \end{aligned} \tag{4.21}$$

$$H_9(\xi,\eta) = (\xi^2-1)(\eta+1)/[4(\alpha^*+1)], \tag{4.22}$$

$$\begin{aligned} H_{10}(\xi,\eta) = {} & (\eta-1)(\eta+1)^2(1-\xi)/[2(3\alpha^*-1)(\alpha^*+1)] \\ & +\xi(\eta+1)(\xi^2-1)(\alpha^*-1)/[4(3\alpha^*-1)], \end{aligned} \tag{4.23}$$

$$H_{11}(\xi,\eta) = (\xi-1)(\eta+1)(\xi+1)^2/[4(\alpha^*+1)], \tag{4.24}$$

$$\begin{aligned} H_{12}(\xi,\eta) = {} & (\eta-1)(\eta+1)^2(\xi+1)/[2(3\alpha^*-1)(\alpha^*+1)] \\ & -\xi(\eta+1)(\xi^2-1)(\alpha^*-1)/[4(3\alpha^*-1)]. \end{aligned} \tag{4.25}$$

In Eqs. (4.14)–(4.25), the parameter α^* $(-1 < \alpha^* < 1, \alpha \neq 1/3)$ denotes the distance between the ξ-axis and the segment B*C*. The features of element (**He$_2$**) are similar to those of the classical two-dimensional Hermite element:

(i) C°—continuity at the boundaries $\xi = \pm 1$ and $\eta = \pm 1$,
(ii) C^1—continuity on the segments $\eta = \pm 1$ of the element and at the geometrical nodes A, B*, C* and D for the partial derivative f'_{η},
(iii) C^1—continuity on the boundaries $\xi = \pm 1$ and at the geometrical nodes A and D for the partial derivative f'_{η}.

Starting from properties related to the classical Hermite element (Béreaux and Clermont [5], Touzot and Dhatt [6]), it proved possible to evaluate the interpolation error e in the domain $V_r \equiv (e_2)$ of the local variables (ξ,η) for a function U to be approximated along the line $(\xi, \alpha)^*$ where the nodes B*$(-1, \alpha^*)$ and C*$(1, \alpha^*)$ are located: only the nodal values f'_{ξ} and f'_{η} are associated with these nodes (Fig. 4.7). The error function $e(\xi,\eta)$ in domain V_r may be estimated by means of the inequality:

$$
\begin{aligned}
|e(\xi,\eta)| \leq & \left\{\left[(\alpha^{*3} + 3\alpha^{*2} - 3\alpha^* + 3) + 2\xi^2)(1-\xi)^2\right] R_H/6 \right.\\
& + (\alpha^* + 1)^3 (\alpha^* - 1)^2 (1-\xi)^2 (1 + -\xi)^2 S_H \\
& \left. + (\alpha^* + 1)^3 (\alpha^* - 1)^2 [T_H/3]\right\} / \left[8(1 - 3\alpha^*)\right],
\end{aligned}
\tag{4.26}
$$

in which the quantities R_H, S_H and T_H are given by the relations:

$$
R_H = \max\left|\partial^4 U/\partial \xi^4\right|_{(\xi,\eta)\in V_r}, \tag{4.27}
$$

$$
S_H = \max\left|\partial^4 U/\partial \xi^2 \partial \eta^2\right|_{(\xi,\eta)\in V_r}, \tag{4.28}
$$

$$
S_H = \max\left|\partial^4 U/\partial \eta^4\right|_{(\xi,\eta)\in V_r}. \tag{4.29}
$$

It may be observed that the error $|e(\xi,\eta)\|$ is found to be smaller when the parameter α^* is close to -1, which corresponds to the node positions of B* and C* shown in Fig. 4.7.

The first and second derivatives of the mapping function f can be obtained from derivatives of the basic functions H_j. To compute the pressure p and tensor components T^{ij}, the four nodal values at points A*, B*, C* and D of each element are chosen as unknowns. As made in previous works with STM [3, 5], derivatives in terms of R and Z are then evaluated on the rectangular mesh by classical finite-difference formulae. Similar results can be obtained in the planar case (Cartesian coordinates).

4.4 Solving the Equations

The aim of STM calculations is the streamline determination, starting from a rectangular computational domain Ω^*. It is obviously expected that small variations of the mapping functions f and g may lead to significant changes of the physical data as the pressure, velocities and stress components, involved in non-linear governing equations. Robust and efficient algorithms are required for the calculations, notably for domains involving singularities for velocities and stresses. In relation to STM characteristics, some elements related to specific features for the governing equations are described in this chapter.

4.4.1 *Consistency and Stability*

When discretizing the equations and unknowns, different approaches are considered in order to ensure consistency and stability of the solving procedure. If a necessary condition of convergence of a numerical scheme is the consistency for linear problems, it should be also of interest to consider numerical errors in non-linear STM situations. Though definitive conclusions on convergence of the programmes cannot be made, it should be of interest to explore effects of successive numerical approximations on the process.

The consistency of a discretizing scheme corresponds to its property to attain an error norm going to zero when the mesh is refined, when this scheme is referred to a known solution to a continuous problem. In other words, if Δx denotes the mesh size corresponding to the discretization in the domain concerned by the solving equations, we get the following property:

$$\lim_{\Delta x \to 0} \|\mathrm{Eq(exact\ solution)}\| - \|\mathrm{Eq(solution\ with\ }\Delta x)\| = 0. \tag{4.30}$$

To verify this equation, it is necessary to consider a problem whose exact (analytical or numerical) solution is known. For example, the reference solution may concern a particular flow as the Poiseuille case. Otherwise, it may be written that the consistency of a scheme is verified, for a problem involving space and time variables, when the truncation error E_{trunc} tends towards zero when the discretization sizes become close to zero. In a problem concerning partial derivatives, this truncation error is defined by the following relation:

$$E_{\mathrm{tronc}} = L_{\mathrm{Taylor}} - L_{\mathrm{Part}}, \tag{4.31}$$

where L_{Taylor} stands for the writing of the discretized scheme the terms of which are replaced by their expressions in terms of Taylor series. The symbol L_{Part} corresponds to the partial derivative operator. In practice, the stability conditions help the rounding errors not to be amplified during the computing progressions.

4.4.2 The Newton–Raphson Algorithm

This method requires determination of the Jacobian matrix of the closed system of N equations of N unknowns:

$$\begin{aligned} &f_1(x_1, x_2, \ldots, x_N) = 0,\\ &f_2(x_1, x_2, \ldots, x_N) = 0,\\ &\ldots\\ &f_M(x_1, x_2, \ldots, x_N) = 0, \end{aligned} \tag{4.32}$$

expressed by

$$\frac{\partial \underline{f}}{\partial \underline{X}}(X_{(i)}) = \left[\frac{\partial f_j}{\partial X_k}\right]. \tag{4.33}$$

The notation $\left[\left.\frac{\partial f_j}{\partial X_k}\right|_{\left(X_1^{(i)}, X_2^{(i)}, \ldots, X_n^{(i)}\right)}\right]$ indicates the partial derivative with respect to the unknown X_k of the jth component of vector $\underline{f}$.

The Newton–Raphson algorithm to solve the matrix equation $f(X) = 0$ may be written as follows:

An initial vector $\underline{X}^{(0)}$ and a small number ε are given.

While $\left\|\Delta \underline{X}\right\| > \varepsilon$, $\underline{X}^{(i+1)}$ is computed from $\underline{X}^{(i)}$, putting $\Delta\underline{X} = \underline{X}^{(i+1)} - \underline{X}^{(i)}$, the following relations are written:

$$\frac{\partial \underline{f}}{\partial \underline{X}}(\underline{X}^{(i)})(\underline{X}^{(i+1)} - X^{(i)}) + \underline{f}(\underline{X}^{(i)}) = 0 \tag{4.34}$$

$$\underline{X}^{(i+1)} = \underline{X}^{(i)} + \Delta\underline{X}$$

with

$$\Delta\underline{X} = \left(\frac{\partial \underline{f}}{\partial \underline{X}}\right)^{-1} \underline{f}. \tag{4.35}$$

If the convergence is obtained, the retained solution $\underline{X}^*$ is the last computed vector $\underline{X}^{(i+1)}$.

4.4.3 Methods Based on Optimization Concepts—Trust Region Algorithm

These methods, based on a closed or over-determined system of equations (more equations than unknowns), are called “Trust Region Optimization algorithms” (e.g.

Ekeland and Teman [7], Fletcher [8], Gastinel [9]). They concern non-linear least squares methods in which one of them corresponds to the Levenberg–Marquardt (LM) computational algorithm. These approaches lead to solve a given system of equations replaced with the study of a function of $\mathbf{R}^n$ in $\mathbf{R}$, the minimum of which is required.

We consider here a closed or over-determined system of M equations of N unknowns, with $M \geq N$:

$$\begin{aligned} & f_1(x_1, x_2, \ldots, x_N) = 0, \\ & f_2(x_1, x_2, \ldots, x_N) = 0, \\ & \ldots \\ & f_M(x_1, x_2, \ldots, x_N) = 0, \end{aligned} \tag{4.36}$$

corresponding to the matrix form:

$$\mathbf{F}(\underline{X}) = \underline{0}. \tag{4.37}$$

We then define the quadratic scalar form, from $\mathbf{R}^m$ to $\mathbf{R}$ (spaces of real numbers vectors), as follows:

$$E(\underline{X}) = \underline{\mathrm{F}}^{\mathrm{T}}(\underline{X}) \cdot \underline{\mathrm{F}}(\underline{X}), \tag{4.38}$$

which is written as

$$E(\mathbf{X}) = \sum_{i=1}^{i=M} (f_i)^2(X_1, X_2, \ldots, X_M), \tag{4.39}$$

where $E(\underline{X})$ denotes the objective function, assumed to be twice differentiable, to be minimized, of gradient $\underline{g}$ required to verify the following relation:

$$\underline{g} = \underline{\nabla E} = \frac{\partial E}{\partial \mathbf{X}} = \left[\frac{\partial E}{\partial X_j}\right] = 0, \tag{4.40}$$

where X_{j} denote the components of the vector $\underline{X}$ of unknowns. The derivation of the objective function $E(\underline{X})$ (Eq. (4.40)) leads to a Jacobian matrix

$$\mathbf{J} = \nabla \underline{E} = \frac{\partial \mathbf{f}}{\partial \mathbf{X}}. \tag{4.41}$$

One also considers the Hessian matrix $H = \underline{\nabla^2 E}$, second derivative of the function E, written as

$$H = \underline{\nabla^2 E} = \left[\frac{\partial^2 E}{\partial X_j \partial X_k}\right]_{1 \leq j,k \leq n} \tag{4.42}$$

also given by

$$\mathbf{H} = \nabla^2 \underline{E} = \mathbf{J}^T \cdot \mathbf{J} + \nabla^2 \underline{E} \cdot \underline{E} \tag{4.43}$$

of components

$$H_{jk} = \nabla^2 \underline{E} = \frac{\partial E_i}{\partial X_j \partial X_k} + \frac{\partial E_i}{\partial X_j \partial X_k} fi, \quad 1 \le j, \ k \le N, \ 1 \le i \le M. \tag{4.44}$$

Trust region methods require evaluations of the gradient of the function E denoted by $\frac{\partial E}{\partial \mathbf{X}} = \left[\frac{\partial E}{\partial X_j} \right]$ of the Hessian or its approximation. To define a trust region optimization algorithm, we make a Taylor development of order 2 for the objective function. Such an approach corresponds to the second-order method leading to the following quadratic local approximation:

$$E(\underline{X} + \underline{\delta}) = E(\underline{X}) + \underline{g}(\underline{X}) \cdot \underline{\delta} + \frac{1}{2} \underline{\delta}^T \mathbf{H}(\underline{X}).\underline{\delta} \tag{4.45}$$

that involves the gradient and the Hessian. Concerning the Test Region, additional details are not provided in this section. However, we underline here that the purpose is to determine a local direction of descent $\underline{\delta}$ towards a solution $\underline{X}$ of the problem under consideration (e.g. Clermont et al. [10]).

In practice, it has been proved that algorithms developed from these methods allow evaluation of iterates towards the numerical solution with an initial estimate rather far from the computed solution [11].

4.4.4 Levenberg–Marquardt (LM) Optimization Algorithm

The Levenberg–Marquardt (LM) method is an iterative and non-stationary optimization algorithm. Its combines the gradient algorithm, slow because it is linear convergence and the faster Newton algorithm of quadratic convergence:

- Initial steps: Gradient algorithm, when the norm of distance of two successive iterates is important, slow but does not diverge.
- Final steps: Newton–Raphson algorithm, when the distance in norm for two successive iterates is reduced.

This method may be considered as a particular case of trust region methods. The iterative algorithm is *non-stationary* in the sense that the matrix operator $G^{(k)}$ that allows the writing of the equation:

$$\underline{X}^{(k+1)} = \underline{G}^{(k)} \left(\underline{X}_{1,}^{(k-1)} \underline{X}^{(k-2)}, \ldots, \underline{X}^{(k-i)} \right), \tag{4.46}$$

for two iterate results at steps (k) and ($k + 1$) is not the same during the iterative process. In the LM algorithm, the Jacobian matrix is generally computed by finite-difference procedures (see, for example, the libraries IMSL and SLATEC). Additional information concerning the Levenberg–Marquardt algorithm are given, for example, in [12].

Numerical works on STM computations have involved the use of the Newton–Raphson method, a trust region optimization algorithm as well as the LM algorithm. The Newton–Raphson approach is retained particularly when the numbers of equations and unknowns are the same.

4.5 Two-Dimensional Flows

APPLICATIONS:

RHEOLOGY, EXTRUSION
CONFINED FLOWS OF POLYMER MELTS
COMPLEX FLUIDS IN CONFINED FLOWS
FREE-SURFACE FLOWS, FIBER SPINNING, TRIBOLOGHY
FLOWS BETWEEN PLATES, TRIBOLOGY

Rheological materials adopted with the governing equations are generally related to Newtonian and non-Newtonian purely viscous fluids, differential and memory-integral viscoelastic models. Obviously, more difficulties arise for materials with non-linear properties often encountered in industry, notably those obeying differential or integral constitutive equations.

- One firstly focus on inelastic materials whose formalism is close to that of the Newtonian model.
- Then, viscoelastic constitutive equations are considered, notably those concerned by a memory-integral forms, expected for a better fit of experimental data.

▼ In the two-dimensional calculations, the primary unknowns (k, p) in planar cases or (f, p) in axisymmetric problems can be considered.
▼ When compared to other methods, the velocity boundary conditions become those related to the mapping functions k (or f).
▼ Specifications for the pressure remain the same than classic velocity–pressure formulations: a pressure value is assumed at a point of the upstream or downstream boundary.

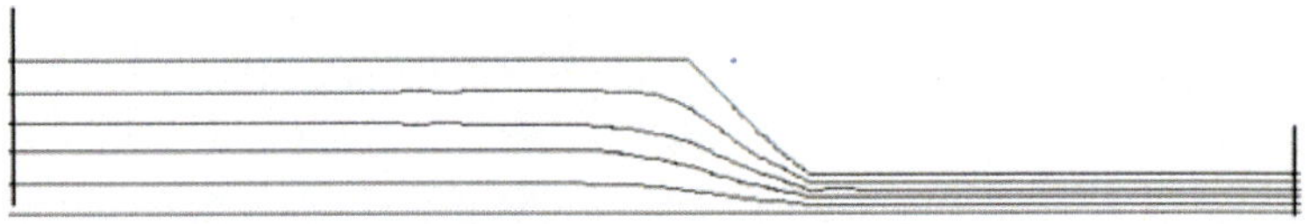

Fig. 4.8 Axisymmetric domain with upstream and downstream sections where the kinematics are known

4.5.1 Flow Rates and Streamlines in a Tube

Let us consider the axisymmetric flow in a non-straight pipe (Fig. 4.8) limited by two upstream and downstream Poiseuille flow zones, corresponding to fully developed flows.

For these respective flow zones, one may refer to cross-sections at $z = z_1$ and $z = z_2$ to write the following equations, corresponding to the flow rate:

$$F_1 = G_1(r) = 2\pi \int_0^r \xi w_1(\xi) d\xi, \tag{4.47}$$

$$F_2 = G_2(r) = 2\pi \int_0^r \xi w_2(\xi) d\xi, \tag{4.48}$$

where w_1 and w_2 are Poiseuille velocities in the upstream and downstream zones.

The abscissae r verifying the relation $F_1 = F_2$ in both relations correspond to the conservation of flow rate for the value of r. Thus, points M_1 and M_2 are located on the same streamline (L) of the flow domain.

4.5.2 Inelastic Models: Newtonian Examples

Such models are often written on the form which is expressed as follows:

$$\underline{\underline{\tau}} = 2\eta(\underline{\underline{D}})\underline{\underline{D}}, \tag{4.49}$$

where $\underline{\underline{D}}$ denotes the rate-of-deformation tensor.

According to Eq. (3.45), for an axisymmetric flow with cylindrical coordinates, the rate-of-deformation tensor $\underline{\underline{D}}$ is expressed by the following matrix:

$$\mathbf{D} = \begin{bmatrix} \frac{\partial u}{\partial r} & 0 & (1/2)\left(\frac{\partial w}{\partial r} + \frac{\partial u}{\partial z}\right) \\ 0 & \frac{u}{r} & 0 \\ (1/2)\left(\frac{\partial w}{\partial r} + \frac{\partial u}{\partial z}\right) & 0 & \frac{\partial w}{\partial z} \end{bmatrix} \tag{4.50}$$

with the following relationships (Eqs. 4.51–4.54). $w_p{'}_R$ denotes the R-derivative of the velocity $w_p(R)$ at the upstream Poiseuille flow section, expressed in terms of the mapping function f:

$$\frac{\partial w}{\partial R} = \frac{w'_{p_R} \cdot R + w_p}{f \cdot f'_R} - \frac{w_p \cdot R \cdot \left(\left(f'_R\right)^2 + f \cdot f''_{RR}\right)}{\left(f \cdot f'_R\right)^2} \tag{4.51}$$

$$\frac{\partial w}{\partial Z} = -\frac{w_p \cdot R \cdot \left(f'_R \cdot f'_z + f \cdot f''_{RZ}\right)}{\left(f \cdot f'_R\right)^2} \tag{4.52}$$

$$\frac{\partial u}{\partial R} = \frac{\partial w}{\partial R} \cdot f'_z + w \cdot f'_{Rz}, \tag{4.53}$$

$$\frac{\partial u}{\partial Z} = \frac{\partial w}{\partial Z} \cdot f'_z + w \cdot f''_{z^2} \tag{4.54}$$

and, along the centreline:

$$\frac{\partial w}{\partial R} = 0, \quad \frac{\partial w}{\partial Z} = -\frac{w_p \cdot f''_{RZ}}{\left(f'_R\right)^3}. \tag{4.55}$$

4.5.2.1 Example 1

Newtonian fluid in a converging flow—Detection of circulation regions (Fig. 4.9).

Secondary flows are generally avoided in ducts as convergent domains of industrial processes. Here, STM simulations consist in detecting the angle α from which

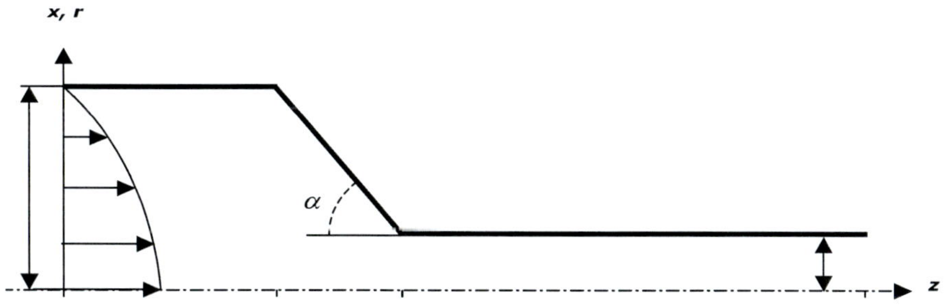

Fig. 4.9 Simple two-dimensional domain involving only open streamlines

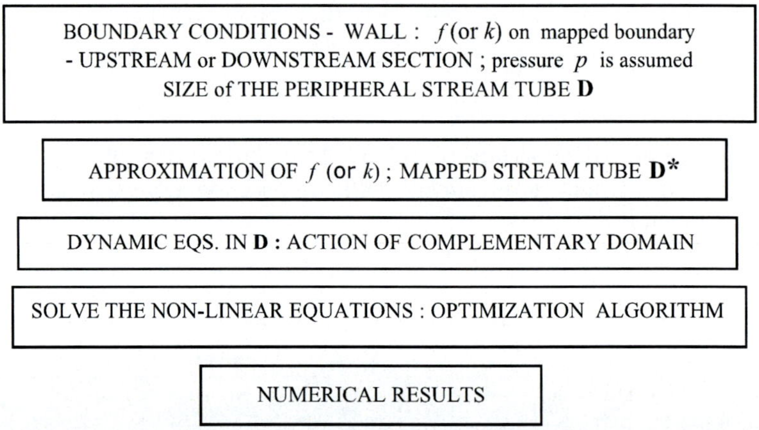

Fig. 4.10 Informations and global steps for computations in the peripheral stream tube: Newtonian fluid in the converging geometry

the algorithm diverges: appearance of eddies. Only the peripheral stream tube is considered while such result requires consideration of the total flow domain with other methods. The computed STM data are considered in relation with predictions of other codes.

According to STM properties, only the peripheral flow domain can be considered for simulations in the converging tube. Figure 4.10 presents different steps for computations in a sub-domain of the converging geometry. The example concerns a Newtonian fluid.

1. *Boundary conditions*

For confined flows with open streamlines (no secondary flows), the boundary conditions for the mapping function k (or f) are expressed as follows:

▼ At the upstream section z_1(fully developed flow), the kinematics are assumed to be known:

$$u = 0, \quad w = w_p(X) \quad \text{or} \quad w = w_p(R). \tag{4.56}$$

In **STM**, such condition corresponds to the following relations:
k (respectively, f) is known at the upstream section such that

$$k = X \quad \text{or} \quad f = R \tag{4.57}$$

and for the derivatives,

$$k'_x = 1 \text{ or } f'_R = 1; \quad k'_z = 0 \text{ or } f'_z = 0. \tag{4.58}$$

This equation indicates that the velocity vector is parallel to the symmetry axis at the upstream section (of fully developed flow) limiting the domain under consideration.

▼ At the downstream section z_2, we write, together with the Dirichlet condition:

$$u = 0, \tag{4.59}$$

a Neumann condition:

$$\frac{\partial w}{\partial z} = 0. \tag{4.60}$$

In STM, the Neumann condition of Eq. (4.61) corresponds to

$$k'_z = 0 \text{ or } f'_z = 0, \tag{4.61}$$

for the formulations (k, p) and (f, p), respectively.

Concerning the pressure p, a value may be assigned at the wall of the downstream Poiseuille section.

▼ the duct wall, the condition for k (or f), is given by the duct shape:

$$k = x_{\text{wall}}(z) \quad (\text{or } f = r_{\text{wall}}(z)). \tag{4.62}$$

At the wall, the stream function Ψ is zero, if adherence of the fluid is assumed, according to the relation:

$$w_p(\text{wall}) = 0. \tag{4.63}$$

2. *Flow domain–Primary formulation–Solving procedure*

An axisymmetric flow domain is considered for the calculations, with only a **peripheral stream tube** (Figs. 4.3, 4.10a) with isothermal conditions. In this stream tube, the elements correspond to those shown in Fig. 4.3, with relations (4.1)–(4.2) for the mapping function f and the pressure p. A primary formulation is associated to a discretization with trapezoidal elements and corresponding relations for the mapping function f and the pressure. Writing the dynamic equations in two-dimensional domains leads to a more complex form than the classic (w, p) equations. For a Newtonian fluid (viscosity μ (Eq. (3.50)), the equations in the axisymmetric isothermal case are written. Using function f and pressure p as primary variables, the non-linear equations are summarized as follows:

$$0 = f^2 (f_R)^4 \frac{\partial \Pi}{\partial R} + A_1 H'(R) + A_2 H'(R) + A_3 H''(R), \tag{4.64}$$

$$0 = f^3 \left(f'_R\right)^4 f'_z \frac{\partial \Pi}{\partial R} - f^3 \left(f'_R\right)^5 \frac{\partial \Pi}{\partial Z} + B_1 H'(R) + B_2 H''(R) + B_3 H'''(R). \tag{4.65}$$

In these equations, $\frac{\partial \Pi}{\partial R}$ and $\frac{\partial \Pi}{\partial Z}$ correspond to the partial derivatives of the variable Π related to the pressure p by the following relation:

$$\Pi = \frac{p}{\mu}. \tag{4.66}$$

▼ The coefficients A_1, A_2, A_3 and B_1, B_2, B_3 are given in Appendix A4.
▼ The action of the complementary domain on the peripheral stream tube under consideration has been written by using Eq. (3.62).
▼ Algorithm to solve the equations: trust region optimization code.

3. *Results*

Figures 4.4, 4.11a and b present, respectively, the peripheral stream tube and the computed streamline limiting the sub-domain under consideration. The results have shown that the trust region algorithm diverges for a converging angle α = 60°, which is the value for which a POLYFLOW-FLUENT code indicates the onset of eddies close to the wall [13].

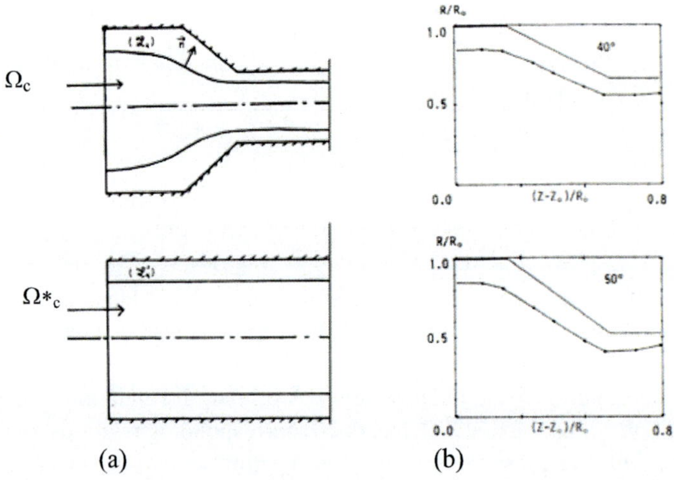

Fig. 4.11 **a** Peripheral stream tube in the physical domain Ω and corresponding stream band in mapped domain Ω*: Ωc is the complementary domain. **b** Examples of a computed streamline for two convergent angles

▼ This numerical study of a peripheral stream tube has allowed to determine the ***appearance of recirculating flow zones*** in a converging flow. Such information may be of interest for industrial processes in converging pipes where secondary flows are avoided.
▼ A similar approach can be applied for fluid material obeying different constitutive models.

4.5.2.2 Example 2: Stick-Slip Problem

This section concerns a numerical Newtonian flow study in a stick-slip (transition from adherence-to-slip problem) in the planar case, with singularities.

The finite element approach allows numerical flow computations without defining special elements or techniques in the vicinity of the transition singularity.

Tribological contacts between plastic or polymer materials can exhibit stick-slip behaviour that generates problems. Numerical studies can be useful for the study of polymer friction leading to developments of new lubricants. Characteristic results are related to adherence-to-slip in a planar surface, using (x, z) Cartesian coordinates. Polymers can have a high static friction coefficient and exhibit stick-slip in tribological contacts.

History: The slip problem has been investigated in several papers (e.g. [14–17]). Several phenomena are associated with slip problems as "shark-skin" in extrusion process (e.g. [18]). Slip at the wall of ducts is considered as a benchmark problem, theoretically (e.g. [19–21]), and widely in numerical simulations ([22–24].

1. *Adherence-to-slip in the planar case—The total flow domain*

The physical domain with boundary conditions is presented in Fig. 4.12. The fluid begins to slip on the wall slip at section $z = z_0$. The mapped computational domain is shown in Fig. 4.13.

The different steps of the simulation process are presented in Fig. 4.14.

2. *Boundary conditions—Governing equations*

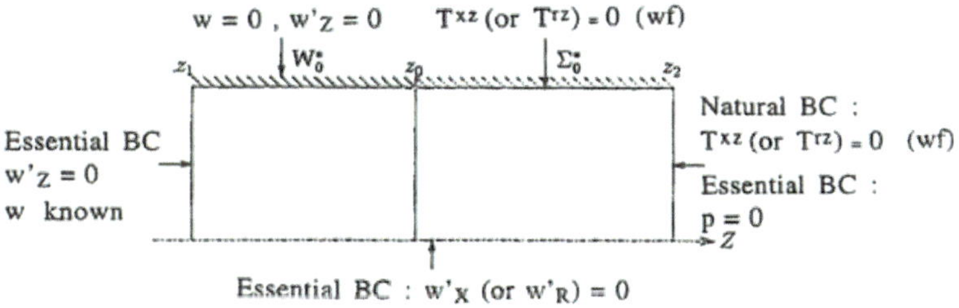

Fig. 4.12 Physical domain Ω with boundary conditions for the stick-slip flow. The natural boundary conditions are written in a weak form for the finite element calculations

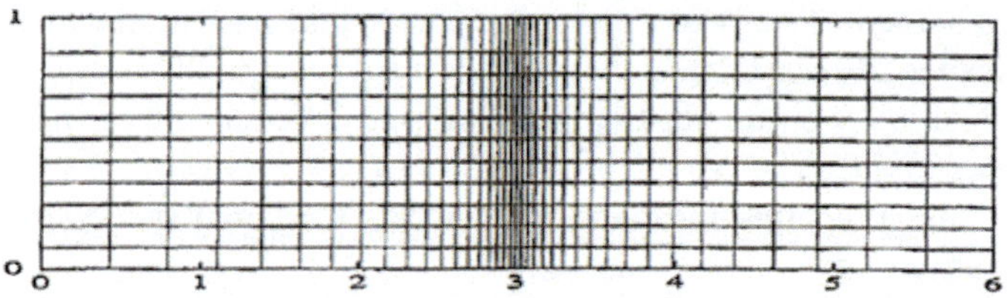

Fig. 4.13 Mapped computational domain (with variables (*x*, *z*)) involving the finite elements with refined mesh in the vicinity of the original slip section

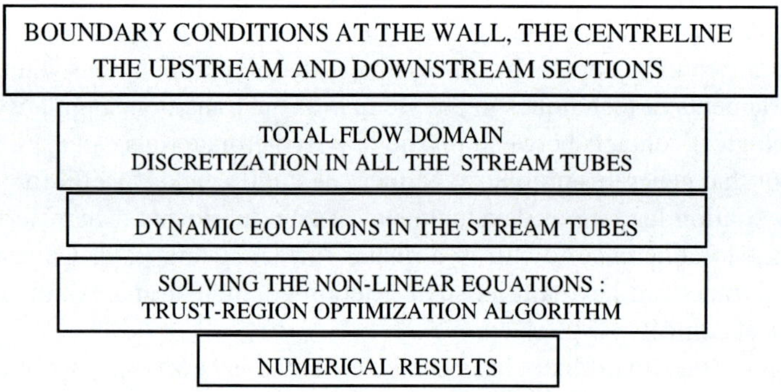

Fig. 4.14 The different steps for the stick-slip problem, solved in the total flow domain by a finite element method

In the planar case, the boundary conditions, summarized in Fig. 4.12, may be expressed as follows, with the mapping function *k*:

▼ At the upstream section z_1:

$$\frac{\partial p(x, z_1)}{\partial z} - h = 0, \tag{4.67}$$

where *h* denotes the head loss. The velocity at the upstream section z_1 is known:

$$w(x, 0) = w_p(x). \tag{4.68}$$

▼ Along the centreline ($X = 0$):

$$k(0, z) = 0, \tag{4.69}$$

$$\frac{\partial p}{\partial X}(X = 0, Z) = 0. \tag{4.70}$$

▼ At the wall, the boundary conditions are expressed as follows:

$$k(X_0, z) = X_0. \tag{4.71}$$

For $z_1 \leq z \leq z_0$,

$$w(X_0, z) = 0. \tag{4.72}$$

For $z_0 \leq z \leq z_2$, the axial velocity component is non-zero and we may write

$$\tau^{yz}(X_0, z) = 0. \tag{4.73}$$

▼ At the downstream section, the following relationships were written, with Neumann conditions, to avoid wiggles:

$$\frac{\partial k(X, z_2)}{\partial z} = 0, \tag{4.74}$$

$$\frac{\partial \tau^{yz}(X, z_2)}{\partial z} = 0. \tag{4.75}$$

The Newtonian fluid is given by $\underline{\underline{T}} = 2\mu\underline{\underline{D}}$, where $\underline{\underline{T}}$ denotes the rheological tensor. Numerical method setups related to partial differential equations generally require information of their mathematical properties. Owing to the third order of derivatives written in the equations, it is not possible, to our knowledge, to determine their type. Nevertheless, one may refer, from a practical viewpoint, to properties relevant to a well-posed problem:

(i) a solution exists;
(ii) this solution is unique;
(iii) the solution depends continuously on the boundary and/or initial conditions.

For reasons of clarity, some elements previously defined in Chap. 3 are recalled here.

Concerning an axisymmetric situation, with variables (r, θ, z) and (R, Θ, Z) in respective domains Ω and Ω*, the transformation involves a function f.

According to previous results of Chap. 3, one may consider an **auxiliary function** F defined, in the axisymmetric case, by

$$F(R, Z) = (1/2) f^2(R, Z). \tag{4.76}$$

This leads to expressing the relationship between F and the w-velocity component as follows:

$$w = \frac{\Psi^*(R)}{F'_R} \quad \text{or} \quad H_1[w, F] = 0. \tag{4.77}$$

In the planar case, with Cartesian coordinates (x, y, z) and (X, Y, Z) for the physical and mapped domains Ω and Ω^*, respectively, the transformation $\Omega^* \rightarrow \Omega$ is given by $x = k(X, Z)$; $y = Y$; $z = Z$. The expression is that the w-component is of the type of Eq. (4.78), with $F = k$. The dynamic Eqs. (3.60) can be formally written in the mapped domain Ω^* as follows:

$$H_2[w, F, p] = 0, \quad H_3[w, F, p] = 0. \tag{4.78}$$

▼ The governing equations are solved in domain Ω^* of boundary $\partial\Omega^*$. We adopt as follows the **Galerkin formulation** with the finite element method:

$L^2(\Omega^*)$ denotes the space of square-integrable functions on Ω^*. The inner product of symbol "<, >" is defined, such that

$$a, b \in L^2(\Omega^*) \quad : < a, b >= \int_{\Omega^a} ab\, \mathrm{d}\omega. \tag{4.79}$$

Defining a partition of diameter h on domain Ω^*, the weak formulation based on the Galerkin method allows us to write an approximate form of problem (P) as follows:

Find $(w_h, p_h) \in W_h \times P_h$ and F that verifies

(i) in Ω^*, the equations: $< H_i[w, p, F, \Phi > = 0\ (i = 2{,}3), \forall\ \Phi \in S_h$

$$H_1[w, F] = 0; \tag{4.80}$$

(ii) the boundary conditions on $\partial\Omega^*$:

$$B_1 - b_1^{(E)} = 0; \quad \left\langle b_2^{(N)}, \Phi^{(N)} \right\rangle = 0, \quad \forall\Phi^{(N)} \in B_h^{(N)}. \tag{4.81}$$

The solution (w, p, F) is as assumed to exist in a space of regular functions $W \times P \times F$. $b_1^{(E)}$ and $b_2^{(N)}$ denote the essential and natural boundary conditions, respectively.

The weak form of the dynamic equations is expressed by using the Green–Gauss theorem, for an element $e^* = [X_1, X_2] \times [Z_1, Z_2]$ of domain Ω^*, in planar and axisymmetric cases. The stress tensor components T^{ij} of the constitutive equation are related to the pressure p and the total stress tensor by the relation $\underline{\underline{\sigma}} = -p\underline{\underline{I}} + \underline{\underline{T}}$. Approximating functions on finite elements for the w-velocity component and the pressure p are considered.

The kinematic relation between w and F, of the type of Eq. (4.57), written at the grid points, is expected to provide data of F in the solving process.

The C^1-continuity at node points, generally difficult to verify because it requires to use prismatic elements (rectangular in two-dimensional cases), is naturally satisfied in STM. The mesh defined on the mapped rectilinear streamlines allows us to adopt Hermitian finite elements of second order, for the w-component. This choice ensures a C^1-continuity of w at the node points [13]. Neumann boundary equations are expressed as essential boundary conditions.

▼ Approximation of the pressure: a classic bilinear interpolation on rectangles.

In the two examples presented in this section (confined and free-surface flows), gravity, surface tension and inertial effects are ignored.

An iterative procedure is defined. Given an iteration number, the following steps related to the problem defined with Eqs. (4.81) are expressed as follows:

(i) w, p and F (f or k) at the boundary (wall and free surface when considered) are estimated from the initialization procedure or the previous iteration step.
(ii) Evaluation of function F in domain Ω^*: the knowledge of estimate values of w at every node of the mesh allows computation of new estimates for the second derivative F''_{X^2}. At each section Z, one computes new values of F by solving with finite element techniques, the one-dimensional Poisson equation:

$$F''_{X^2} = \mathcal{F}(Z)\,, \tag{4.82}$$

where $\mathcal{F}$ denotes a function whose estimate is known.
(iii) Computation of new estimates of the unknowns (w, p) by solving the set of governing equations and boundary conditions by the Levenberg–Marquardt (LM) algorithm: new estimates are provided for w and p.

Boundary conditions corresponding to the two-dimensional stick-slip problem have been shown in Fig. 4.12. Boundaries of the physical and mapped computational domains Ω and Ω^* are identical:

▼ A Poiseuille flow profile is assumed at section $Z_1 = z_1$;
▼ Solid flow conditions are written at the downstream section $Z_2 = z_2$.

Hermite elements retained for the w-velocity component allow to take into account the kinematic singularity at section Z_0.

The evolution norm of the **objective function** is of the order of 10^{-5}.

3. *Results*

Satisfactory comparisons between STM results for computed streamlines and those from Georgiou et al. [15] are provided in the planar stick-slip case (Fig. 4.15).

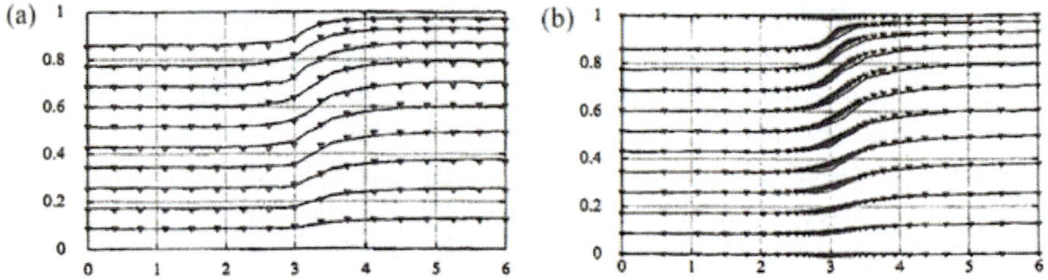

Fig. 4.15 The STM results [14] are consistent with literature data and numerical predictions of the FLUENT code. Comparisons for computed streamlines in the planar stick-slip problem: **a** (∇): results from Georgiou et al. [15], (—): STM results, **b** (•): results from the Fluent code, (—): STM results*

The same STM approach, applied to the axisymmetric swell problem for a fluid at the exit of a cylindrical tube, has also provided consistent results.

4.5.3 Viscoelastic Models in STM Problems

In this section, the stream-tube method is applied to viscoelastic two-dimensional flow problems with differential and memory-integral constitutive equations.

4.5.3.1 Differential Constitutive Equations—Deborah and Weissenberg Numbers

Many papers involving flow computations of viscoelastic materials obeying differential equations have been considered in the literature (e.g. [13]). Those particularly concern finite-difference methods, mostly finite volume and finite element approaches. The complexity of the adopted constitutive equations led to limits of numerical convergence beyond flow rates encountered in experimental or industrial situations. In viscoelasticity, benchmark problems, still considered in the literature, are generally referred to the dimensionless Deborah and Weissenberg numbers, as defined by Poole [16].

– The Deborah number De is defined as

$$De = \lambda / T, \tag{4.83}$$

where T is a characteristic time for the deformation process and λ the corresponding relaxation time.

– The Weissenberg number *We* can be considered as the following ratio of elastic forces and viscous forces for a given material:

$$We \approx \frac{\text{elastic forces}}{\text{viscous forces}}. \tag{4.84}$$

Based on the flow of a viscoelastic fluid in a tube, with the wall shear rate γ_w and the characteristic relaxation time λ of the material, the Weissenberg number *We* is also defined by the following relation:

$$We = \lambda \cdot \gamma_w. \tag{4.85}$$

Among viscoelastic differential equations, the Oldroyd-B model (Chap. 2) has been largely investigated in numerical simulations, notably in benchmark problems (4:1 axisymmetric contraction, swelling problem (e.g. Crochet et al. [13], Owens and Phillips [17]). Using the total stress tensor $\underline{\underline{\sigma}}$, the Oldroyd-B model may be written by the following equations:

$$\underline{\underline{\sigma}} = -\mathrm{p}\underline{\underline{I}} + 2\eta_s\underline{\underline{D}} + \underline{\underline{T}}, \tag{4.86}$$

$$\lambda\frac{\Delta\underline{\underline{T}}}{\Delta t} - +\underline{\underline{T}} = 2\eta_m\underline{\underline{D}}, \tag{4.87}$$

where $\underline{\underline{D}}$ is the rate-of-deformation tensor. Equation (4.87) involves the elastic extra-stress tensor $\underline{\underline{T}}$. $\frac{\Delta}{\Delta t}$ denotes the upper convected derivative operator. By using a classical physical interpretation, η_m and η_s are called the "Newtonian solvent viscosity" and the "polymer viscosity", respectively.

The "viscosity fraction", as defined by Bush [18], is given by the following equation:

$$\beta = \frac{\eta_m}{\eta_m + \eta_s}. \tag{4.88}$$

The viscosity of the Oldroyd-B model is a constant: $\eta_m + \eta_s$. The respective Weissenberg and recoverable shear numbers (*We* and S_R) are considered. For this model, the so-called recoverable shear rate S_R is defined by the following relation:

$$S_\mathrm{R} = \beta\, We. \tag{4.89}$$

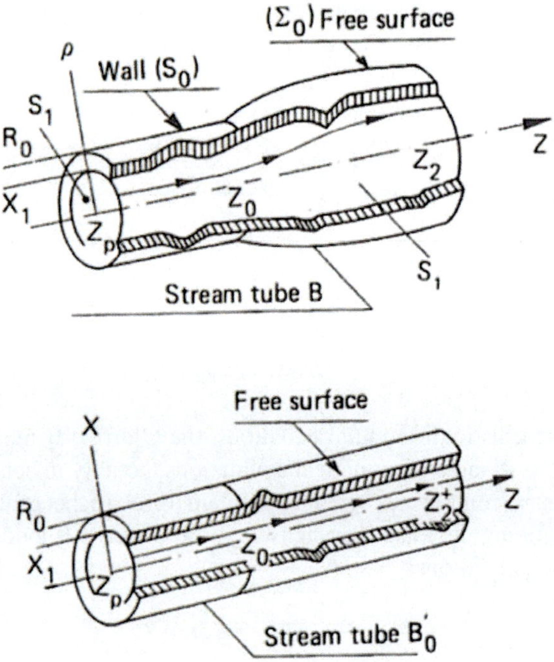

Fig. 4.16 Peripheral stream tube in the swelling problem: physical and mapped domains*

4.5.3.2 Example 3: Extrudate Swell with an Oldroyd-B Fluid

This example concerns STM numerical simulation of extrudate swell in an axisymmetric domain for differential Oldroyd-B fluids, with a streamline approximation. The shape of the free surface, the velocity and pressure are unknowns.

Figure 4.16 presents a three-dimensional view of the swelling problem which concerns a peripheral stream tube limited by

- ▼ the wall and free surface,
- ▼ an inner stream surface,
- ▼ an upstream Poiseuille annular sub-domain and
- ▼ a solid annular flow region of constant axial velocity.

Figure 4.17 presents a planar view of the annular domains corresponding to the peripheral stream tube and its transformed. Unit normal vectors to the free surface and to the upstream and downstream annular sections of the stream-tube sub-domain are also depicted.

Computations with the peripheral stream tube

1. *Boundary conditions—Dynamic equations*

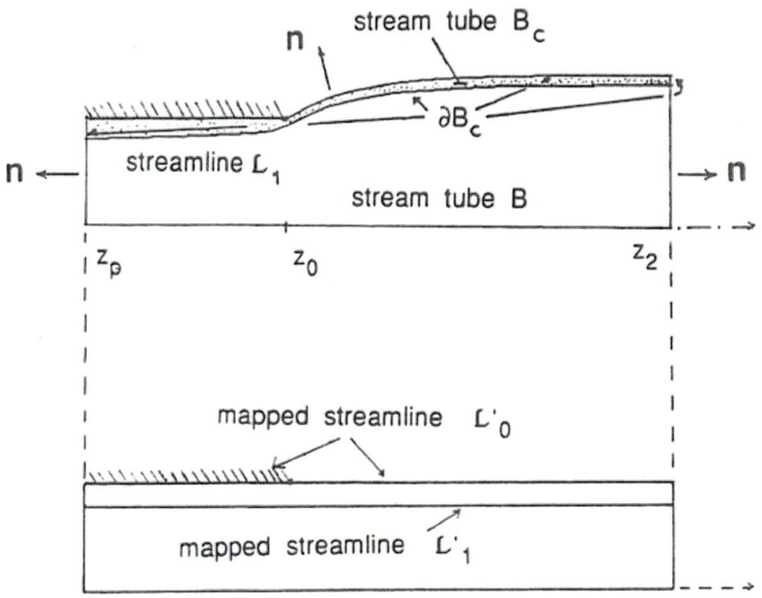

Fig. 4.17 Peripheral stream tube limited by the streamline L_1

Boundary conditions for an axisymmetric flow (variables (r, z)):

▼ At section $z = z_1$: Poiseuille flow. The conditions correspond to those of the axisymmetric stick-slip problem, with Eqs. (4.66)–(4.67).
▼ Along the centreline, same equations than (4.69) and (4.70) for the stick-slip problem.
▼ Along the tube wall, for $z_1 \leq z \leq z_0$, the velocity is zero, from the adherence condition, expressed by Eq. (4.72).
▼ Along the unknown free surface ($z_0 \leq z \leq z_2$) of non-zero velocity, the pressure $p(R_0, z)$ is assumed to be negligible: the surface force components, denoted by F_r and F_z, are zero. At the free surface, the velocity and pressure are unknown. Using n_r and n_z, the two components of the outward unit surface normal vector $\underline{n}$ to the free surface, one may write the following equations at the free surface:

$$u\, n_r + w\, n_z = 0, \tag{4.90}$$

$$\tau_r n_r + \tau_2 n_z = \sigma_t \left(\frac{1}{\rho_1} + \frac{1}{\rho_2} \right), \tag{4.91}$$

projection of the stress vector on the unit normal vector to the surface:

$$\tau_r n_z - \tau_z n_r = 0. \tag{4.92}$$

In Eq. (4.91), ρ_1 and ρ_2 denote the principal radii of curvature of the free surface. σ_t is the surface tension coefficient.

Referring to the cylindrical basis ($\underline{e}_r$, $\underline{e}$, $\underline{e}_z$), the components of the outer normal vector $\underline{n}$ to the free surface are given by the following relations, in terms of the mapping function f:

$$n_r = \frac{l}{\sqrt{1 + (f'z)^2}}, n_\theta = 0, \quad n_r = -\frac{f_z}{\sqrt{1 + (f'z)^2}}. \tag{4.93}$$

In this case, the dynamic equations in cylindrical coordinates are given by **relations (3.61) of Chap.** 3.

The following steps for determining the swell surface by considering the peripheral stream tube are presented in Fig. 4.20 (Fig. 4.18).

2. *Analytical equations for the streamlines: Batchelor–Horsfall approximation*

The physical and mapped domains, shown in Figs. 4.16 and 4.17, are limited:

▼ in the tube of radius r_0, by the last upstream Poiseuille section $z_{p.}$
▼ in the solid flow section z_2 of the jet: R_0 is the final jet radius.

Inertia, surface tension and body forces are ignored.

The flow involves only open streamlines and, for $z > z_2$ (solid flow section), the streamlines become parallel to the z-direction.

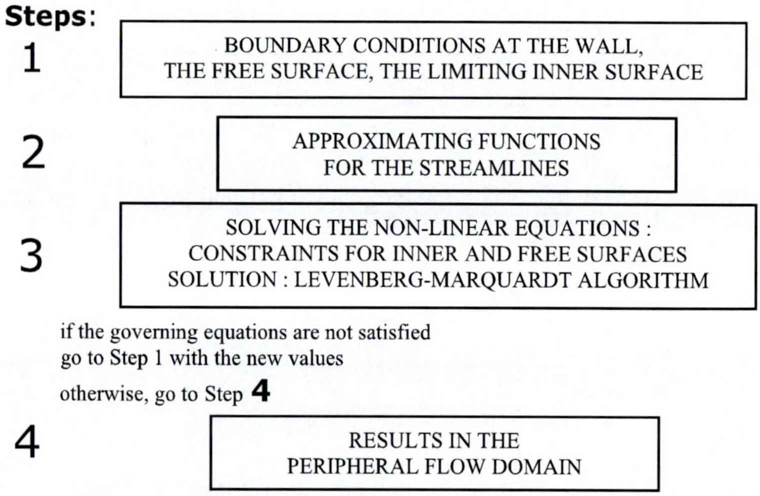

Fig. 4.18 Different steps for determining the unknown free surface—computations for determining the free surface and obtain results for the peripheral stream

To compute the flow field, an analytical approximation of the streamlines is adopted. The following equation:

$$R(z) = r_0[1 + A\exp(-Bz)], \tag{4.94}$$

where A and B are constants, proposed by Batchelor and Horsfall [19, 22] fits well with the jet surface for several fluids [21, 22]. Thus, approximating streamlines in the tube and the jet, using (ρ, Z) as variables in the mapped domain Ω^*, are written as follows:

$$f(\rho, Z) = C_1(\rho, Z) + A(\rho)q(\rho)\{1 - \exp[-(Z - z_0)B(\rho)]\}, \text{ for } z_p \le Z \le z_0 \tag{4.95}$$

$$f(\rho, Z) = R(\rho) + A(\rho)\{\exp[-(z_2 - z_0)B(\rho)] - \exp[-(Z - z_0)B(\rho)]\}, \quad \text{for } z_0 \le Z \le z. \tag{4.96}$$

The function $R(\rho)$ is determined by the conservation of flow rate between the two limiting sections z_p and z_2.

The final jet radius of the free surface is given by

$$R(\rho_0) = R_0. \tag{4.97}$$

Functions A(ρ), B(ρ), $C_1(\rho, Z)$ and B(Z) involve unknown coefficients. Equations (4.95) and (4.96) have provided a good approximation of the streamlines in swell problems [10].

Analytical relations

Analytical streamline relations are considered in the peripheral stream tube limited by

- streamlines L_0 and L_1 in the physical flow domain Ω;
- the upstream and downstream cross-sections at z_p and z_2, respectively.

The following concerns analytical expressions:

Let x_0 and x_1 be the original abscissae of streamlines L_0 and L_1, respectively, at section z_p. The distance $|r_0 - r_1|$ being small, the functions $A(\rho)$ and $B(\rho)$, for r $\in[r_1, r_0]$, can be written as Taylor expansion series:

$$\rho \in [r_1, r_0] : A(\rho) = \alpha_0 + \alpha_1(\rho - r_1) + \alpha_2(\rho - r_1)^2 + O\left[(r_0 - r_1)^3\right], \tag{4.98}$$

$$\rho \in [r_1, r_0] : B(\rho) = \beta_0 + \beta_1(\rho - r_1) + \beta_2(\rho - r_1)^2 + O\left[(r_0 - r_1)^3\right]. \tag{4.99}$$

The function $A(\rho)$, considered as the swelling law *versus* the variable ρ, verifies the boundary condition:

$$A(\rho)|_{\rho=\rho_0=r_0} = A^*. \tag{4.100}$$

A* depends on the final jet radius R_0, according to the following equation:

$$A^* = R_0 - r_0 = \chi\, r_0, \qquad (4.101)$$

where χ is the swell ratio R_0/r_0.

Functions $C_1(\rho, Z)$ and $q(\rho)$ are similar to those given in [19].

Coefficients α_0, α_1, α_2, α_0 and α_1 should be determined. The peripheral stream tube B_c in Ω and its mapped domain B'_c in Ω^* are shown in Fig. 4.20. A mesh is built on the stream band B'_c in Ω^* to compute the unknowns.

3. *Procedure for determining swelling—Constraints*

The complementary domain B_c and functions A, B, C lead to write a non-linear boundary condition equation. Denoting by Σ_0 the unknown free surface, we write the two following constraints (C_1) and (C_2) related to determination of the swell ratio χ:

$$(C_1) : \left[\int_{\Sigma_0} (-p\underline{\underline{I}} + \underline{\underline{T}})\underline{n}\,\mathrm{ds}\right] \cdot \underline{e}_z \text{ is minimum} \qquad (4.102)$$

$$(C_2) : \left[\int_{\partial B_c} \left(-p\underline{\underline{I}} + \underline{\underline{T}}\right)\underline{n}\,\mathrm{ds}\right] \cdot \underline{e}_z \text{ is minimum.} \qquad (4.103)$$

$\underline{\underline{I}}$ denotes the unit tensor and $\underline{n}$ the unit outward vector normal to the boundary surface ∂B_c. The pressure p is required to satisfy the following boundary condition:

$$p(r, z2) = 0 \qquad (4.104)$$

at the solid flow section z_2. The known values of f at this section are obtained from the volume conservation rate.

Using mapping function approximations and stress expressions from the Oldroyd-B equations versus the kinematics, the dynamic equations may be expressed formally as non-linear forms:

$$E_1\{[\alpha_0, \alpha_1, \alpha_2, \beta_0, \beta_1, \beta_2], p_{(M)}\} = 0, \qquad (4.105)$$

$$E_2\{[\alpha_0, \alpha_1, \alpha_2, \beta_0, \beta_1, \beta_2], p_{(M)}\} = 0, \qquad (4.106)$$

for points M of the line L'_1 which limits stream tube B' in the transformed domain Ω^* (Fig. 4.17). The pressure p is an unknown in the computational domain.

The general form of the momentum conservation equations for the stream tube B under consideration, of boundary ∂B, is also written as

$$\left[\int_{\partial B}(-p\underline{\underline{I}} = \underline{\underline{T}})\underline{n}\mathrm{d}s\right]\cdot\underline{e}_z. \tag{4.107}$$

Equation (4.107) involves known boundary values of stresses and pressures at sections z_p and z_2 and may be expressed formally as

$$E_3\{[\alpha_0, \alpha_1, \alpha_2, \beta_0, \beta_1, \beta_2], p_{(M)}\} = 0. \tag{4.108}$$

The constraint C_1 (Eq. 4.102) to be written for the boundary ∂B_c of stream tube B_c requires evaluation of the following quantities:

▼ the unit normal vector $\underline{n}$, by using the functions involved in Eqs. (4.95) and (4.96);
▼ the pressure p;
▼ the stress components T^{ij} at the limiting sections and along streamlines L_0 and L_1 (Figs. 4.16 and 4.17).

Thus, the primary equations and constraints may be expressed in terms of the unknowns $\{\alpha_0, \alpha_1, \alpha_2, \alpha_0, \alpha_1, \alpha_2\}$ and the pressure values of mesh points of the mapped streamline L'_1.

4. *Equations for the peripheral stream tube: Levenberg–Marquardt optimization algorithm*

Equations defined by (4.105)–(4.107) and simple boundary condition equations do not define a closed system with the unknowns. They are written as

$$A(X_1, X_2, \ldots, X_N) = 0, \quad i = 1, 2, \ldots, M, M \neq N, \tag{4.109}$$

where $\underline{X} = (X_1, X_2, \ldots, X_N) \in \mathbf{R}^N$ denotes the vector of unknowns.

Such systems are solved by minimization techniques. They may concern one of the following mathematical problems (P_1) or (P_2), with objective functions F or G.

– **Problem ($\mathbf{P_1}$)** may be expressed as follows:

$$(P_1) : \min\{F(\underline{X}) : \underline{X} \in R^N\} \tag{4.110}$$

with the constraints:

$$C_k(\underline{X}) = 0, \quad \text{with } k = 1, 2. \tag{4.111}$$

Here, the constraints correspond to Eqs. (4.103)–(4.104) related to the swell ratio. F is a function from $\mathbf{R}^N$ to $\mathbf{R}$ and denotes the following quadratic function:

$$F(\underline{X}) = A^T(\underline{X}) \cdot A(\underline{X}) = \sum_{i=1}^{i=M} A_i^2(X_1, X_2, \ldots, X_N). \tag{4.112}$$

– **Equivalent problem (P$_2$):**

$$(P_2) : \min\{G(\underline{X}) : \underline{X} \in R^N\}, \tag{4.113}$$

where G is a quadratic function from $\mathbf{R}^N$ to $\mathbf{R}$ given by the equation:

$$\begin{aligned} G(\underline{X}) &= A^T(\underline{X}) \cdot A(\underline{X}) + C_1(\underline{X})^2 + C_2(\underline{X})^2 \\ &= F(\underline{X}) + C_1(\underline{X})^2 + C_2(\underline{X})^2. \end{aligned} \tag{4.114}$$

Thus, the quadratic function G involves the constraints related to Eqs. (4.102) and (4.103). Non-linear constraint problems are generally "hard cases" (Moré [25]), particularly for non-convex situations. For problems (P_1) and (P_2), the Levenberg–Marquardt (LM) algorithm [12] is used. This solver is economic in terms of time consumption, since the Hessian of the objective function is not explicitly calculated at each iteration step.

When applied to problem (P_1), the LM algorithm provides a satisfactory solution of Eqs. (4.111).

5. *Computations in the total flow domain*

Figure 4.19 presents positions of inner stream tubes in the flow domain as also a typical discretization of elements.

Successive stream tubes T_i $(2 < i <+ \infty)$ (T_2, the first one, is limited by the computed streamline L_1) are considered up to the centreline $\rho = 0$. The procedure for determining the solution for stream tubes T_i, limited by flow streamlines L_{i-1} and L_i, is summarized below. Stream bands B'$_i$ are considered in the computational domain Ω* (Fig. 4.19).

Equations (4.95)–(4.96) provide approximations for the mapping function f. In domain Ω*, the transformed streamlines of L_{i-1} and L_i originated from points ρ_{i-1} and ρ_i, respectively (Fig. 4.20), at section z_p. Similar equations to (4.99) and (4.100) are used for the functions A(ρ) and B(ρ), where the variable ρ is defined on intervals $[\rho_i, \rho_{i-1}]$ for the stream tube T_i (or stream band B_i).

The unknown coefficients $\{\alpha_0, \alpha_1, \alpha_2, \alpha_0, \alpha_1, \alpha_2\}$ related to the expansions of A(ρ) and B(ρ) in $[\rho_i, \rho_{i-1}]$ are introduced in equations similar to (4.105)–(4.107). Data of f and p on the streamline L_{i-1} provide known boundary conditions for the relevant equations, solved with the LM algorithm. The inner stream tubes are calculated up to the centreline.

6. *Example of numerical results*

In the simulations, the Oldroyd-B model is considered with $\lambda = 1$ and several values of the parameter β (Eq. (4.89), from $\beta = 0.01$ to $\beta = 1$, including the case $\beta = 8/9$,

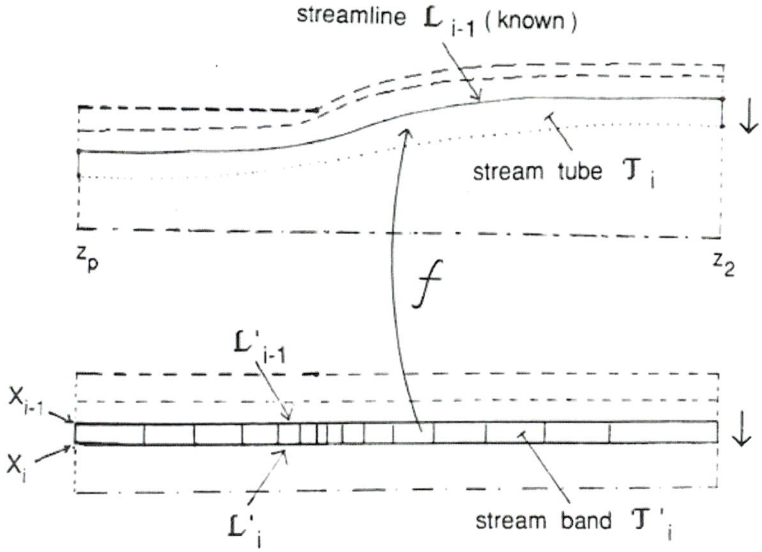

Fig. 4.19 Stream tubes and stream bands in the physical and mapped domains: typical mesh in a stream band. Reference to a stream tube T_i and mapped tube T^*_i

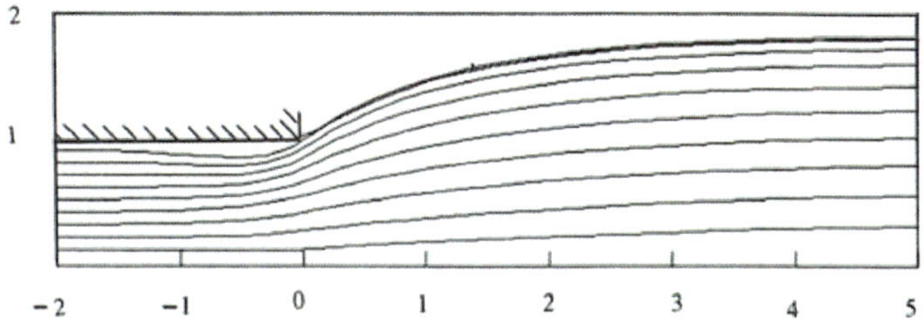

Fig. 4.20 Computed streamlines in a swelling flow with an Oldroyd-B fluid, β = 0.89 for a Weissenberg number of 6. STM swell results have been found to be in agreement with literature data*

largely used in non-Newtonian calculations [13]. The computational domain in the tube and the jet is chosen such that

$$(z_0 - z_p)/r_0 = 2, \quad (z_2 - z_0)/r = 5. \tag{4.115}$$

Little change in the results is observed for the mesh where $(z_2 - z_0)/r_0 = 4$. z_2 corresponds to the downstream solid flow section.

*Reprinted from Ref. [21]. Copyright 1993, with permission from Elsevier

4.5.3.3 Memory-Integral Constitutive Equations

Corotational Memory-Integral Models—the Goddard–Miller Fluid

This model, which requires determination of the corotational frame at every time, is considered for flows in a four-to-one circular contraction. In non-Newtonian fluid mechanics, such geometry is retained for what is called a benchmark problem [13]. In the literature, numerical studies are generally associated to an axisymmetric geometry, particularly for the Oldroyd-B differential constitutive equation [13]. Theoretical works have led to propose robust and efficient numerical schemes. Convergence has been obtained at high Weissenberg numbers W_e defined by [26]

$$W_e = \lambda \frac{U}{L}, \tag{4.116}$$

where λ is a characteristic time, U and L denote a characteristic velocity and a reference length, respectively. In relation to the large potential for fluids with memory to serve as models for industrial polymers, several numerical studies using memory-integral equations have been performed.

Early attempts to solve the governing equations for memory fluids with integral models have involved approximations for the kinematics (Crochet et al. [13]). Convergence was obtained for a limit value of W_e. Relating to problems arising with particle-tracking requirements, some authors have used a Protean coordinate system introduced by Adachi [27, 28], as pointed out in Chap. 3.

Example 4: Flow in an Abrupt Axisymmetric Contraction (Corotational Model)

In the axisymmetric flow in a contraction geometry, a Goddard–Miller constitutive equation, written in a corotational frame, is considered. This model does not reveal the problem of existence of Poiseuille flow beyond a critical Weissenberg number [3]. It should be expected to represent at least qualitatively the behaviour of some real fluids in complex flows as in the present case. The momentum conservation equations and non-linear boundary conditions specific to STM are considered in the study. Convergence is obtained up to a Weissenberg number of 31.

1. *Transformation of the contraction flow domain—Basic equations*

A cylindrical coordinate system $x^1 = r, x^2 = \theta, x^3 = z$ is associated with the axisymmetric flow domain of the 4/1 contraction (Fig. 4.21). The limiting sections z_0 and z_{max}, corresponding to upstream and downstream Poiseuille flows, respectively, are assumed to be known, as also their corresponding velocity profiles. The stream function $\Psi(r, z)$, for $z = z_0$, is then given by

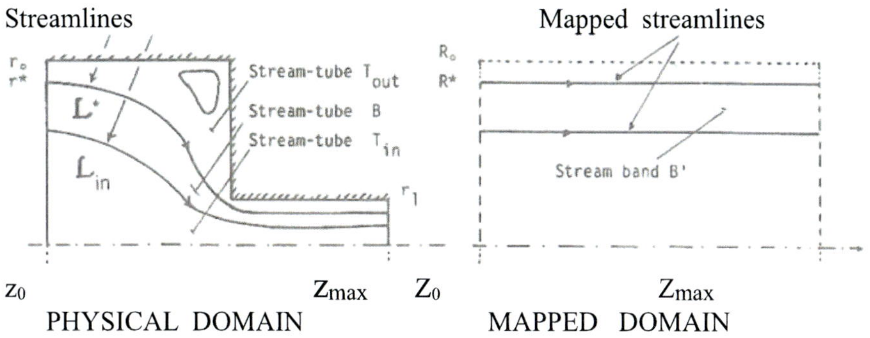

Fig. 4.21 Mapping of the four-to-one contraction geometry*

$$\Psi(r, z_0) = \int_0^r w(s, z_0) s \, ds. \tag{4.117}$$

For the main flow in the physical domain Ω, a one-to-one transformation between the physical domain $\Omega_1 \subset \Omega$ and its transformed domain Ω^*_1 of variables (R, Θ, Z) may be defined, for a mapping function f. The set of streamlines is a simply connected region (non-singular Jacobian of the transformation). In the contraction geometry, STM results should concern open streamlines as shown in Fig. 4.23, for a 4/1 axisymmetric contraction.

According to experimental and computational literature results [13], a recirculating flow zone is expected at the corner (Fig. 4.21). This zone cannot be investigated by STM. The approach concerns a sub-domain of the total flow domain limited by a streamline originated at a point close to the wall, at the upstream section, as shown in Chap. 3 (Fig. 3.12 of Sect. 3.3.7.4). Ω^* is the computational domain where the mapped rectilinear streamlines are parallel to the z-axis (Fig. 4.22). The peripheral stream tube is the first sub-domain limited by the limit streamline.

The main flow elements can be considered in the following cases:

▼ only the peripheral stream band in the mapped domain Ω^*;
▼ all the stream tubes of the mapped domain, starting from the peripheral sub-domain of Fig. 4.22.

Figure 4.23 provides an example of computed open streamlines.

In **Example 4**, STM concerns the global non-circulating flow region, with the Goddard–Miller model. Kinematic variables related to memory-integral equations are evaluated by considering the motion of particles versus time on rectilinear pathlines of the transformed flow domain Ω^* (Fig. 4.22). The time τ corresponding to the position ζ of the particle on the streamline is given by

*Reprint from Ref. [3]. Copyright 1993, with permission from Elsevier

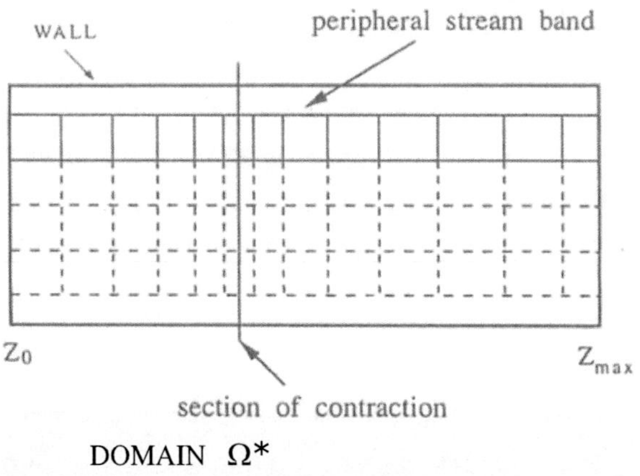

Fig. 4.22 Mapped computational sub-domains for the flow in an abrupt contraction

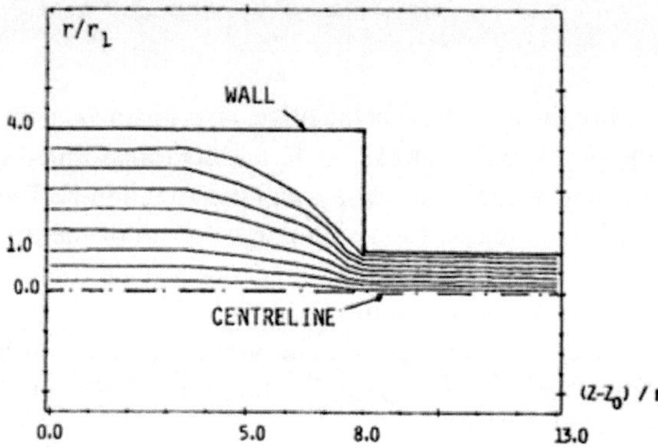

Fig. 4.23 Example of computed streamlines from STM computations, for a 4/1 contraction geometry

$$\tau - \tau_0 = \int_0^{\varsigma} ds/w(R, s), \quad R \leq R^*, \tag{4.118}$$

where τ_0 denotes a reference time related to the position Z_0 of the particle.

A kinematic growing function $\Phi(\mathrm{II}_D)$ of the shear rate is considered. $\mathrm{II_D}$ denotes the second invariant of the rate-of-deformation tensor. Thus, the stress tensor corresponding to the rheological equation is written as follows in a corotational frame:

$$T^*(\mathrm{t}) = \int_{\infty}^{t} \Phi(\mathrm{II}_D) \exp[-\lambda(\mathrm{t} - \tau)\underline{\underline{D}} * (t)\mathrm{d}\tau. \tag{4.119}$$

The asterisk indicates that the equation is related to a corotational frame at time t. λ is related to an average characteristic time, $\underline{\underline{D}}$ the rate-of-deformation tensor. Φ is a kinematic function of the second invariant $\mathrm{II_D}$ of $\underline{\underline{D}}$ [3]. The non-zero stress

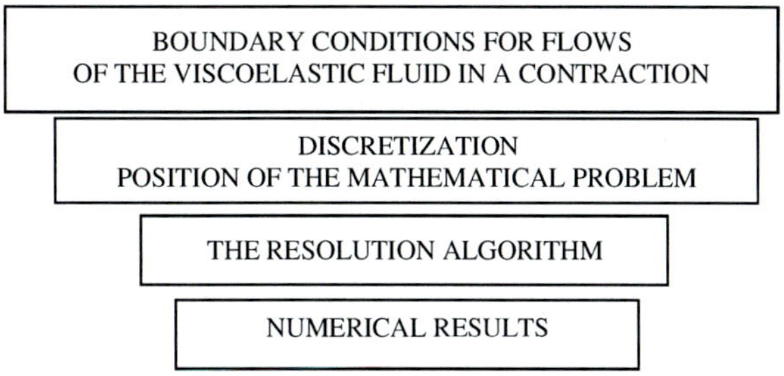

Fig. 4.24 Methodology for the computations with memory-integral equations

components of the extra-stress tensor are

$$
\begin{aligned}
T^{*11}(t) =& \frac{1}{2}\int_{-\infty}^{t} \Phi(\mathrm{II}_D)\exp[-\lambda(\mathrm{t}-\tau)] \\
&\times \big[(\partial u/\partial r + \partial w/\partial z) - (\partial w/\partial r + \partial u/\partial z)\sin \mathrm{A}^*(x,t,\tau) \\
&+(\partial u/\partial r - \partial w/\partial z)\cos \mathrm{A}^*(x,t,\tau)\big]\mathrm{d}\tau,
\end{aligned}
\tag{4.120}
$$

$$
T^{*22}(t) = \frac{1}{2}\int_{-\infty}^{t} \Phi(\mathrm{II}_D)\exp[-\lambda(\mathrm{t}-\tau)](u/r)\mathrm{d}\tau,
\tag{4.121}
$$

$$
\begin{aligned}
T^{*13}(t) =T^{*31}(t) = & \frac{1}{2}\int_{-\infty}^{t} \Phi(\mathrm{II}_D)\exp[-\lambda(\mathrm{t}-\tau)] \\
&\times \big[(\partial u/\partial r - \partial w/\partial z)\sin \mathrm{A}^*(x,t,\tau) \\
&+(\partial w/\partial r + \partial u/\partial z)\cos \mathrm{A}^*(x,t,\tau)\big]\mathrm{d}\tau,
\end{aligned}
\tag{4.122}
$$

$$
T^{*33}(t) = -\Big[\underline{\underline{T}}^{*1}(t) + \underline{\underline{T}}^{*22}(t)\Big].
\tag{4.123}
$$

$\mathrm{A}^*(x, t, \tau)$ is related to the rotation of the corotational frame at time t [3].

Computation of these stress components implies evaluation of integrals of the type of

$$
I_k(t) = \frac{1}{2}\int_{-\infty}^{t} K(x,t,\tau)\exp[-\lambda(\mathrm{t}-\tau)]\mathrm{d}\tau.
\tag{4.124}
$$

Figure 4.24 summarizes the steps defined for STM computations in domain Ω^*.

2. *Boundary condition equations*

Successive stream tubes are considered from the wall to the central flow region. Figure 4.24 illustrates features of the computed streamlines in the main flow zone, determined with successive open stream tubes. The limiting streamline L^* of the first (peripheral) stream tube B (Figs. 4.22, 4.23) investigated in the present analysis starts from the radial abscissa r^* at the upstream section z_0. Recirculating vortices are completely imbedded in the outer stream tube T_{out}. The boundary, not explicitly taken into account in the dynamic equations (existence of secondary flows close to the wall), provides boundary conditions for the function f.

On the inside, the complementary flow domain T_{in} is limited by the streamline L_{in} originating at T_{in} (at $z = z_0$). The limiting open streamlines L^* and L_{in} (of original abscissae r^* and r_{in}, respectively) are unknowns, when omitting the arrival points at the downstream flow section $z_{\max}$, determined by conservation of the flow rate.

The action of the complementary domains T_{in} and T_{out} of the stream tube B implies consideration of non-linear boundary condition equations, expressing the global form of the momentum conservation equation:

$$\int_S \underline{\underline{\sigma}} \cdot \underline{\text{n}}\, \text{ds} = \int_S \left[-p\underline{\underline{I}} + \underline{\underline{T}}\right]\underline{n}\, \text{ds} = \underline{0}, \tag{4.125}$$

where S denotes a closed surface, bounded by streamlines, limiting a volume V, and $\underline{n}$ the unit outward vector normal to the surface S. Inertia band body forces are ignored. The limiting wall provides known boundary conditions on the function f. This equation is applied to the two complementary regions in a form presented in Chap. 3 (Sect. 3.3.7.4) and, from the numerical point of view, was proved to be a sufficient condition for calculations on elementary stream tubes. The components n_1, n_2 and n_3 of $\underline{\boldsymbol{n}}$, unit vector normal to the surface, may be expressed in terms of the transformation function f by

$$n_1 = 1/\left(1 + f_z'^2\right)^{1/2}; \quad n_2 = 0, \quad n_3 = \text{f}_z'/\left(1 + f_z'^2\right)^{1/2}. \tag{4.126}$$

In the axisymmetric case, the previous boundary equation may be reduced into a scalar equation where $\underline{e}'_z$ denotes the unit vector along the z-axis:

$$\left(\int_S \underline{\underline{\sigma}} \cdot \underline{\text{n}}\, \text{ds}\right) \cdot \underline{e}'\text{z} = \left[\int_S \left[-p\underline{\underline{I}} + \underline{\underline{T}}\right]\underline{n}\, \text{ds}\right] \cdot \underline{e}'\text{z} = 0. \tag{4.127}$$

These non-linear boundary condition equations involve the transformation function f, the pressure p and components of the extra-stress tensor. These relations should be written in addition to the simple boundary condition equations.

3. *Discretization—Position of the mathematical problem*

Restricting computations to a stream tube B of the main flow region, we write in a mixed formulation:

▼ the dynamic and constitutive equations;
▼ the simple boundary condition equations related to data of unknowns at the boundaries;
▼ two boundary condition equations, previously proposed, concerning the action of the outer and inner complementary stream tubes T_{in} and T_{out}. These equations are constraints for the governing equations and the ordinary boundary conditions.

The unknowns are considered on a stream band, using Eqs. (4.4)–(4.10) related to the specific rectangular six-point element (Sect. 3.2.3.2.1). To compute the pressure p and tensor components T^{ij}, the four nodal values at points 1, 2, 3, 4 (corner of the elements) of Fig. 4.5 are chosen as unknowns. Using finite-difference formulae, derivatives in terms of R and Z are evaluated on the rectangular mesh, with refinements in the vicinity of the contraction section.

4. *Solving algorithm*

In STM, a previous approximation of the transformation f and the pressure p leads to use second-order computational methods [11] with the Levenberg–Marquardt (LM) algorithm [3, 8]. This particular situation is related to great sensitivity of the equations versus variations of the transformation function. The algorithm has proved its robustness and efficiency. One may recall here the following elements.

The system of the governing equations is written as follows:

$$A_i(X_1, X_2, \ldots, X_N) = 0, \quad i = 1, \ldots, N. \tag{4.128}$$

The numerical computation of the unknowns may be related to the following problem:

$$(\mathrm{P}) : \min\left\{\mathrm{F}(\underline{\mathrm{X}}) : \underline{\mathrm{X}} \in \Re^{\mathrm{N}}\right\}, \tag{4.129}$$

under the constraints

$$C_k(X) = 0, \quad \text{with } k = 1, 2, \tag{4.130}$$

which represent the action of the two complementary domains of the stream tube under consideration.

In Eq. (4.130), F denotes the quadratic function expressed as follows:

$$F(\underline{X}) = A^T(\underline{X}) \cdot A(\underline{X}) = \sum_{i=1}^{i=N} A^2(X_1, X_2, \ldots, X_N). \tag{4.131}$$

In optimization, non-linear constraint problems are generally "hard" to solve [26], particularly for a priori non-convex constraints. To overcome this difficulty, Eqs. (4.131) of problem (P) are formulated as a problem (P'), defined as

$$(P') : \min\left\{G(\underline{X}) : \underline{X} \in R^N\right\}. \tag{4.132}$$

To this problem corresponds the following set of equations:

$$\chi_i(X_1, X_2, \ldots, X_N) = 0, \quad i = 1, 2, \ldots, M, \tag{4.133}$$

in which the initial closed system and the additional equations of constraints are considered in such a way that the quadratic form of F is given by

$$G(\underline{X}) = \chi^T(\underline{X}) \cdot \chi(\underline{X}) = \sum_{i=1}^{i=M} \chi_i^2(X_1, X_2, \ldots, X_N). \tag{4.134}$$

Boundary conditions at the upstream, the downstream sections and at the wall are also considered.

5. *Numerical results for the Goddard–Miller fluid*

Numerical experiments are performed at different flow rates, corresponding to various Weissenberg numbers W_e defined as $W_e = t_M \cdot \gamma_w$, where γ_w is the wall shear rate at the downstream fully developed flow section. Eight stream tubes were considered. t_M denotes the average characteristic time set at 0.48 s and the average time is related to the coefficient λ of the Goddard–Miller equation by the following relation:

$$t_M = 1/\lambda. \tag{4.135}$$

For the [0–22.0] Weissenberg number range, the LM procedure converges in less than 15 iterations. A distinguishing feature of the algorithm is the possibility of determining the solution by direct computation without performing step-by-step calculation at lower Weissenberg numbers. For $W_e > 10$, the length of the mesh had to be increased (Figs. 4.25 and 4.26).

Codeformational Memory-Integral Equations—the K-BKZ Fluid

An axisymmetric case is considered for a codeformational model, with the respective Cauchy and Finder objective strain tensors $\underline{\underline{C}}_t(t')$ and $\underline{\underline{C}}_t^{-1}(t')$ (Eqs. 3.92–3.93 of Chap. 3). Cylindrical coordinates (r, θ, z) and (R, Θ, Z) are adopted. In the frame corresponding to (R, Θ, Z), the Cauchy and

Finger tensor components can be obtained from those of $\underline{\underline{F}}_t$ with Eqs. (3.99)–(3.104). The K-BKZ model (Papanastasiou et al. [29]) is written as

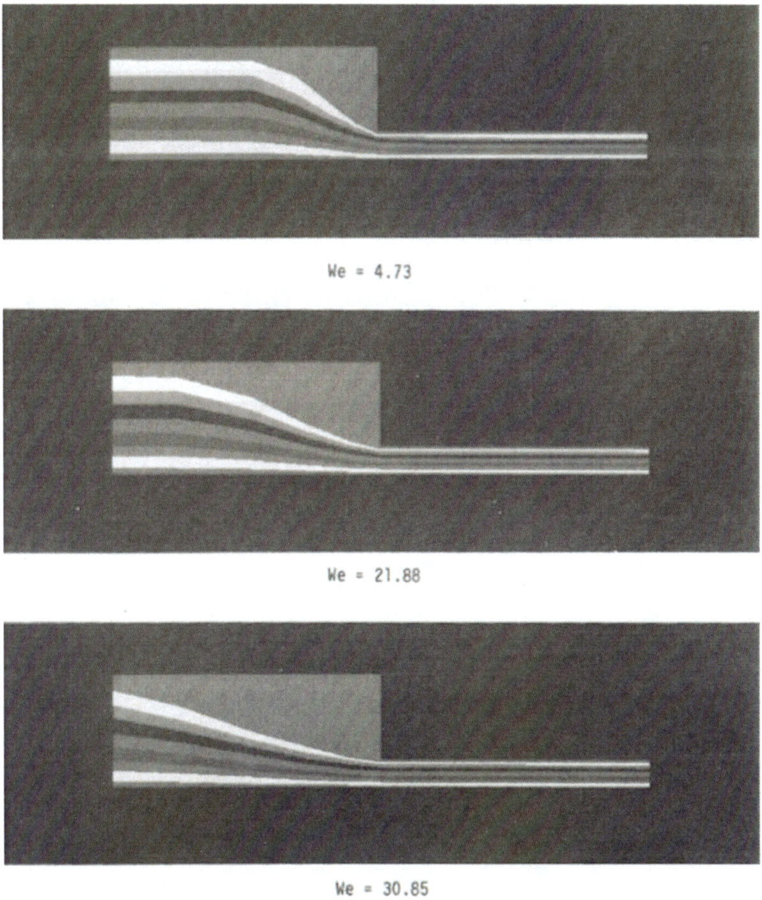

Fig. 4.25 Computed stream tubes for $W_e = 4.73$, $W_e = 21.88$ and $W_e = 30.85$*

$$\underline{\underline{T}}(t) = \int_{\tau=-\infty}^{\tau-t} \sum_{p=1}^{p=8} \left(a_p/\lambda_p\right)\overline{\Psi}(I_C, I_{C-1}) \exp\left[-(t-\tau)/\lambda_p\right] \underline{\underline{C}}_t^{-1}(\tau)d\tau. \qquad (4.136)$$

$\overline{\Psi}(I_C, I_{C-1})$ denotes a kinematic function given by

$$\overline{\Psi}(I_C, I_{C-1}) = \frac{\alpha}{\alpha - 3 + \beta I_{\underline{C}^{-1}} + (1-\beta) I_{\underline{C}}}, \qquad (4.137)$$

where λ_p and a_p are the respective relaxation times and modulus coefficients, α and β material constants, and $I_{\underline{C}}$ and $I_{\underline{C}^{-1}}$ are first invariants of Cauchy tensors $\underline{\underline{C}}_t(\tau)$ and $\underline{\underline{C}}_t^{-1}(\tau)$, respectively.

---

*Reprinted from Ref. [3]. Copyright 1993, with permission from Elsevier

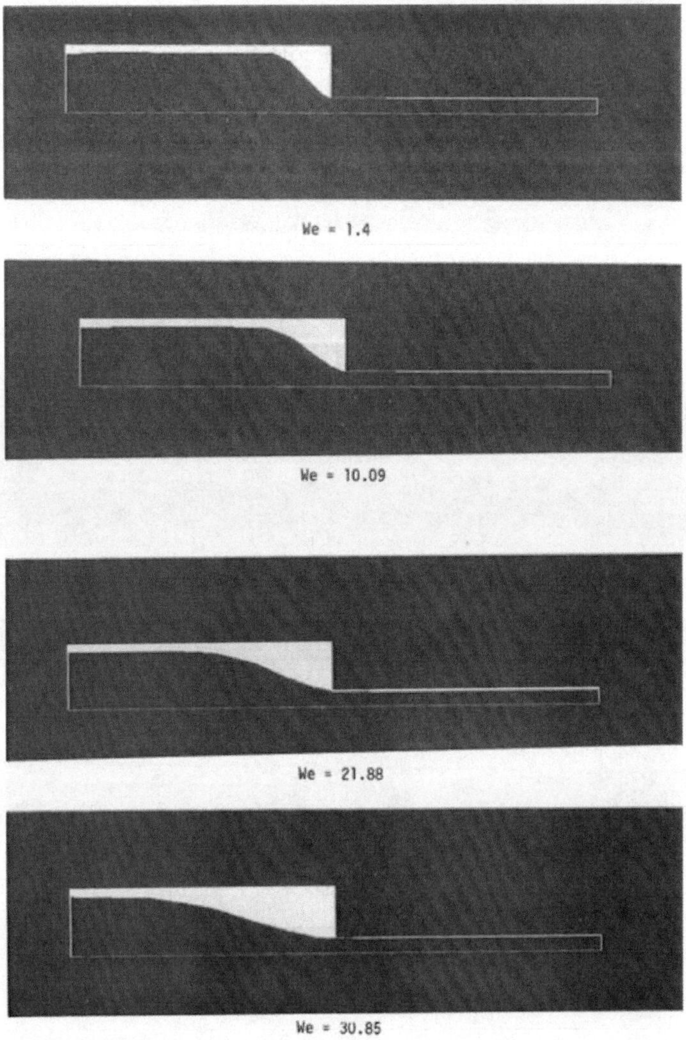

Fig. 4.26 Evolution of the recirculating flow (white) zone (4:1 axisymmetric contraction) at different Weissenberg numbers*

Example 5: Flow in an Axisymmetric Contraction (Codeformational Model)

A contraction geometry is considered with the K-BKZ constitutive equation. Calculations are made by considering the **peripheral stream tube**.

1. *Governing equations and unknowns*

In isothermal cases, only the dynamic governing equations are considered, for the main flow domain. As already pointed out, this interesting property entails taking into account the action of the complementary domain of the stream tube under consideration. In the axisymmetric problem, the scalar integral equation

$$\left(\int_S \underline{\underline{\sigma}} \cdot \underline{n}\, ds\right) \cdot \underline{e}'_z = \left[\int_S \left[-p\underline{\underline{I}} + \underline{\underline{T}}\right]\underline{n}\, \mathrm{d}s\right] \cdot \underline{e}'_z = 0 \tag{4.138}$$

must therefore be written. A mixed system of unknowns is considered, with primary unknowns f, p as well as stress tensor components, evaluated at points of a mapped rectilinear streamline passing through the position M(R, Θ, Z_0) of the upstream Poiseuille flow section (Figs. 4.20 and 4.22).

For computations in a peripheral stream tube B of the main flow region, the governing equations to be solved are

▼ the dynamic equations;
▼ the constitutive equation (mixed formulation);
▼ the simple boundary condition equations related to data concerning the unknowns at the upstream, the downstream sections and the wall;
▼ two boundary condition equations of the same type of those used in the problem considered previously (*Example 5*), concerning the action of the outer and inner complementary stream tubes. These relations are written as constraints for the set of governing equations.

2. *Approximation of the unknowns—Resolution algorithm*

The unknowns concern the simple computational peripheral sub-domain of Fig. 4.21. The stream band is divided into rectangular elements built on the limiting rectilinear mapped streamlines with refinements close to the contraction section. As for the Goddard–Miller model, the boundary is not explicitly taken into account because of secondary flows close to the wall. Known boundary conditions for the mapping function f are provided by the limiting wall data.

Elements (**He$_1$**) and (**He$_2$**) presented in the respective Sects. 4.3.3.2.2 and 4.3.3.2.3 have been tested for the calculations. The modified Hermite element (**He$_2$**) is found to be satisfactory for K-BKZ calculations. So, the same technique than that adopted for the Goddard–Miller case has been set up for the optimization process, with the LM algorithm.

3. *Results with the K-BKZ equation*

For the computations, the following flow parameters are considered:

▼ Apparent shear rate Γ, defined in the downstream tube of radius r

$$\Gamma = 4\, w_1 / r_1, \tag{4.139}$$

where w_1 denotes the average axial velocity in the downstream tube.

▼ A dimensionless number S_R given by [30]:

$$S_R = \left[N_1/(2T^{13})\right]_w, \tag{4.140}$$

where N_1 stands for the so-called “normal stress difference” [30]. The subscript “w” indicates that the ratio is considered at the wall of a Poiseuille flow in the smaller downstream tube. Slow variations are observed at high shear rates.

The numerical results presented here are related to the peripheral stream tube. For the flow rates considered up to $S_R = 2.67$, no convergence problems were encountered with the computational procedure.

Figure 4.27 illustrates the computed limiting streamline L^* of the peripheral stream tube with the K-BKZ model for different values of S_R. The shape of the computed limiting streamlines clearly indicates the relative importance of the recirculating flow zone.

4.6 Concluding Remarks

- To compute the mapping function, different types of discretization and mesh elements are used in the stream bands of the rectangular mapped domains. Finite elements of the Hermite type are generally considered.
- In the stream-tube formulation, the governing dynamic equations are non-linear. To determine the unknowns in the peripheral stream tube, computations may lead to write boundary integral equations at the wall and the inner limiting streamline. Such conditions are constraints taken into account by mean square methods.
- The equations are solved with optimization methods. The Levenberg–Marquardt algorithm proves to be appropriate and accurate for solving STM relations.
- STM simulations highlight interest of computing flows with Newtonian, viscoelastic differential and integral models in various situations.
- In the example of a converging geometry, the limit angle before divergence of the algorithm for a peripheral stream tube is found to be in agreement with data resulting from computations in the total flow domain (FLUENT code).
- For STM flow calculations of viscoelastic memory-integral fluids in abrupt contractions, convergence is obtained at large flow rates (high Weissenberg numbers). The results are in agreement with physical characteristics and literature data (Fig. 4.27).
- It should be underlined that, for flows in abrupt contraction, the streamline close to the vortex zone can be determined by considering a peripheral stream tube. An example is provided with the K-BKZ memory-integral fluid.
- With the adopted optimization algorithms, STM numerical solutions in the rectangular mapped computational domains are obtained with reduced CPU time

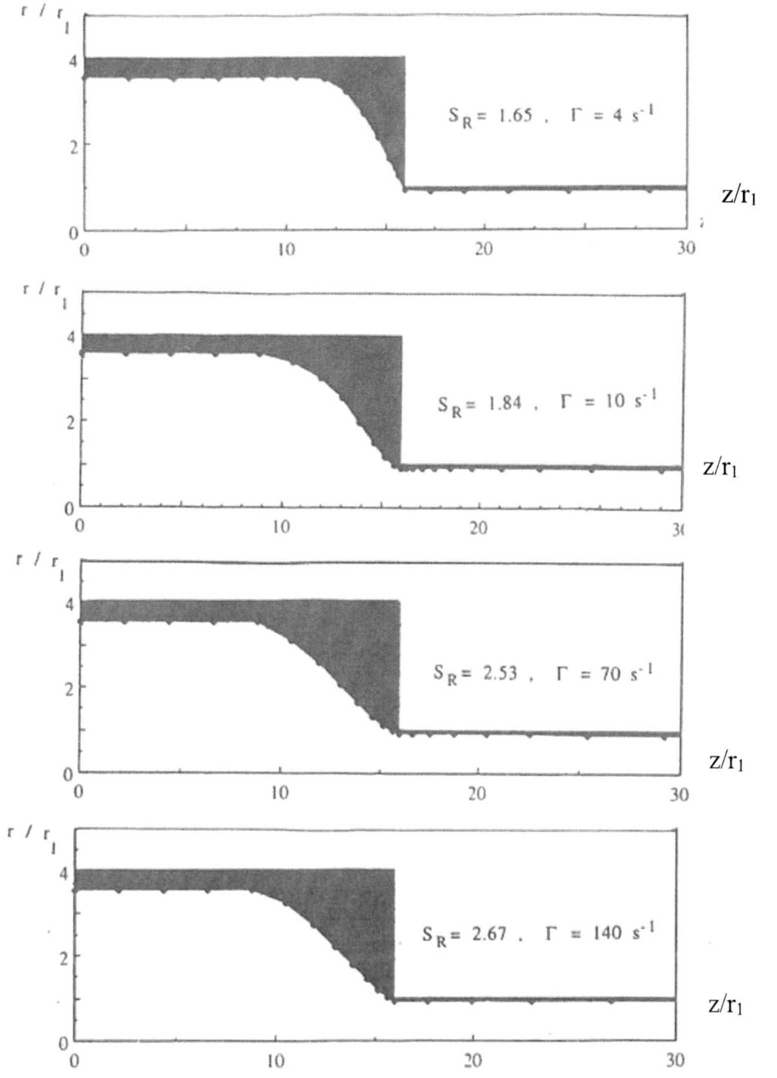

Fig. 4.27 Computed limiting streamlines L^*, showing the evolution of the recirculating flow zones at various S_R values (K-BKZ fluid). © 1995 by John Wiley & Sons, Ltd: reproduced from Ref. [5] by permission of John Wiley & Sons Ltd

and a limited number of unknowns. This should be underlined notably for memory-integral models since particle-tracking problems are avoided.

4.7 Examples of Two-Dimensional Flow Situations for STM

The goal of the examples presented in Figs. 4.27, 4.28 and 4.29 is to propose some elements leading to solve two-dimensional steady flow problems by the STM approach. The three practical examples concern planar and axisymmetric cases (Fig. 4.30).

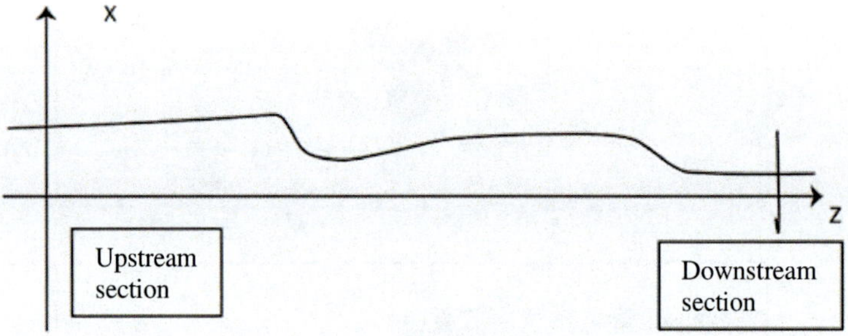

Fig. 4.28 Flow between two plates: Cartesian coordinates (x, z), the flow characteristics do not depend on the y-direction. Poiseuille flows at the upstream and downstream sections. The computations are made in the plane of Fig. 4.28 with kinematic and dynamic equations with variables (x, z) (see Chap. 3). Different constitutive equations can be selected for the calculations

Fig. 4.29 Flow of an axisymmetric duct. Dimensions of the duct may be selected for the calculations. Equations should be written with cylindrical coordinates. A constitutive equation (Newtonian) should be chosen. Choose a flow rate (which leads to the head loss conditions). Write Poiseuille flow conditions at the upstream and downstream sections to obtain the velocities and thus the corresponding streamline positions at these sections

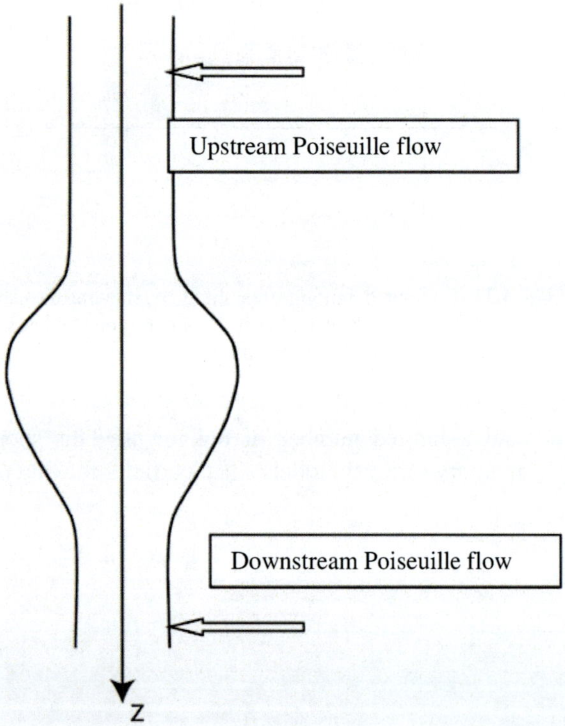

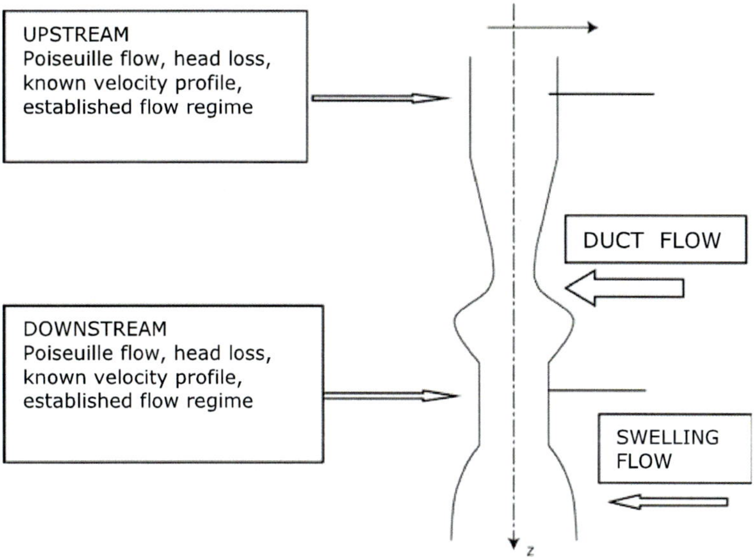

Fig. 4.30 Flow in an axisymmetric tube with swelling at the exit section—choose a flow rate and a head loss at the upstream section. Computations on a rectangular mapped domain—according to the shape of the boundary, a peripheral stream tube may be considered a priori or all the stream tubes, simultaneously, if the swelling region is not concerned. When angles at the boundary are far too important at the wall to ensure only open streamlines, a close limiting line should be considered as shown in Fig. 4.1 and for Examples 4 and 5 of the main text. Different constitutive models can be used. For upstream and downstream Poiseuille sections, Eqs. (4.47) and (4.48) allow a correspondence between streamline points. With more complicated models, upstream and downstream profiles should be computed by using a more complex equation. The downstream head loss can also be determined (equations for Poiseuille conditions)

References

1. Chine A, Ammar A, Clermont J-R (2015) Simulations of two-dimensional steady isothermal and non-isothermal steady flows with slip for a viscoelastic memory-integral fluid. Eng. Comput 32(8):2318–2342
2. Smith GD (1985) Numerical solution of partial differential equations: finite difference methods, 3rd edn. Oxford University Press
3. Clermont J-R, de la Lande ME (1993) Calculation of main flows of a memory integral fluid in an axisymmetric contraction at high Weissenberg numbers. J Non-Newtonian Fluid Mech 46:89–110
4. Chung TJ (1978) Finite element analysis in fluid dynamics. McGraw-Hill Publishers, New-York
5. Bereaux Y, Clermont JR (1995) Numerical simulation of complex flows of non-Newtonian fluids using the stream-tube method and memory-integral constitutive equations. Int J Numer Methods Fluids 21:371–389
6. Touzot G, Dhatt G (1984) Une présentation de la Méthode des Eléments Finis. Editions Maloine, Paris
7. Ekeland I, Temam R (1974) Analyse Convexe et Problèmes Variationnels. Dunod, Gauthier-Villars
8. Fletcher R (1980) Practical methods in optimization, vol 1. Wiley, New-York
9. Gastinel N (1966) Analyse numérique linéaire. Hermann, Paris

10. André P, Clermont J-R (1990) Experimental and numerical study of the swelling of a viscoelastic liquid using the Stream-Tube Method and a kinematic singularity approximation. J Non-Newtonian Fluid Mech 38:1–29
11. Clermont JR, de la Lande ME, Pham Dinh T, Yassine A (1991) Analysis of plane and axisymmetric flows of incompressible fluids with the Stream-Tube Method. Numerical simulation by Trust Region optimization algorithm. Int J Numer Methods Fluids 13:371–399
12. Gourdin A, Boumahrat M (1989) Méthodes numériques appliquées. Technique et Documentation, Editions Lavoisier, Paris
13. Crochet MJ, Davies AR, Walters K (1984) Numerical simulation of non-newtonian flow. Elsevier, New York
14. Normandin M, Radu D, Clermont JR, Mahmoud A (2002) Finite-element and stream tube formulations for computing flows with slip or free surfaces. Math Comput Simul 60(1–2):129–134
15. Georgiou GC, Olson LG, Schultz WW (1991) The integrated singular basis function method for the stick-slip and the die-swell problem. Int J Numer Methods Fluids 13:1253–1265
16. Poole RJ (2012) The Deborah and Weissenberg numbers. Rheol Bull 53(2):32–39
17. Owens RG, Phillips T (2002) Computational rheology. Imperial College Press
18. Bush MB (1990) A numerical study of extrudate swell in very dilute polymer solutions represented by the Oldroyd-B model. J Non-Newtonian Fluid Mech 1:15–24
19. Bachelor J, Horsfall H (1971) Die-swell in elastic and viscous fluids. Rubber and Plastics Research Association of Great Britain. Rep. 189
20. Butler CW, Bush MB (1989) Extrudate swell in some dilute elastic solutions. Rheol Acta 28:294–301
21. Clermont JR, Normandin M (1993) Numerical simulation of extrudate swell for Oldroyd-B fluids by using the stream-tube analysis and a streamline approximation. J Non-Newtonian Fluid Mech 50:193–215
22. Normandin M, Clermont JR (1996)Three-dimensional extrudate swell: formulation with the stream-tube method and numerical results for a Newtonian fluid. Int J Numer Methods Fluids 23:937–952
23. Moré JJ (1983) Recent developments in algorithm and software for trust region methods. Mathematics Programming. The state of the art. Springer, Berlin, pp 258–287
24. Crochet MJ, Keunings R (1982) Finite element analysis of a highly elastic liquid. J Non-Newtonian Fluid Mech 10:339–356
25. Clermont JR, de la Lande ME (1991) On the use of the stream tube method in relation to apparition of secondary flows in plan and axisymmetric geometries. Mech Res Commun 18(5):303–310
26. Auslender A (1976) Optimisation. Méthodes Numériques Appliquées. Editions Masson, Paris
27. Adachi K (1983) Calculation of strain histories with Protean coordinates systems. Rheol Acta 22:326–335
28. Adachi K (1986) A note on the calculation of strain histories in orthogonal streamline coordinate systems. Rheol Acta 25:555–563
29. Keunings R (2003) Finite element methods for integral viscoelastic fluids. In: Binding DM, Walters K (eds) Rheology reviews. pp 167–195
30. Luo XL, Mitsoulis E (1990) An efficient algorithm for strain-history tracking infinite element computation of non-Newtonian fluids with integral constitutive equations. Int J Numer Methods Fluids 11:1015–1031

Chapter 5
Stream-Tube Method in Three-Dimensional Problems

5.1 Introduction

In Chaps. 3 and 4, theory and applications of the stream-tube method are devoted to planar and axisymmetric cases, using one transformation function. A formulation allowing STM studies in three dimensions, presented here, extends computing possibilities to more general cases than those already explored. Streamlines are a priori warping curves. The analysis still refers to the concept of stream tubes in relation with a transformation of the physical domain Ω. The three-dimensional study still allows a streamline L of Ω to be transformed into a straight line parallel to the main flow direction. In relation to three-dimensional flow calculations, a specific method is presented to calculate contour curves in fully developed flow zones of ducts of complex shapes. Such a technique allows the determination of constant velocity curves to be considered in reference domains adopted for the calculations. An example related to a three-dimensional simply connected duct is presented.

The idea of transforming a complicated two- or three-dimensional geometry into a simple region in order to solve boundary value problems of partial differential equations has been developed in numerous cases. This has concerned particularly grid generation techniques in computational fluid dynamics, notably for solving Navier–Stokes equations (e.g. Thompson et al. [3], Eiseman [4]). The set of transformations concerns here three-dimensional flow studies.

5.2 Analysis of Three-Dimensional Flows

5.2.1 Basic Equations

For a three-dimensional flow (Sect. 3.3.3.1) in a domain Ω, the velocity vector $\underline{V}$ is expressed as

J.-R. Clermont and A. Ammar, *Stream-Tube Method*,
https://doi.org/10.1007/978-3-030-65470-2_5

$$\underline{V} = u(x, y, z)\underline{\varepsilon}_1 + v(x, y, z)\underline{\varepsilon}_2 + w(x, y, z)\underline{\varepsilon} \tag{5.1}$$

in a Cartesian orthonormal basis ($\underline{\varepsilon}_i$). As in the previous cases considered, the z-axis is assumed to be related to the main flow direction.

In the present analysis, one considers a limiting upstream section $z_{0,}$ where the kinematics are known [1].

From the incompressibility condition $\nabla \cdot \underline{V} = 0$ of Eq. (2.25), a pair of stream functions $\Psi_1(x, y, z)$ and $\Psi_2(x, y, z)$ can be defined, leading to express the velocity vector $\underline{V}$ by the following vector product:

$$\underline{V} = \underline{\nabla}\Psi_1 \times \underline{\nabla}\Psi_2. \tag{5.2}$$

According to this equation, a streamline L of the physical domain Ω can be considered as the line of intersection of the two surfaces $\Psi_1(x, y, z)$ and $\Psi_2(x, y, z)$, according to the following equations for the velocity components:

$$u = \frac{\partial \Psi_1}{\partial y}\frac{\partial \Psi_2}{\partial z} - \frac{\partial \Psi_1}{\partial z}\frac{\partial \Psi_2}{\partial y}, \tag{5.3}$$

$$v = \frac{\partial \Psi_1}{\partial z}\frac{\partial \Psi_2}{\partial x} - \frac{\partial \Psi_1}{\partial x}\frac{\partial \Psi_2}{\partial z}, \tag{5.4}$$

$$w = \frac{\partial \Psi_1}{\partial x}\frac{\partial \Psi_2}{\partial y} - \frac{\partial \Psi_1}{\partial y}\frac{\partial \Psi_2}{\partial x}. \tag{5.5}$$

Let $x^i(x^1 = x, x^2 = y, x^3 = z)$ be the Cartesian coordinate system related to the physical domain Ω. We consider the one-to-one transformation T which maps a domain Ω^* into Ω^* as follows:

$$T : \Omega^* \to \Omega$$
$$\underline{X} \to \underline{x}$$

with

$$x^i = T^i\left(X^j\right). \tag{5.6}$$

X^j $(j = 1, 2, 3)$ denote the coordinates in the transformed domain Ω^* as follows:

$$X^1 = X, X^2 = Y, X^3 = Z. \tag{5.7}$$

The one-to-one transformation $x^i = T^i(X^j)$ implies the equations:

$$x = f(X, Y, Z), y = g(X, Y, Z), z = Z, \tag{5.8}$$

where the functions f and g are unknowns.

The following assumptions are made:

(i) There exists a section z_0 such that the kinematics is known for $z \leq s_0$.
(ii) The flow does not involve secondary motions.
From assumption (ii), Eq. (5.8) must verify the initial conditions:

$$x = f(X, Y, z_0) = X, y = g(X, Y, z_0) = Y, z = Z = z_0. \tag{5.9}$$

(iii) The mapped streamlines in domain Ω^* are parallel to the Z (or z) axis.

When considering the section z_0 of the duct, the knowledge of the kinematics at $z = z_0$ allows determination of the contour lines Γ of constant velocity w, such that (Fig. 5.1):

$$\Gamma = \{(x, y, z) \in \Omega : w(x, y, z_0) = \text{constant}\}. \tag{5.10}$$

These lines are assumed here to define simply connected regions as shown in Fig. 5.2. The curve $\overline{\Gamma}$ shown in Fig. 5.1 limits the flow domain for $z = z_0$. Determination of the lines of equal velocity is considered in the next section.

According to Eqs. (5.8)–(5.10), the transformed domain Ω^* is a cylinder of basis limited by the curve $\overline{\Gamma}$, of generants parallel to the Z-axis (Fig. 5.3). Concerning the w-velocity component given in terms of functions $\Psi_1(x, y, z)$ and $\Psi_2(x, y, z)$ (Eqs. (5.3)–(5.5)), a function $\overline{\Psi}(X, Y)$ can be defined as follows by the means of Eq. (5.9):

$$\overline{\Psi}(X, Y) = \frac{\partial \Psi_1(x, y, z_0)}{\partial x} \frac{\partial \Psi_2(x, y, z_0)}{\partial y} - \frac{\partial \Psi_1(x, y, z_0)}{\partial y} \frac{\partial \Psi_2(x, y, z_0)}{\partial x}. \tag{5.11}$$

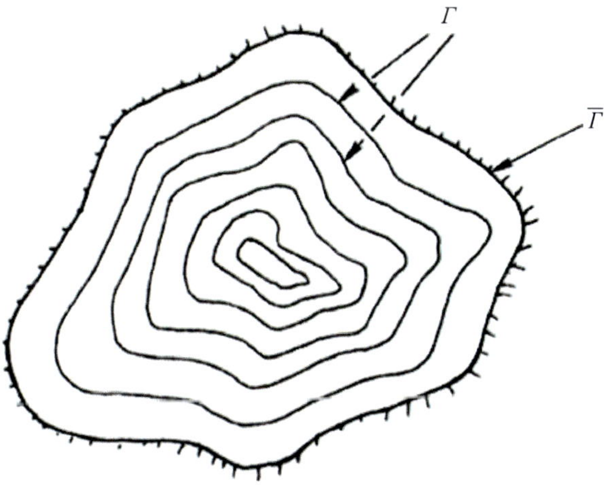

Fig. 5.1 Lines Γ of equal velocity and boundary $\overline{\Gamma}$ in the z_0-plane of the physical domain Ω. Reprinted by permission from Springer Nature Customer Service Centre GmbH from [1], Copyright 1988

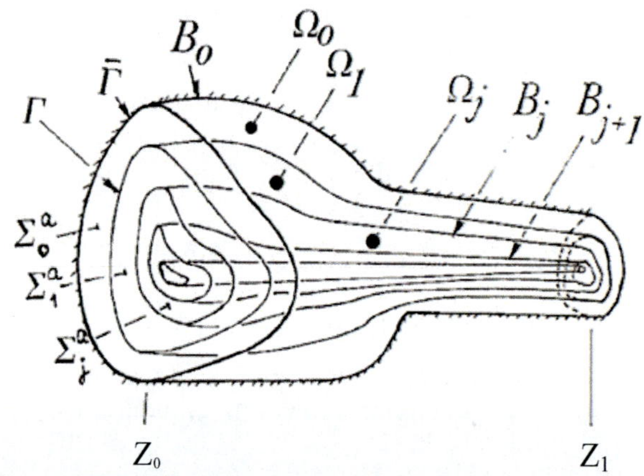

Fig. 5.2 Stream tubes in the physical domain Ω. Reprinted by permission from Springer Nature Customer Service Centre GmbH from [1], Copyright 1988

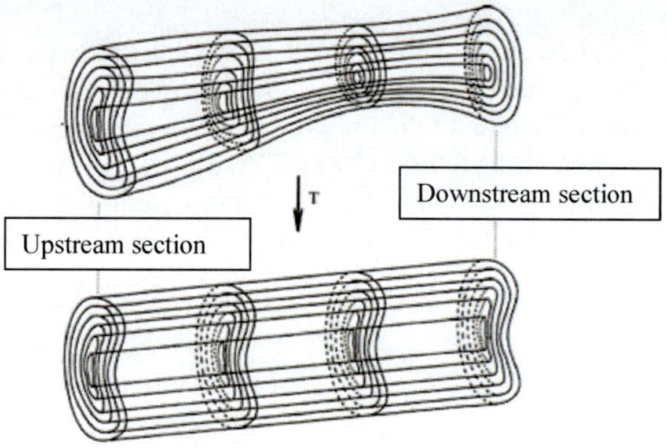

Fig. 5.3 Transformation of the physical domain Ω in a three-dimensional case

The function $\overline{\Psi}(X, Y)$ is known according to assumption (i). From Eq. (5.10), $\overline{\Psi}$ is a constant on every contour curve $\mathbf{\Gamma}$ of the plane $Z = z_0$. A stream tube of the physical domain is mapped into a cylindrical tube the cross-section of which is limited by two curves $\mathbf{\Gamma}$ (Fig. 5.2). The mapped streamlines are parallel and straight. A streamline of domain Ω may be considered as the intersection curve of two surfaces f and g. In domain Ω*, the transformed line of the original streamline is the intersection line of a straight cylinder built on a curve $\mathbf{\Gamma}$ with a plane parallel to the Z-axis. The boundary of domain Ω determines the correspondence between the wall of the physical spatial domain and the boundary of the transformed domain Ω* (Fig. 5.3). The zero-velocity may correspond to the wall. Otherwise, the boundary may stands for a free surface of the flow, as for example, in swelling problems (Fig. 5.4).

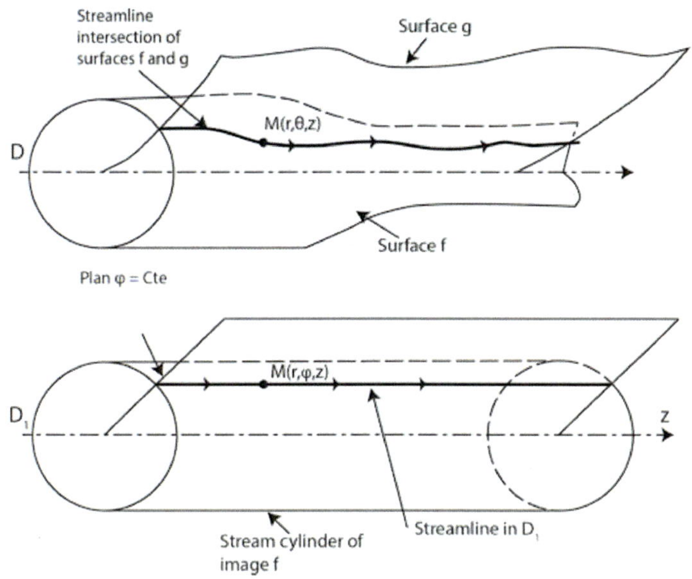

Fig. 5.4 Streamline, surfaces f, g, and their transformed in a mapped domain. The streamline is a warping curve

5.2.2 *Determination of Velocity Contour Curves in Poiseuille Flows*

Lines of equal velocity are of interest for three-dimensional calculations with the Stream-tube method to provide boundary conditions at the upstream and downstream sections, where fully developed flows are assumed. On the other hand, calculations of laminar flow through and around boundaries of complex shapes is related to many practical situations. Geometrical singularities as sharp angles can lead to numerical difficulties arising from the discretization of the governing equations and unknowns (e.g. Roache [2]) particularly when non-linear constitutive fluid models are adopted in the simulations.

This section concerns an original determination of velocity contour values in a complex flow zone corresponding to Poiseuille flow conditions.

The assumption of fully developed flows in ducts of a constant cross-section has been discussed in the literature (see for example [3, 4]). Here, the assumption of only one non-zero velocity component, corresponding to the flow direction, is adopted as follows:

$$\overline{V}(x, y) = \boldsymbol{w}(x, y)\underline{\boldsymbol{\varepsilon}_3} \tag{5.12}$$

or

$$\underline{V}(r, \theta) = \boldsymbol{w}(r, \theta)\underline{\boldsymbol{\varepsilon}_3} \tag{5.13}$$

when using a cylindrical coordinate system ($r = 1, \theta = 2, z = 3$). The zero-velocity assumption is made at the wall.

The existence of a laminar flow (inertia and body forces are neglected) implies a constant gradient for the pressure p along the z-direction, as

$$\frac{\partial p}{\partial z} = -h = \text{constant} \quad (h > 0). \tag{5.14}$$

In such flow conditions, it may be readily shown that the momentum conservation equations are expressed by the following vector equation:

$$\nabla \cdot \underline{\underline{\sigma}} = \nabla \cdot (-p\underline{\underline{I}} + \underline{\underline{T}}) = 0, \tag{5.15}$$

where $\underline{\underline{\sigma}}$ and $\underline{\underline{T}}$ denote the total stress and the rheological stress tensors, respectively. For a Newtonian fluid of constant viscosity η, the relation (5.15) may be written in the form of a Poisson equation:

$$\nabla^2 \cdot w = -\left(\frac{1}{\eta}\right)h. \tag{5.16}$$

where ∇^2 is the Laplacian operator. It is known that Eq. (5.15) also describes the torsion of an elastic rod of uniform and simply connected cross-section (e.g. [5]). Thus, the significant literature on this problem (e.g. [5, 6]) can be transferable to the laminar flow of Eq. (5.15).

The formulation may be adapted to the three following types of cross-sections [3]:

1. Simply connected sections of type A, already investigated [2], where a global one-to-one transformation T_A can be defined (Fig. 5.5);
2. Sections of type B, still simply connected, the shape of which does not allow to define a one-to-one transformation with regard to the velocity contour values (Fig. 5.5a);
3. Sections of type C, doubly connected, close in shape to annular cross-sections as shown in Fig. 5.5b, where a global one-to-one transformation cannot generally be determined.

The approach considered here is specific to ducts B and C:

1. We investigate local sub-domains W_i allowing one-to-one local transformations T_ι: $W*_i \rightarrow W$ to be defined, notably sub-regions close to the boundary ∂W of zero velocity. Simple circular lines are used as contour velocity curves and mesh lines in the mapped computational sub-domains $W*_i$. This limits singularity problems when evaluating the derivative operators defined in the governing equations.

The following equations may be defined as

$$T_A : (R, \phi) \rightarrow (r, \varphi), \tag{5.17}$$

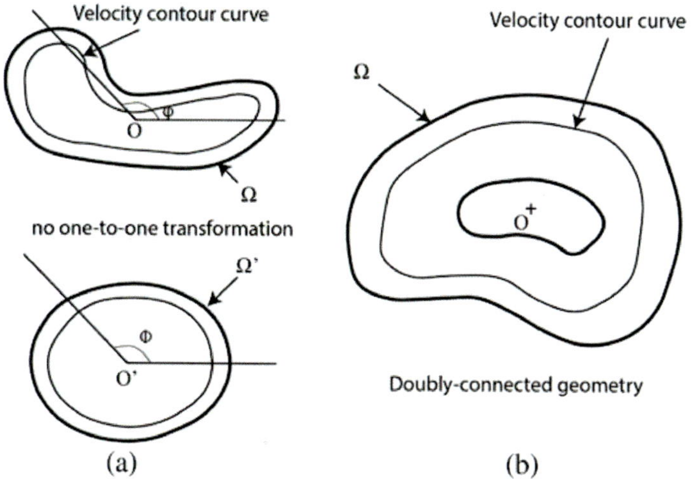

Fig. 5.5 Different shapes of ducts for local transformations and decomposition of domains. **a** Ducts of type B; **b** ducts of type C

$$R = \lambda(R,\phi); \quad \varphi = \phi. \tag{5.18}$$

The central flow region, denoted by W_c, is expected to involve more than local maximum for the velocities. It may be also conjectured that the boundary ∂B_c does not contain geometrical singularities. In this sub-domain, the governing equations are solved directly, using the variables (r, φ) of the sub-domain W.

We have found it of interest to provide information related to details from a paper on the subject [3].

- *Ducts of type B*

In this case, the original physical domain W is subdivided into three overlapping sub-domains B_i ($i = 1, 2, 3$) and B_c according to the following relation (Fig. 5.6a):

$$W = \left[\bigcup_{i=1,2,3} W_i\right] \bigcup Wc. \tag{5.19}$$

The central sub-domain W_c may involve closed velocity contour values. The local transformations relate sub-domains W_i ($i = 1, 2, 3$), involving parts of the boundary ∂W, into sub-domains W^*_i ($i = 1, 2, 3$), where the velocity contours are assumed to be concentric circles. Sub-domains W^*_2 and W^*_3 are located at the ends of the physical domain W; as shown in Fig. 5.6a.

Similar equations to (5.17) and (5.18) relate the geometrical mapped sub-domains W^* used in the computations (Fig. 5.6b) to sub-domains W_i. Local cylindrical coordinate systems (r_i, φ_i) and (R_i, ϕ_i) are adopted, respectively for sub-domains W_i and

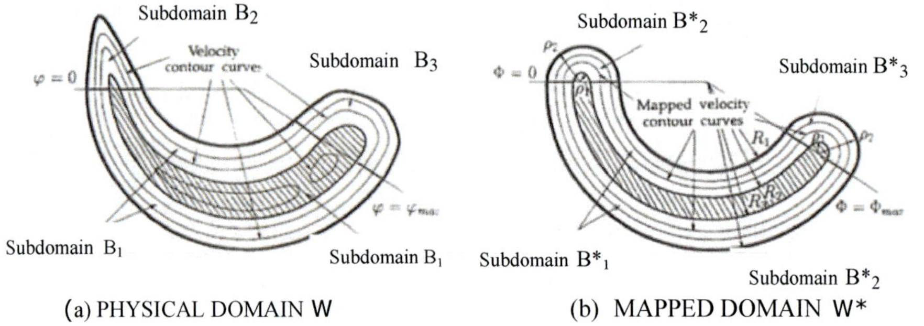

Fig. 5.6 Transformation of domains for cross-sections of type B: **a** physical domain W involving the hachured region W_c non-concerned by the transformation; **b** mapped sub-domains W^*_1, W^*_2 and W_c[*]

W^*_i ($i = 1, 2, 3$). A one-to-one transformation T_i may be expressed as follows:

$$i = 1, 2, 3: \quad r_i = \lambda_i(R_i, \phi_i), \quad \varphi_I = \phi_I \tag{5.20}$$

with the specific relations (Fig. 5.6b):

$$R_1 \leq R \leq R_2, R_3 \leq R \leq R_4, (0 < R_1 < R_2 < R_3 < R_4), \quad 0 \leq \phi \leq \phi_{\max}. \tag{5.21}$$

For sub-domain W^*_1 and for sub-domains W^*_2, W^*_3 (Fig. 5.7):

$$\rho_1 \leq R_i \leq \rho_2, 0 \leq \Phi \leq \pi.$$

- *Conservation equations*

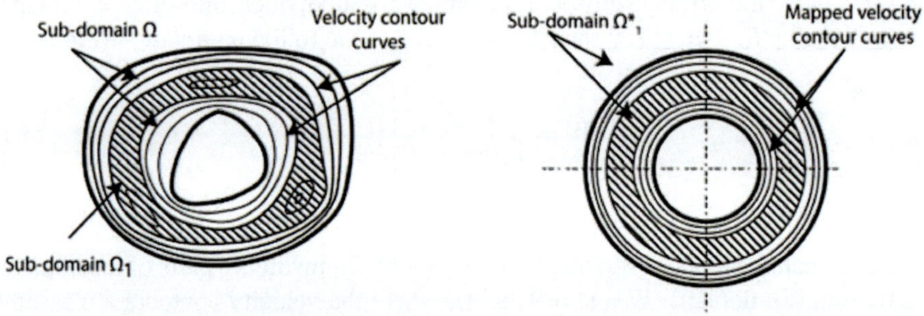

Fig. 5.7 Transformation of domains for cross-sections of type C

[*]Reprinted from Ref. [3], Copyright 2006, with permission from Elsevier

The governing equations are reduced into the dynamic equations and boundary conditions. The stress tensor components are obtained from the constitutive equation for the fluid under consideration.

In cylindrical coordinates ($r = 1$, $\varphi = 2$, $z = 3$), we get, using a pressure drop h, the following dynamic equations in terms of the stress components as

$$\frac{\partial p}{\partial r} - \frac{\partial T^{rr}}{\partial r} = 0, \tag{5.22}$$

$$\frac{\partial p}{\partial \varphi} - \frac{\partial T^{\varphi\varphi}}{\partial \varphi} = 0, \tag{5.23}$$

$$\frac{\partial T^{rz}}{\partial \varphi} + \frac{T^{rz}}{r} + \left(\frac{1}{r}\right)\frac{\partial T^{\varphi z}}{\partial \varphi} - h = 0. \tag{5.24}$$

These equations are written for direct calculations in the central sub-domain region W_c in the original domain W.

Using the unknown function λ for a mapped sub-domain W^* of variables (R, ϕ) and the derivative operators of Eq. (5.18), the dynamic Eqs. (5.22)–(5.24) become

$$\frac{\partial p}{\partial R} - \frac{\partial T^{rr}}{\partial R} = 0, \tag{5.25}$$

$$\frac{\partial p}{\partial \Phi} - \frac{\partial T^{\varphi\varphi}}{\partial \Phi} = 0, \tag{5.26}$$

$$\lambda'_R T^{rz} + \lambda'_R \frac{\partial T^{rz}}{\partial R} - \lambda'_\varphi \frac{\partial T^{\varphi z}}{\partial R} + \lambda'_R \frac{\partial T^{\varphi z}}{\partial \phi} = -\lambda \lambda'_R h, \tag{5.27}$$

where the superscripts correspond to coordinates $r = 1$, $\varphi = 2$ and $z = 3$.

Equations (5.22), (5.23), (5.25), (5.26) vanish for Newtonian and purely viscous fluids where only dynamic Eqs. (5.26) and (5.27) are concerned [3]. When using viscoelastic constitutive equations and denoting by z_0 a reference abscissa, the pressure may be written as [7]

$$p(r, \varphi, z) = -h(z - z_0) + \mathrm{K}(r, \varphi). \tag{5.28}$$

Together with the dynamic and constitutive equations, we write.

- the boundary equations for the unknowns,
- the symmetry conditions,
- the compatibility equations at the boundaries of adjacent sub-domains.

- *Discretization and numerical procedure*

The computationally mapped sub-domains W^*_i, referred to cylindrical coordinates (R, ϕ), allow a rectangular grid to be defined for discretizing the equations and unknowns. The iterative procedure used for solving the equations leads to consider, in domain W at each step [k], the updated contour curve of constant velocity defined

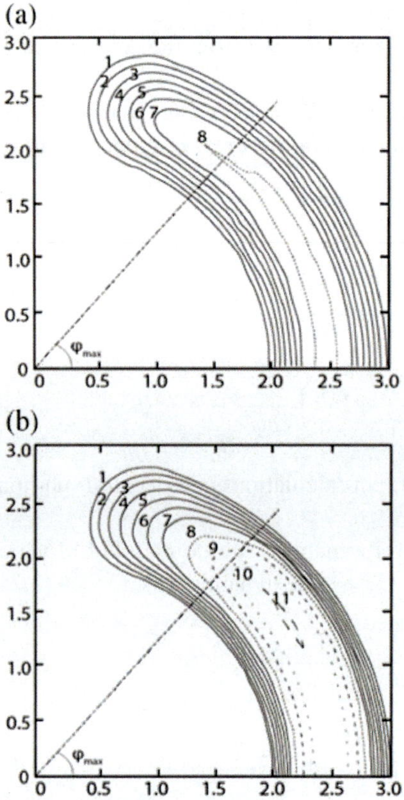

(a)

No. of contour value	w [cm/s]
1	0.00
2	1.60
3	3.05
4	4.35
5	5.48
6	6.47
7	7.29
8	8.60

(b)

No. of contour value	w [cm/s]
1	0.00
2	1.07
3	2.03
4	2.90
5	3.66
6	4.31
7	4.86
8	5.73
9	6.61
10	7.48
11	8.35

Fig. 5.8 Computed velocity contour values in duct of type α_1 **a** Newtonian fluid; **b** non-Newtonian purely viscous (Carreau) fluid

from the use of sub-domains W^*_i. The equations are solved by the LM (Levenberg–Marquardt) algorithm [3]. Some examples are provided in Figs. 5.8 and 5.9.

The method allows the determination of contour curves of constant velocity, in relation to flow rates corresponding to upstream and downstream sections of established flow regimes.

5.2.3 Computations of Kinematics

According to Eq. (5.9), the derivative operators in terms of x, y, z can be written as follows:

$$\frac{\partial}{\partial x} = \frac{1}{\Delta}\left[g'_Y \frac{\partial}{\partial x} - g'_X \frac{\partial}{\partial y} \right], \tag{5.29}$$

Fig. 5.9 Computed velocity contour values in ducts with the viscoelastic Wagner fluid*

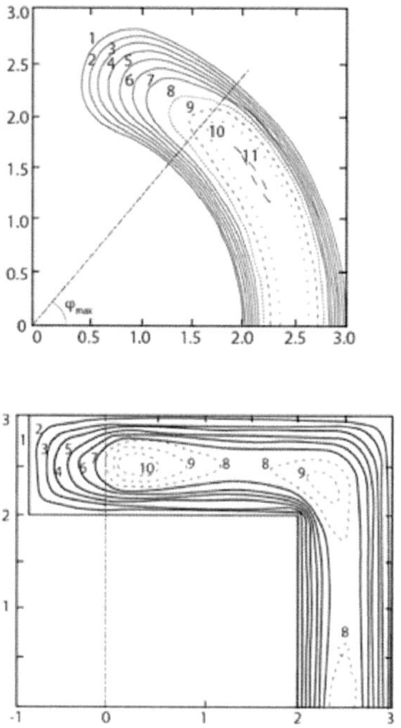

No. of contour value	w [cm/s]
1	0.00
2	1.07
3	2.03
4	2.90
5	3.66
6	4.31
7	4.86
8	5.73
9	6.61
10	7.48
11	8.35

No. of contour value	w [cm/s]
1	0.00
2	1.61
3	3.07
4	4.38
5	5.53
6	6.51
7	7.34
8	8.07
9	8.81
10	9.54

$$\frac{\partial}{\partial y} = \frac{1}{\Delta}\left[f'_{Y} \frac{\partial}{\partial X} + f'_{X} \frac{\partial}{\partial Y} \right], \tag{5.30}$$

$$\frac{\partial}{\partial z} = \frac{1}{\Delta}\left[\left(f'_{Y} g'_{Z} - f'_{Z} g'_{Y}\right) \frac{\partial}{\partial X} + \left(f'_{Z} g'_{X} - f'_{X} g'_{Z}\right) \frac{\partial}{\partial Y} \right] + \frac{\partial}{\partial Z}, \tag{5.31}$$

where Δ is the Jacobian of the transformation T:

$$\Delta = \frac{\partial(x, y, z)}{\partial(X, Y, Z)} \tag{5.32}$$

of inverse

$$\Delta^{-1} = \frac{\partial(X, Y, Z)}{\partial(x, y, z)} \tag{5.33}$$

is given by the relation

$$\Delta(x, y, z) = f'_{X} g'_{Y} - g'_{X} f'_{Y} \,. \tag{5.34}$$

The non-singularity of Jacobian implies that the transformation is one-to-one which, in turn, implies that secondary flows are not taken into account. This condition

*Reprinted from Ref. [3], Copyright 2006, with permission from Elsevier

also means restricting the computation only to main flows, as in two-dimensional cases considered in Chaps. 3 and 4.

The condition $\Delta \neq 0$ is related to the assumption that the tangent to the streamline curve in the physical domain has no inflexion point in the plane (x, y).

In relation with Eq. (5.29), two functions H and K are defined according to the following expressions:

$$H = f''_{\varphi} \cdot g'_{Z} - f''_{Z} \cdot g'_{\varphi}, \tag{5.35}$$

$$K = f'_{Z} \cdot g'_{R} - f''_{R} \cdot g'_{Z}. \tag{5.36}$$

From Eqs. (5.3)–(5.5), (5.29)–(5.31), the computation of the velocity components in terms of f,g and $\overline{\Psi}(X, Y)$ leads to

$$u = \frac{f'z}{\Delta}\overline{\psi}(X, \mathrm{Y}), \tag{5.37}$$

$$v = \frac{g'z}{\Delta}\overline{\psi}(X, \mathrm{Y}), \tag{5.38}$$

$$w = \frac{1}{\Delta}\overline{\psi}(X, \mathrm{Y}). \tag{5.39}$$

The natural basis $(\underline{B}_i)$ related to the new coordinate system (X^j), for a material point of vector $\underline{M}$ is given by

$$\underline{B}_1 = \frac{\partial \underline{M}}{\partial X} = f'_X\, \underline{\varepsilon}_1 + g'_X\, \underline{\varepsilon}_2, \tag{5.40}$$

$$\underline{B}_2 = \frac{\partial \underline{M}}{\partial X} = f'_Y\, \underline{\varepsilon}_1 + g'_Y\, \underline{\varepsilon}_2, \tag{5.41}$$

$$\underline{B}_3 = \frac{\partial \underline{M}}{\partial X} = f'_Z\, \underline{\varepsilon}_1 + g'_Z\, \underline{\varepsilon}_2 + \underline{\varepsilon}_3. \tag{5.42}$$

The derivative operators of a function A, for expressions in terms of variables (X, Y, Z), are written as follows:

$$\frac{\partial A(x, y, z)}{\partial X} = \frac{1}{\Delta}\left[g'_Y \frac{\partial A(X, Y, Z)}{\partial X} - g'_X \frac{\partial A(X, Y, Z)}{\partial Y}\right], \tag{5.43}$$

$$\frac{\partial A(x, y, z)}{\partial Y} = \frac{1}{\Delta}\left[f'_Y \frac{\partial A(X, Y, Z)}{\partial X} - f'_X \frac{\partial A(X, Y, Z)}{\partial Y}\right], \tag{5.44}$$

$$\frac{\partial A(x, y, z)}{\partial Z} = \frac{1}{\Delta}\left[H \frac{\partial A(X, Y, Z)}{\partial X} + K \frac{\partial A(X, Y, Z)}{\partial Y} + \frac{\partial A(X, Y, Z)}{\partial Y}\right], \tag{5.45}$$

using relations (5.33)–(5.34).

From Eqs. (5.20)–(5.22), the velocity vector, in the natural basis, is obtained as follows:

$$\underline{V} = \frac{\overline{\Psi}(R,\varphi)}{f\Delta(R,\varphi,Z)}\underline{E}_3. \tag{5.46}$$

As already pointed out for two-dimensional cases, the flow description is particularly well adapted to the use of materials with memory as those described by integral equations. Indeed, since the computation is carried out in the mapped domain Ω^*, where the transformed streamlines are straight, each line L^* of Ω^* can discretize into time intervals. From the velocity Eq. (5.44), the time evolution of a particle is given by

$$t - t_0 = \frac{1}{\overline{\Psi}(X,Y)}\int_{Z_0}^{Z}\left(f'_{Z}\,g'_{Y} - g'_{Z}\,f'_{Y}\right)_{(X,Y,Z)}\mathrm{d}Z, \tag{5.47}$$

where t_0 denotes the corresponding time to the position (X, Y, Z_0).

5.2.4 Conservation Laws and Boundary Conditions

Consider a steady, isothermal, three-dimensional incompressible flow under assumptions (i) and (ii) of the previous section. For a constant fluid density, the mass conservation of mass is automatically verified. Only momentum equations should be written with boundary condition equations [8]. In Cartesian coordinates, the three-dimensional momentum conservation equations are written as

$$\frac{\partial p}{\partial x} + \frac{\partial T^{xx}}{\partial x} + \frac{\partial T^{xy}}{\partial y} + \frac{\partial T^{xz}}{\partial z} = 0, \tag{5.48}$$

$$\frac{\partial p}{\partial y} + \frac{\partial T^{xy}}{\partial x} + \frac{\partial T^{yy}}{\partial y} + \frac{\partial T^{yz}}{\partial z} = 0, \tag{5.49}$$

$$\frac{\partial p}{\partial z} + \frac{\partial T^{xz}}{\partial x} + \frac{\partial T^{yz}}{\partial y} + \frac{\partial T^{zz}}{\partial z} = 0. \tag{5.50}$$

In relations corresponding to (X, Y, Z) in mapped domain Ω^*, the non-linear differential system may be written as

$$\frac{1}{\Delta}\cdot g'_{Y}\cdot\frac{\partial p}{\partial X} - \frac{1}{\Delta}\cdot g'_{X}\cdot\frac{\partial p}{\partial Y} + \frac{1}{\Delta}g'_{Y}\frac{\partial T^{xx}}{\partial X} - \frac{1}{\Delta}\cdot g'_{X}\cdot\frac{\partial T^{xx}}{\partial Y}$$

$$
\begin{aligned}
&-\frac{1}{\Delta}f'_{Y}\frac{\partial T^{XY}}{\partial X}+\frac{1}{\Delta}f'_{X}\frac{\partial T^{XY}}{\partial Y}\\
&+\frac{1}{\Delta}(f'_{Y}g'_{Z}-f'_{Z}g'_{Y})\cdot\frac{\partial T^{xz}}{\partial X}+\frac{1}{\Delta}(f'_{Z}g'_{X}-f'_{X}g'_{Z})\cdot\frac{\partial T^{XZ}}{\partial Y}+\frac{\partial T^{XZ}}{\partial Z}=0.
\end{aligned}
\tag{5.51}
$$

$$
\begin{aligned}
&-\tfrac{1}{\Delta}\cdot f'_{Y}\cdot\tfrac{\partial p}{\partial X}+\tfrac{1}{\Delta}\cdot f'_{X}\cdot\tfrac{\partial p}{\partial Y}+\tfrac{1}{\Delta}\cdot g'_{Y}\cdot\tfrac{\partial T^{XY}}{\partial X}\\
&-\tfrac{1}{\Delta}f'_{Y}\tfrac{\partial(fT^{yy})}{\partial X}-\tfrac{1}{\Delta}\cdot f'_{Y}\cdot\tfrac{\partial T^{yy}}{\partial X}+\tfrac{1}{\Delta}\cdot f'_{X}\cdot\tfrac{\partial(fT^{yy})}{\partial Y}\\
&+\tfrac{1}{\Delta}(f'_{Y}g'_{Z}-f'_{Z}g'_{Y})\cdot\tfrac{\partial T^{yz}}{\partial X}+\tfrac{1}{\Delta}(f'_{Z}g'_{X}-f'_{X}g'_{Z})\tfrac{\partial T^{yz})}{\partial Y}+\tfrac{\partial T^{yz}}{\partial X}=0.
\end{aligned}
\tag{5.52}
$$

$$
\begin{aligned}
&\frac{1}{\Delta}(f'_{Y}g'_{Z}-f'_{Z}g'_{Y})\cdot\frac{\partial p}{\partial X}+\frac{1}{\Delta}(f'_{Z}g'_{X}-f'_{X}g'_{Z})\cdot\frac{\partial p}{\partial Y}+\frac{\partial p}{\partial Z}\\
&+\frac{1}{\Delta}g'_{Y}\frac{\partial T^{xz}}{\partial X}-\frac{1}{\Delta}g'_{X}\frac{\partial(T^{xz}}{\partial Y}-\frac{1}{\Delta}\cdot f'_{Y}\cdot\frac{\partial T^{yz}}{\partial X}+\frac{1}{\Delta}\cdot f'_{X}\cdot\frac{\partial(fT^{yz})}{\partial Y}\\
&+\frac{1}{\Delta}(f'_{Y}g'_{Z}-f'_{Z}g'_{Y})\cdot\frac{\partial T^{zz}}{\partial X}+\frac{1}{\Delta}(f'_{Z}g'_{X}-f'_{X}g'_{Z})\frac{\partial T^{zz}}{\partial Y}+\frac{\partial T^{zz}}{\partial Z}=0.
\end{aligned}
\tag{5.53}
$$

The stress components expressions from a given constitutive equation can be inserted in the dynamic Eqs. (5.37)–(5.39) for a primary formulation in terms of the unknown mapping functions f and g. For Newtonian and some purely viscous constitutive models, third order of differential equations may be written formally as

$$
\mathrm{F_i}\big[f,f',f'',f''',g,g',g'',g''',p\big],\quad i=1,2,3. \tag{5.54}
$$

The symbols ($'$), ($''$) and ($'''$) of this equation, associated with the mapping functions f and g, stand for first-, second- and third-order spatial derivatives, respectively. So far, it has not been possible to determine the type of this set of third-order equations.

To solve the governing equations in three-dimensional situations, a mixed formulation is retained as defined in Chap. 4 (two-dimensional case). An integrodifferential form of the dynamic equations should be written for memory-integral constitutive equations. However, it is underlined that a better accuracy of stress calculations on rectilinear mapped streamlines is verified in comparison with usual approximations of three-dimensional solving techniques. A primary formulation could be adopted for Newtonian and purely viscous fluids but, in practice, a *mixed formulation* is generally retained for solving the equations, as reported in Chap. 4.

5.2.5 Boundary Condition Equations

Three-dimensional stream tubes have been shown in Fig. 5.3. An elementary stream tube B of complementary domain B_c is now considered in the physical domain Ω. The primary unknowns (transformation functions and pressure) and tensor components involved by the mixed formulation should be determined. Computing unknowns in the stream tube B of complementary domain B_c in Ω leads to consider the action of domain B_c (Fig. 5.10), as done in the two-dimensional case (Chap. 4). Thus, two different types of boundary condition equations may be concerned [1, 9].

1. Simple boundary condition equations related to the primary unknowns and the stresses at the upstream flow section, at the wall and the downstream flow section, in a confined flow.
2. Additional equations should be written when a free surface is considered.
3. Boundary conditions related to the action of the complementary domain B_c of the stream B under consideration. This is done by writing the global form of the momentum conservation equations. If $\underline{R}$ and $\underline{M_0}$ denote the resultant and moment vectors at a given point O, the following equations may be written:

$$\underline{R} = \int_{\partial B_c} \underline{\underline{\sigma}}\,\underline{n}\mathrm{d}s = \int_{\partial B_c} \left(-p\underline{\underline{I}} + \underline{\underline{\tau}}\right) \cdot \underline{n}\mathrm{d}s = \underline{0}, \tag{5.55}$$

$$\underline{M_0} = \int_{\partial B_c} \underline{OM} \times \underline{\sigma}^{v}(M)\mathrm{d}s = \underline{0}. \tag{5.56}$$

In these equations, ∂B_c denotes the limiting surface of the complementary domain $B_c = S_1 \cup S_2 \cup S_3$. $\underline{\underline{\sigma}}$ and $\underline{\sigma}^v$ are the total stress tensor and vector, respectively, $\underline{n}$ is the outward unit vector normal to the surface, $\underline{\underline{I}}$ the identity tensor and $\underline{\underline{\tau}}$ the extra-stress tensor such that

$$\underline{\underline{\sigma}} = -p\underline{\underline{I}} = \underline{\underline{T}}. \tag{5.57}$$

The plane perpendicular to the axis Oz passing through the point M of surface ∂B_c intersects the axis at a point O' and Eq. (5.54) becomes

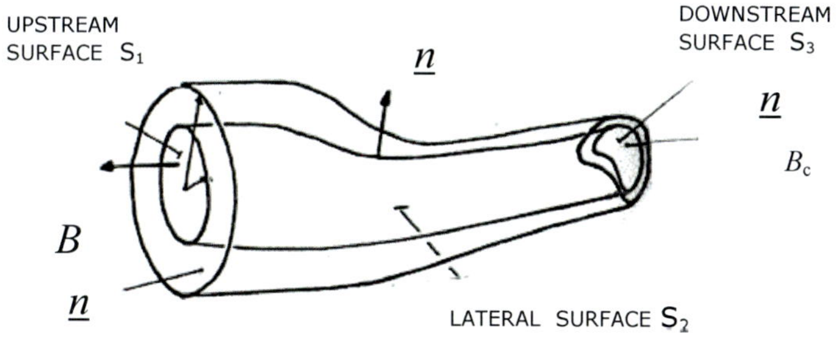

Fig. 5.10 Stream tube B and its complementary domain B_c

$$\underline{M}_0 = \int_{\partial B_c} \underline{OO'} + \underline{O'M} \times \underline{\sigma}^v(M)\mathrm{d}s = \underline{0}. \tag{5.58}$$

5.2.6 The Transformation in Cylindrical Coordinates

Transformation functions are now considered in terms of cylindrical coordinates $r = 1, \theta = 2, z = 3$ in the physical domain Ω. With coordinates $R = 1, \varphi = 2, Z = 3$ in the transformed flow domain Ω^*, one writes

$$r = f(R, \varphi, Z), \quad \theta = g(R, \varphi, Z), \quad z = Z \tag{5.59}$$

in a cylindrical orthonormal basis ($\underline{e}_i$). Similarly to the Cartesian case, the same notations "f"" (r) and "g" (θ) for the two functions depending on (R, φ, Z) are adopted. With the selected notations, such transformation is illustrated in Fig. 5.10.

As in the Cartesian case, the z-axis is assumed to be related to the main direction of flow. The upstream section at $z_0 = Z_0$ corresponds to the initial boundary conditions in a cylindrical orthonormal basis ($\underline{e}_i$) written as

$$r_0 = f(R, \varphi, Z_0), \quad \theta_0 = g(R, \varphi, Z_0), \quad z_0 = Z_0. \tag{5.60}$$

At section z_0, one writes

$$r = R, \quad \varphi = \theta. \tag{5.61}$$

The incompressibility condition leads to the relation $\nabla \cdot \underline{V} = 0$. In domains Ω and Ω^*, the initial sections at $z = z_0$ and $Z = Z_0$ are identical. According to the simply connected assumption for an upstream section, one defines a stream function such that

$$w(R, \varphi, Z_0) = R\Psi(R, \varphi), R \neq 0. \tag{5.62}$$

This leads to the following velocity components of $\underline{V}$:

$$u = \frac{f'_z}{f \cdot \Delta}\Psi(R, \varphi), \tag{5.63}$$

$$v = \frac{g'_z}{\Delta}\Psi(R, \varphi), \tag{5.64}$$

$$w = \frac{1}{f \cdot \Delta}\Psi(R, \varphi). \tag{5.65}$$

The Jacobian Δ of the transformation from Ω^* to Ω, assumed to be non-singular, may be written as

$$\Delta = \frac{\partial(z, \theta, z)}{\partial(R, \varphi, Z)} = f'_R g'_\varphi - f'_\varphi g'_R . \tag{5.66}$$

The natural basis $(\underline{E}_i)$, related to coordinates $R = 1$, $\varphi = 2$, $Z = 3$, for a material point of vector $\underline{M}$, is given by

$$\underline{E}_R = \frac{\partial \underline{M}}{\partial R} = f_R{}' \underline{e}_r + f g'{}_R \underline{e}_\varphi, \tag{5.67}$$

$$\underline{E}_\varphi = \frac{\partial \underline{M}}{\partial R} = f_\varphi{}' \underline{e}_r + f g'{}_\varphi \underline{e}_\varphi, \tag{5.68}$$

$$\underline{E}_3 = \frac{\partial \underline{M}}{\partial R} = f'{}_z \underline{e}_1 + f g''{}_Z + \underline{e}_z. \tag{5.69}$$

This non-singularity implies a one-to-one transformation. Only the main flows are considered explicitly.

To compute the temporal particle evolution, one selects t_0 as a reference time corresponding to a position at section Z_0. A relationship close to Eq. (5.47) is written as

$$t - t_0 = \frac{1}{\Psi(R, \varphi)} \int_{Z_0}^{Z} f(R, \varphi, \zeta) \Delta(R, \varphi, \varsigma) \mathrm{d}\zeta. \tag{5.70}$$

In this equation, t_0 denotes the corresponding time for the position (X, Y, Z_0) of the particle.

5.2.7 *Dynamic Equations with Cylindrical Coordinates*

As in the Cartesian case, the relevant equations are considered here under isothermal and steady conditions. Inertia and body forces are ignored. Mass conservation is automatically verified from the formulation. The stream tubes are considered either separately or in the global flow domain, as in Chap. 4. With cylindrical coordinates, the three equilibrium equations are given by

$$-\frac{\partial p}{\partial r} + \frac{\partial T^{rr}}{\partial r} + \frac{1}{r}\frac{\partial T^{r\theta}}{\partial \theta} + \frac{\partial T^{xz}}{\partial z} + \frac{T^{rr} - T^{\theta\theta}}{r} = 0, \tag{5.71}$$

$$\frac{\partial T^{r\theta}}{\partial r} - \frac{1}{r}\frac{\partial p}{\partial \theta} + \frac{1}{r}\frac{\partial T^{\vartheta\vartheta z}}{\partial \theta} + \frac{\partial T^{\theta z}}{\partial z} + 2\frac{T^{r\theta}}{r} = 0, \tag{5.72}$$

$$\frac{\partial T^{rz}}{\partial r}+\frac{1}{r}\frac{\partial T^{\theta z}}{\partial \theta}-\frac{\partial p}{\partial z}+\frac{\partial T^{zz}}{\partial z}+\frac{T^{rz}}{r}=0. \tag{5.73}$$

In these equations, T^{ij} denote the components of the stress tensor $\underline{\underline{T}}$, while the superscripts are related to $r = 1$, $\theta = 2$ and $z = 3$. Using the derivative operators (5.41)–(5.43), the governing Eqs. (5.69)–(5.71) become, in terms of variables (R, φ, Z),

$$\begin{aligned}
&\frac{1}{\Delta}\cdot g'_{\varphi}\cdot\frac{\partial p}{\partial R}-\frac{1}{\Delta}g'_{R}\cdot\frac{\partial p}{\partial \theta}+\frac{1}{f\Delta}g'_{\varphi}\frac{\partial (fT^{rr})}{\partial R}-\frac{1}{f\Delta}\cdot g'_{R}\cdot\frac{\partial (f\cdot T^{rr})}{\partial \theta}\\
&-\frac{1}{f\Delta}f'_{\varphi}\frac{\partial (fT^{r\varphi})}{\partial R}+\frac{1}{f\Delta}f'_{R}\frac{\partial (fT^{r\varphi})}{\partial \varphi}+\frac{1}{\Delta}\left(f'_{\varphi}g'_{Z}-f'_{Z}g'_{\varphi}\right)\cdot\frac{\partial T^{rz}}{\partial R}\\
&+\frac{1}{\Delta}\left(f'_{Z}g'_{R}-f'_{R}g'_{Z}\right)\cdot\frac{\partial T^{rZ}}{\partial \varphi}+\frac{\partial T^{rZ}}{\partial Z}=0,
\end{aligned} \tag{5.74}$$

$$\begin{aligned}
&\frac{1}{f\Delta}\cdot f'_{\varphi}\cdot\frac{\partial p}{\partial R}+\frac{1}{f\Delta}\cdot f'_{R}\cdot\frac{\partial p}{\partial \varphi}+\frac{1}{f\Delta}\cdot g'_{\varphi}\cdot\frac{\partial (f\cdot T^{r\varphi})}{\partial R}\\
&-\frac{1}{f\Delta}g'_{R}\frac{\partial (fT^{r\varphi})}{\partial \varphi}-\frac{1}{f\Delta}\cdot f'_{\varphi}\cdot\frac{\partial (fT^{\varphi\varphi})}{\partial R}+\frac{1}{f\Delta}\cdot f'_{R}\cdot\frac{\partial (fT^{\varphi\varphi})}{\partial \varphi}\\
&+\frac{1}{\Delta}\left(f'_{\varphi}g'_{Z}-f'_{Z}g'_{\varphi}\right)\cdot\frac{\partial (fT^{\varphi z})}{\partial R}+\frac{1}{\Delta}\left(f'_{Z}g'_{R}-f'_{R}g'_{Z}\right)\frac{\partial T^{\varphi z}}{\partial \varphi}+\frac{\partial T^{\varphi z}}{\partial \varphi}=0,
\end{aligned} \tag{5.75}$$

$$\begin{aligned}
&\frac{1}{\Delta}\left(f'_{\varphi}g'_{Z}-f'_{Z}g'_{\varphi}\right)\cdot\frac{\partial p}{\partial R}+\frac{1}{\Delta}\left(f'_{Z}g'_{R}-f'_{R}g'_{Z}\right)\cdot\frac{\partial p}{\partial \varphi}+\frac{\partial p}{\partial Z}\\
&+\frac{1}{f\Delta}g'_{\varphi}\frac{\partial (fT^{rz})}{\partial R}-\frac{1}{f\Delta}g'_{R}\frac{\partial (fT^{rz})}{\partial \varphi}-\frac{1}{f\Delta}\cdot f'_{\varphi}\cdot\frac{\partial (fT^{\varphi z})}{\partial R}\\
&+\frac{1}{f\Delta}\cdot f'_{R}\cdot\frac{\partial (fT^{\varphi z})}{\partial \varphi}+\frac{1}{\Delta}\left(f'_{\varphi}g'_{Z}-f'_{Z}g'_{\varphi}\right)\cdot\frac{\partial T^{zz}}{\partial R}\\
&+\frac{1}{\Delta}\left(f'_{Z}g'_{R}-f'_{R}g'_{Z}\right)\cdot\frac{\partial T^{zz}}{\partial \varphi}+\frac{\partial T^{zz}}{\partial Z}=0.
\end{aligned} \tag{5.76}$$

Numerous problems involving symmetries are frequently encountered in three-dimensional cases. We choose here to provide examples related to cylindrical variables.

The rate-of-deformation tensor is adopted in flow studies for Newtonian fluids and materials obeying purely viscous equations. Examples concerning viscoelastic materials with memory-integral constitutive equations are also considered. Cauchy and Finger strain tensors, related to codeformational models, are adopted in the STM calculations.

5.2.8 Kinematic Tensors for Codeformational Models

In order to evaluate the Cauchy and Finger tensors in the context of Stream-tube method, the elements developed in [3] and [6] are adopted, by starting from Adachi's work [4, 5]. Accordingly, the deformation gradient tensor $\underline{\underline{F}}(\tau)(R, \varphi, \zeta)$ is expressed in the frame corresponding to (*R*, φ, Z) coordinates by the following matrix:

$$\mathbf{F}_t(\tau) = \begin{bmatrix} 1 & 0 & 0 \\ 0 & 1 & 1 \\ \partial Z_\tau/\partial R_t & \partial Z_\tau/\partial \varphi_t & \partial Z_\tau/\partial Z_t \end{bmatrix}, \tag{5.77}$$

where the subscripts indicate reference times. The unknown terms of this matrix may be written as

$$\frac{\partial Z_\tau}{\partial R_t} = w(R, \varphi, Z_\tau) \int_{Z_t}^{Z_t} \frac{[\partial w(R, \varphi, \zeta)/\partial R_t]}{w^2 R, \varphi, \zeta)} \mathrm{d}\zeta, \tag{5.78}$$

$$\frac{\partial Z_\tau}{\partial \varphi_t} = w(R, \varphi, Z_\tau) \int_{Z_t}^{Z_t} \frac{[\partial w(R, \varphi, \zeta)/\partial \varphi_t]}{w^2 R, \varphi, \zeta)} \mathrm{d}\zeta, \tag{5.79}$$

$$\frac{\partial Z_\tau}{\partial Z_t} = \frac{w(R, \varphi, Z_\tau)}{w(R, \varphi, Z_\tau)}. \tag{5.80}$$

- Evaluating the respective Cauchy and Finger tensors $\underline{\underline{C}}_t$ and $\underline{\underline{C}}_t^{-1}$ involves consideration of the respective natural and reciprocal basis ($\underline{e}_\mathrm{i}$) and $\left(\underline{\chi}^j\right)$ related to the orthonormal frame ($\underline{e}_\mathrm{i}$), with variables (r, θ, z). According to the expressions of the Jacobian Δ and of the following relations:

$$H = f'_\varphi \cdot g'_Z - f'_Z \cdot g'_\varphi, \tag{5.81}$$

$$K = f'_Z \cdot g'_\mathrm{R} - f'_R \cdot g'_Z, \tag{5.82}$$

these bases can be written as

$$\begin{aligned} &\underline{b}_1 = f'_\mathrm{R}\, \underline{c}_1 + f\, g'_\mathrm{R} \underline{c}_2, \underline{b}_2 = f'_\varphi\, \underline{c}_1 + f\, g'_\varphi \underline{c}_2, \\ &\underline{b}_3 = f'_\mathrm{Z}\, \underline{c}_1 + f\, g'_\mathrm{Z} \underline{c}_2 + \underline{c}_3 \end{aligned} \tag{5.83}$$

and

$$\underline{\chi}^1 = \frac{1}{f\Delta}\left(f\, g'_\varphi\, \underline{c}_1 - f'_\varphi\, \underline{c}_2 + f\, H \underline{c}_3\right), \tag{5.84}$$

$$\underline{\chi}^2 = \frac{1}{f\Delta}(-f\, g'_R \underline{c}_1 - f'_R \underline{c}_2 + f\, K \underline{c}_3), \tag{5.85}$$

$$\underline{\chi}^3 = \underline{c}_3. \tag{5.86}$$

- The corresponding metric tensor $\underline{\underline{G}}^{-1}$ is given by the following matrix of components expressed as follows in Eq. (5.87):

$$\begin{aligned}
&(G^{-1})^{11} = (f'_R)^2 + f^2 (g'_Z)^2 \\
&(G^{-1})^{12} = (G^{-1})^{21} = f'_R\, f'_\varphi + f^2\, g'_R\, g'_\varphi \\
&(G^{-1})^{13} = (G^{-1})^{31} = f'_R\, f'_Z + f^2\, g'_R\, g'_Z \\
&(G^{-1})^{22} = f'_\varphi + f^2 (g'_\varphi)^2 \\
&(G^{-1})^{23} = (G^{-1})^{32} = f'_\varphi\, f'_Z + f^2\, g'_\varphi\, g'_Z \\
&(G^{-1})^{33} = 1 + (f'_Z)^2 + f^2 (g'_Z)^2.
\end{aligned} \tag{5.87}$$

- The components of the metric tensor $\underline{\underline{G}}$ are written as follows in Eq. (5.88):

$$\begin{aligned}
&G^{11} = \frac{\left(f'_R / f'^2_\varphi\right)^2 + (g'_\varphi)^2 + \mathrm{H}^2}{\Delta^2} \\
&G^{13} = G^{31} = \frac{\mathrm{H}}{\Delta} \\
&G^{22} = \frac{(f'_R / f)^2 + (g'_R)^2 + \mathrm{K}^2}{\Delta^2} \\
&G^{23} = G^{32} = \frac{\mathrm{K}}{\Delta} \\
&G^{33} = 1.
\end{aligned} \tag{5.88}$$

- The Finger tensor $\underline{\underline{B}}(\tau)$ is expressed by the equation (where the superscript "T" denotes the transposition of a tensor):

$$\underline{\underline{C}}_t(\tau)^{-1} = \left[\underline{\underline{F}}_t(\tau)^{-1}\right]_j^{\cdot} = \left(\left[\underline{\underline{F}}_t(\tau)^{-1}\right]\right)^{\mathrm{T}}. \tag{5.89}$$

In the natural basis $(\underline{b}_i)$, one writes the Finger tensor as follows:

$$\underline{\underline{B}}(\tau)^{-1} = \left[\underline{\underline{F}}_t^{-1}\right]_j^i \cdot G^{jh} \cdot \left[\underline{\underline{F}}_t^{-1}\right]_h^k \underline{b}_j \otimes \underline{b}_k. \tag{5.90}$$

Equation (5.89) points out that the Finger tensor is a two-contravariant tensor in the natural basis of the reference configuration at time t. Similarly, it can be shown that the Cauchy tensor is a two-covariant tensor in the reciprocal basis $(\underline{\chi}^j)$ such that

$$\underline{\underline{C_t}}(\tau) = \left[\underline{\underline{F_t}}^T\right]^i_j \cdot G_{ih} \cdot \left[\underline{\underline{F_t}}^{-1}\right]^h_k \underline{\chi}^j \otimes \underline{\chi}^k. \tag{5.91}$$

- The expression of the Finger tensor in cylindrical coordinates is obtained from Eq. (5.82), using the matrix $\boldsymbol{\mu}$ that ensures the passage from the natural basis $(\underline{E}_i)$ in the reference configuration to the cylindrical basis $(\underline{e}_r, \underline{e}_\varphi, \underline{e}_Z)$ in the same configuration, such that

$$\underline{\underline{B}}(\tau)\Big|_{(e_i)} = \boldsymbol{\mu} \cdot \underline{\underline{B}}(\tau)\Big|_{(Ej)} \cdot \boldsymbol{\mu}^{\mathrm{T}}. \tag{5.92}$$

  In the cylindrical basis, the matrix $\boldsymbol{\mu}$ is given by

$$\boldsymbol{\mu} = \begin{bmatrix} f'_{\mathrm{R}} & f & g'_{\mathrm{R}} & 0 \\ f'_\varphi & f & g'_\varphi & 0 \\ f'_{\mathrm{Z}} & f & g'_{\mathrm{Z}} & 1 \end{bmatrix}. \tag{5.93}$$

- Similar calculations can be performed to evaluate the Cauchy tensor, inverse of the Finger tensor. According to Eq. (5.89), the Cauchy tensor can be evaluated by the relation:

$$\underline{\underline{C_t}}(\tau)\Big|_{(ei)} = \left[\boldsymbol{\mu} \cdot \underline{\underline{C_t}}(\tau)^{-1}\Big|_{(Ej)} \cdot \boldsymbol{\mu}^{\mathrm{T}}\right]^{-1}. \tag{5.94}$$

Relations (5.91) and (5.94) provide accurate computations with memory-integral constitutive equations.

5.3 STM Applications

Examples provided in this section may be related to practical engineering problems. Three-dimensional isothermal cases are considered in cylindrical coordinates, ignoring inertia and body forces. It is found of interest to provide examples of confined and free surface flows, using Newtonian and viscoelastic constitutive equations:

(i) flows in a converging geometry involving a threefold rotational symmetry,
(ii) free surface flows at the exit of a tube of the squared cross-section.

5.3.1 Newtonian Fluid in a Converging Domain

The physical and mapped domains are presented in Fig. 5.11. This geometry close to processing flows, as those of fibre industry, concerns.

- at upstream, a circular cylinder;
- an axisymmetric converging domain;
- a triangular equilateral domain in the downstream region.

Fully developed flow regions are assumed at upstream and downstream the converging domain. According to Fig. 5.12, the flow domain involves an upstream tube of circular cross-section and a downstream domain with a threefold rotational symmetry. The prism of triangular cross-section intersects the truncated cone which joins the upstream circular cylinder.

Computations of flow characteristics are achieved by taking into account geometrical specifications and features of STM. The form of the dynamic equations has been given in relations (5.74)–(5.76). For a Newtonian fluid of viscosity η_0, the stress tensor components are given by

$$T^{11} = 2\eta_0 \frac{\partial u}{\partial r}, \tag{5.95}$$

$$T^{22} = 2\eta_0 \left[\frac{1}{r}\frac{\partial v}{\partial \theta} + \frac{u}{r} \right], \tag{5.96}$$

$$T^{33} = -\left(T^{11} + T^{22}\right), \tag{5.97}$$

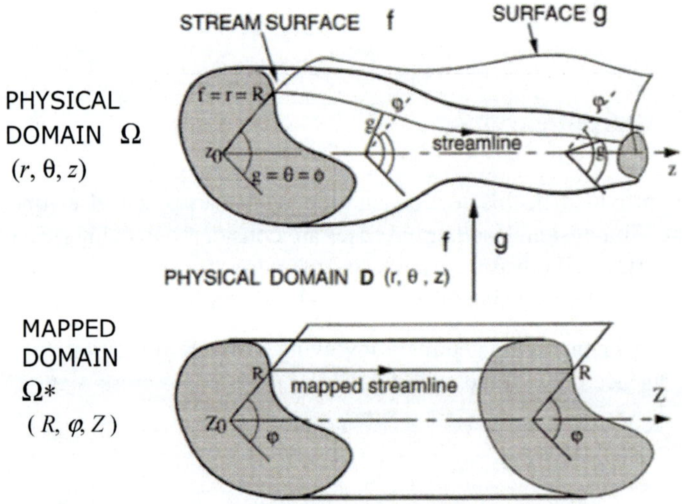

Fig. 5.11 Physical and mapped domains Ω and Ω* in cylindrical coordinates: $f = r = R$ and $g = \theta = \varphi$ at the upstream section

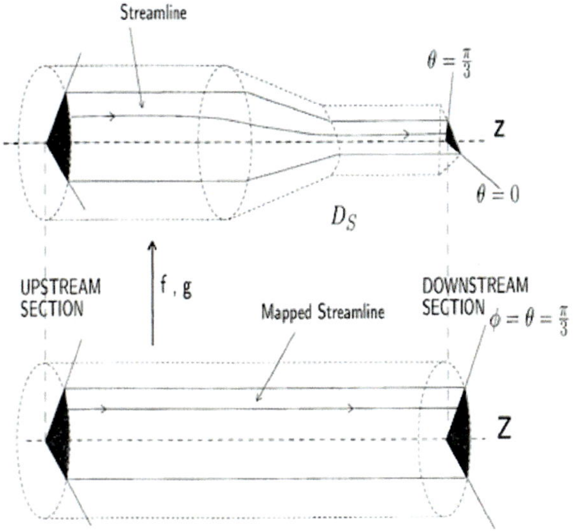

Fig. 5.12 Physical domain Ω involving a triangular downstream section and its mapped domain Ω^*

$$T^{12} = \eta_0 \left[\frac{\partial v}{\partial r} - \frac{v}{r} + \frac{1}{r} \frac{\partial u}{\partial \theta} \right], \tag{5.98}$$

$$T^{13} = \eta_0 \left[\frac{\partial u}{\partial z} + \frac{\partial w}{\partial r} \right], \tag{5.99}$$

$$\mathrm{T}^{23} = \eta_0 \left[\frac{1}{r} \frac{\partial w}{\partial \theta} + \frac{\partial v}{\partial z} \right]. \tag{5.100}$$

The expressions of the kinematic components are related to Eqs. (5.95)–(5.100) in terms of the unknown transformation functions f and g.

Owing to symmetry planes, only a 60° sector model may be considered for computing the flow. Convergent sections formed by planes passing through the axis 0z are shown in Fig. 5.13, from $\varphi = \theta = 0$ to $\varphi = \theta = \pi/3$. Minimum and maximum restriction ratios are 2 (for $\varphi = \theta = 0$) and 4 (for $\varphi = \theta = \pi/3$).

This example is of interest, mainly for the following reasons:

- Applications to flows where an n-fold ($n > 1$) rotational symmetry are assumed.
- The duct geometry is close to those which may be encountered in processing flows of rheology, notably in the fibre industry.
- As mentioned previously, azimuthal sections of the wall surface define the shape of a converging geometry profile. Hence, numerical results from the present analysis may provide predictions on streamlines, pressures and stresses.

1. *Governing equations*

According to the STM theory, the equations to be considered are the differential Eqs. (5.74)–(5.76), the rheological stress Eqs. (5.95)–(5.100) and boundary

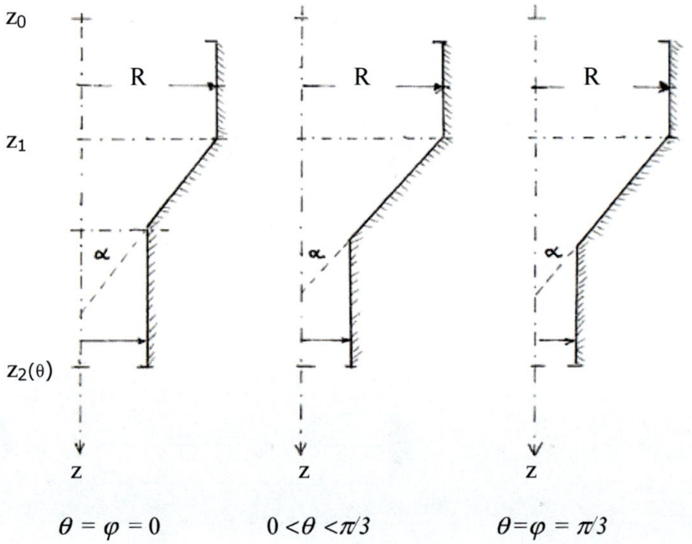

Fig. 5.13 Convergent profiles of the flow domain Ω for different sections about the axis *0z*

conditions. As stated earlier (Sect. (5.2.4)), these last equations involve the integral boundary condition Eqs. (5.55)–(5.56) when a stream tube is considered in the flow domain. It is shown below that these equations reduce into one scalar equation when the geometry involves an *n*-fold rotational symmetry.

The case, where $n = 3$, is considered. When projecting the vector equation on the three axes corresponding to cylindrical coordinates, it can easily be seen that the terms related to vectors $\underline{e}'_r$ and $\underline{e}'_\varphi$ vanish three by three, owing to the threefold symmetry. Thus, only the following scalar equation, obtained by projection on the z-axis,

$$\left[\int_{\partial B_c} (-p\underline{\underline{I}} + \underline{\underline{\tau}}) \cdot \underline{n}\mathrm{d}s\right] \cdot \underline{e}'_z = 0, \tag{5.101}$$

should be considered. A similar argument may be used with the form (5.56) recalled here:

$$\underline{M}_0 = \int_{\partial B_c} (\underline{OO'} + \underline{O'M}) \times \sigma^v(M)\mathrm{d}s = \underline{0}.$$

According to the transformation T between domains D and D': $(r, \theta, z) \to (R, \varphi, Z)$, a vector $\underline{N}$ normal to a stream surface can be expressed in terms of the transformation functions f and g by

$$\underline{N}\left(1+\frac{g'_R f'_\varphi}{\Delta}, \frac{f'_R f'_\varphi}{f, \Delta}, -\frac{f'_\varphi \mathrm{K}}{\Delta} - f'_Z\right). \tag{5.102}$$

The unit vectors $\underline{n}_1$ and $\underline{n}_3$ (see Fig. 5.4) normal to the upstream and downstream surfaces S_1 and S_3 are, respectively,

$$\begin{aligned} &\underline{n}_1(0, 0, -1) \text{ for } S_1, \\ &\underline{n}_3(0, 0, 1) \text{ for } S_3. \end{aligned} \tag{5.103}$$

From Eq. (5.56), the third scalar equation

$$\underline{M}_0 \cdot \underline{e}'_z = 0 \tag{5.104}$$

is automatically verified. Considering again the flow domain of Fig. 5.4, Eq. (5.101) may be expressed as

$$\underbrace{\left[\int_{S_1} \underline{\underline{\sigma}} \cdot \underline{n} \mathrm{d}s\right]}_{I_1} \cdot \underline{e}'_z + \underbrace{\left[\int_{S_2} \underline{\underline{\sigma}} \cdot \underline{n} \mathrm{d}s\right]}_{I_2} \cdot \underline{e}'_z + \underbrace{\left[\int_{S_3} \underline{\underline{\sigma}} \cdot \underline{n} \mathrm{d}s\right]}_{I_3} \cdot \underline{e}'_z = 0. \tag{5.105}$$

The integral I_1 corresponds to the upstream surface $S_{1,}$ I_2 is related to the lateral surface S_2 and I_3, to the downstream surface S_3 (Fig. 5.4).

At the upstream Poiseuille flow section, for purely viscous fluids, Eqs. (5.95)–(5.100) lead to

$$T^{11} = T^{22} = T^{33} = 0. \tag{5.106}$$

Hence, it writes

$$I_1 = \pi (R_1)^2 p_0, \tag{5.107}$$

where R_1 denotes the upstream radius of the complementary domain B_c of the stream tube B. p_0 is the pressure assumed to be a constant at the upstream section z_0. A similar relation is obtained for the downstream Poiseuille flow (at z_{max}) of area S_3, such that

$$I_3 = -S_3 p_1. \tag{5.108}$$

p_1 denotes the constant pressure at the downstream section.

Concerning the integral I_2, it is necessary to consider the unit normal vector $\underline{\mathrm{n}}$ related to the vector $\underline{\mathrm{N}}$, according to Eq. (5.104). Using the symmetry property of the flow domain *D*, it may be shown that the contribution to the lateral surface to Eq. (5.105) is given by

$$I_2 = 3 \int_0^{2\pi/3} d\varphi \int_z^z \max \operatorname{sign}\left(\frac{\Delta}{f'_R}\right)\left[f\, g'_\varphi T^{13} - f'_\varphi T^{23} - Hf\left(p + T^{11} + T^{22}\right)\right] d\zeta, \tag{5.109}$$

where Δ denotes the Jacobian of the domain transformation. H is defined by Eq. (5.81).

2. *Discretization—Approximation of unknowns*

We firstly recall elements for STM simulations of the problem:

- The governing equations are non-linear and may be written in a matrix form as

$$E(\underline{X}) = 0. \tag{5.110}$$

$\underline{\mathbf{X}}$ denotes the vector of unknowns X_1, X_2, …, X_n, related to mapping functions f and g, pressure p and stress components T^{11}, T^{22}, T^{12}, T^{13}, T^{23} of the mixed formulation for the dynamic Eqs. (5.74)–(5.76) and stress Eqs. (5.95)–(5.100).

- The computational domain D' is simple: a parallelepiped is obtained by using variables (R, φ, Z) and the rectilinear mapped streamlines. Figure 5.14a shows a sub-domain corresponding to an elementary mapped stream tube, on which a rectangular mesh is constructed.
- Stream bands: Sections of the mesh defined by planes φ = constant determine stream bands, denoted by the symbol **C**, from $\mathbf{C} = 1$ for $\varphi = 0$ to $\mathbf{C} = \mathbf{C}_{\max}$ for $\varphi = \pi/3$. Each stream band is divided into elements (e), presented in Chap. 3 (Sect. 3.3.3.1) (Fig. 3.7).
- Function f

  The transformation function is assumed to be quadratic and expressed by the following equation, for a given stream band **C**:

$$M \in (e) : f_M = \sum_{m=1}^{m=6} \Phi_m(x, y) f_m^{(e)}. \tag{5.111}$$

  Figure 5.14b shows the element retained for the approximations of the function f. (x, y) denotes the local coordinates of the point M of global coordinates ($R^{(e)}$, $Z^{(e)}$).
  Writing the relations

$$\Delta R = R_{(1)}^{(e)} - R_{(4)}^{(e)}, \quad \Delta Z = Z_{(2)}^{(e)} - Z_{(1)}^{(e)}, \tag{5.112}$$

the basis functions are expressed by Eq. (5.111).

Thus, the first and second derivatives of f can be readily obtained from those of the basic functions ϕ_m, taking into account the following relations:

$$f'_R = f'_X X'_R = -f'_X, \; f'_Z = f'_Y Y'_Z = f'_Y. \tag{5.113}$$

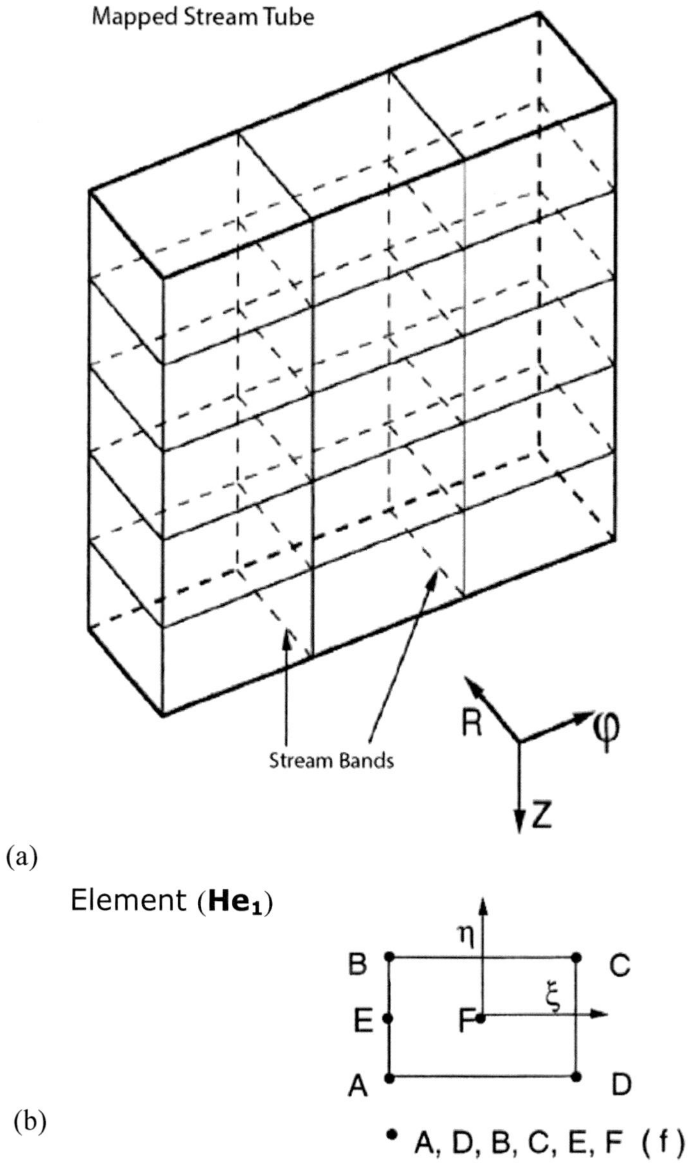

Fig. 5.14 **a** Computational sub-domain (three-dimensional mapped stream tube). **b** Quadrilateral element (**He1**) in a stream band

- Function g

To compute the mapping function g and tensor components T^{ij}, the nodal values at points 1, 2, 3, 4 of each element are chosen as unknowns. Their derivatives, together with those of the function f in terms of the variable φ are then evaluated on the rectangular mesh by finite-difference formulae.

3. *Solving the equations*

The governing equations and *simple* boundary conditions may be written as a closed set of equations, as

$$A_{\mathrm{i}}(X_1, X_2, \ldots, X_{\mathrm{n}}) = 0, i = 1, \ldots, n. \tag{5.114}$$

When considering an elementary stream tube of the flow domain, the action of one or two complementary flow domains must be involved by writing a non-linear boundary condition equation of the form of (5.105). Then, computation of the unknowns of a stream tube may be related to the following problem:

$$(P)\ \min\{E(\mathbf{X}) : \mathbf{X} \epsilon R^n\}, \tag{5.115}$$

under one of two constraints:

$$\begin{aligned} &C_1(X) = 0(\text{peripheral stream tube}) \\ \text{or}\quad &C_1(X) = 0 \text{ and } C_2(X) = 0(\text{inner stream tube}). \end{aligned} \tag{5.116}$$

These constraints correspond to the action of one or two complementary domains of the stream tube under consideration.

Q corresponds to the following quadratic function:

$$Q(X) = A^{\mathrm{T}}(X) \cdot A(X) = \sum_{i=1}^{i=n} A_i^2(X_1, X_2, \ldots, X_{\mathrm{n}}). \tag{5.117}$$

Equations of problem (P) are formulated as a problem (P') defined as

$$(P')\ \min\{G(\mathbf{X}) : \mathbf{X} \epsilon R^n\}. \tag{5.118}$$

To this problem corresponds the following set of equations:

$$\chi_{\mathrm{i}}(X_1, X_2, \ldots, X_{\mathrm{n}}) = 0, i = 1, \ldots, M, n < \mathrm{M}, \tag{5.119}$$

which represents (5.116) and (5.118) in such a way that the quadratic form defined in (5.117) is given by

$$G(X) = \chi^{\mathrm{T}}(X) \cdot \chi(X) = \sum_{i=1}^{i=M} A_i^2(X_1, X_2, \ldots, X_n). \tag{5.120}$$

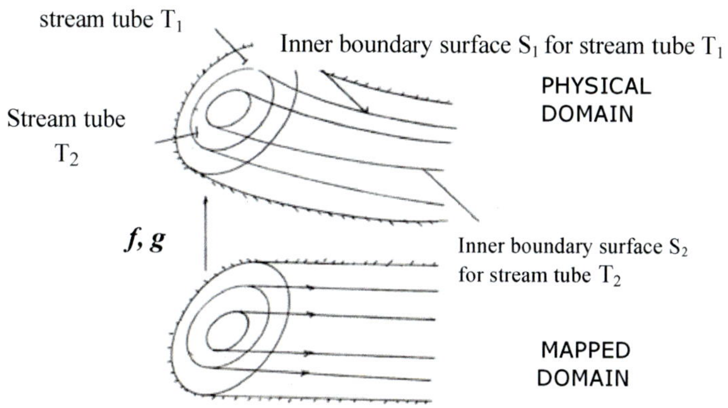

Fig. 5.15 Elementary stream tubes in the physical and mapped domains

The numerical solution $\boldsymbol{X}^*$ of the equations related to problem (P') is obtained by the LM algorithm [10].

4. *Procedure for determining unknowns—Mesh, sub-domains, velocity contour values*

To compute the unknowns, successive stream tubes from the wall to the central flow region are considered. Figure 5.15 shows an example for elementary stream tubes and stream surfaces concerned in the calculations. For each stream tube, the same procedure is adopted to solve the governing equations. The computational flow domain is the cylinder involving rectilinear streamlines (Fig. 5.15).

The specific computing approach may be summarized as follows:

(1) Peripheral stream tube T_1:

In the computational domain, surface $\overline{S}_0$ (the surface f) corresponding to the wall is known. This concerns:

- simple boundary condition equations,
- a function $\overline{\Psi}(X, Y)$ the dynamic Eqs. (5.51)–(5.53)
- the constitutive Eqs. (5.95)–(5.100),

for a stream surface embedded in the stream tube (Figs. 5.15 and 5.16), and for the surface $\overline{S'}_1$ limiting the inside of the tube T_1.

One also writes, as constraints C_k, the boundary integral equation for the complementary stream tube:

– at the outside of the stream surface $\overline{S}_1^*$
– at the inner complementary stream tube limited by the unknown stream surface $\overline{S}_1$.

All the unknowns are computed simultaneously with the LM algorithm. Thus, in the stream tube, the velocities are obtained from Eqs. (5.51) to (5.53).

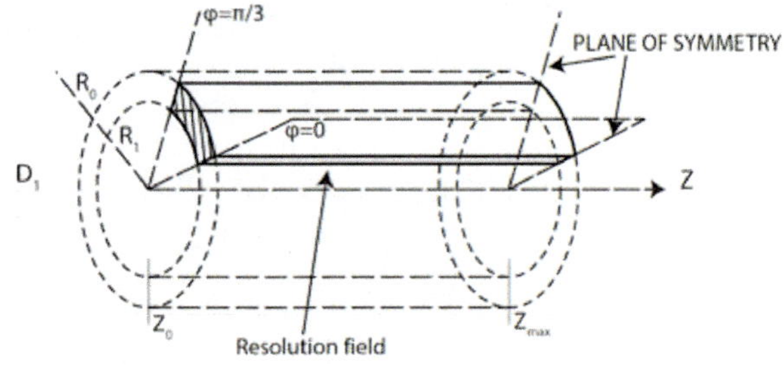

Fig. 5.16 View of a mapped sub-domain for the computations

(2) Stream-tube T_2:

This stream tube is limited by

- the known surface $\overline{S}_1$,
- the surface $\overline{S}_2$ to be determined.

The surface $\overline{S}_1$ provides simple boundary condition equations. Similarly, the governing equations are written for points of a mapped stream surface $\overline{S}_2^*$(Figs. 5.15 and 5.16).

The boundary integral equations are written for the surfaces $\overline{S}_1^*$ and $\overline{S}_2$. The computational procedure is then used to determine the surface $\overline{S}_2$, the other unknowns and the corresponding velocities.

The same technique is applied for the stream tubes T_i ($i > 2$), up to the central flow region. In this computing approach for determining the solution in successive stream tubes, it is of interest to specify the following points:

1. The outer wall of each stream tube T_i under consideration is known and only provide boundary values for the unknowns of the problem, not for their derivatives. The computed solution is not assumed to ensure continuity of the derivatives for the unknowns corresponding to the stream surface limiting two adjacent stream tubes.
2. The inflow and outflow boundary conditions are taken into account by writing the simple boundary condition equations. Those are also globally involved in the non-linear boundary condition equations of the form of (5.105).

The traces of the stream tubes T_i correspond to the contour values of the velocity w, at upstream and downstream Poiseuille flow sections z_0 and z_{max}. For a Newtonian fluid of viscosity η_0, the contour values are circles at section z_0 and, at the downstream section z_{max}, may be computed numerically from the analytical Poiseuille flow solution in a duct of triangular cross-section [9]:

$$w = h_2[x + a][2a - x + \sqrt{3}y][2a - x - \sqrt{3}y](12a\eta_0) \tag{5.121}$$

in a Cartesian coordinate system (x, y). h_2 denotes the pressure drop, a is the distance of the side parallel to the y-axis from the origin, as shown in Fig. 5.12a. The downstream contour values at z_{max} (which statisfy the conservation of mass) corresponding to the upstream velocity contours w are illustrated in Fig. 5.17b.

3. In the computations, a mesh similar to that shown in Fig. 5.18a has been used for each stream tube. The number of unknowns is approximately 1000. A satisfactory convergence of the algorithm is obtained in about 15 iterations. For Newtonian flow results, the "warping" of the computed streamlines has been found to be weak. However, this result does not allow flow in a converging section of constant φ to be predicted by considering separately the computational results related to the same flow section in a corresponding planar or axisymmetric geometry. The predicted stress components T^{13}, versus Z, close to the wall, for ducts with $\alpha =$

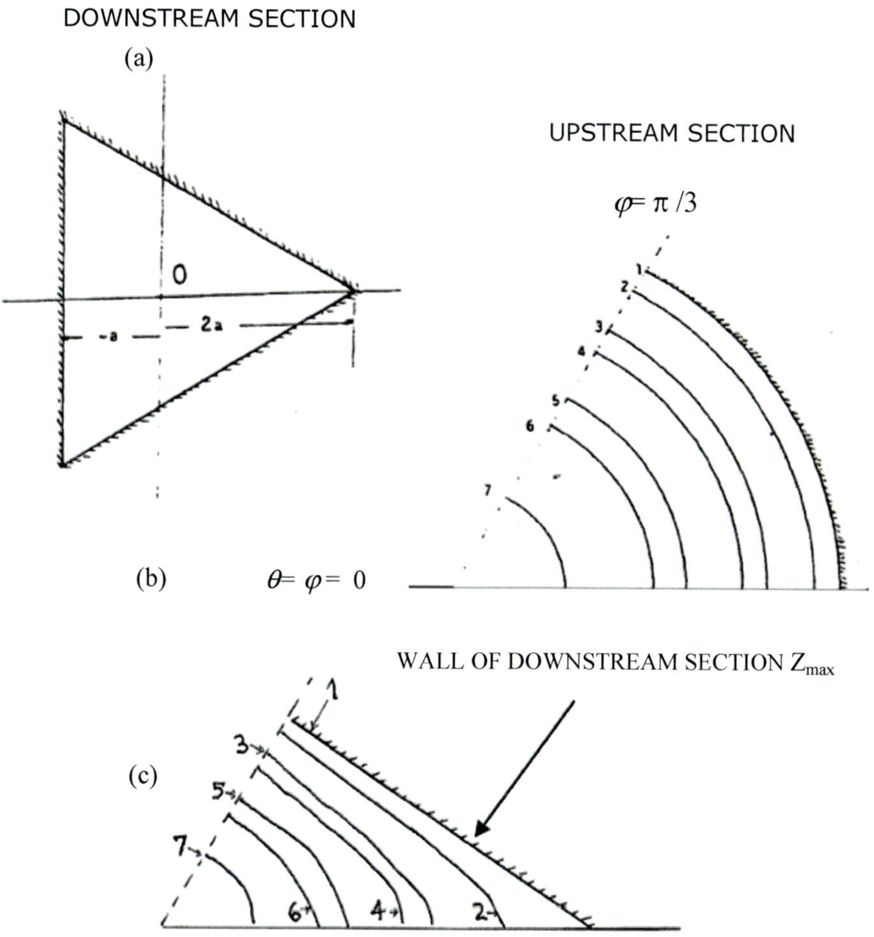

Fig. 5.17 **a** The triangular cross-section of the duct, downstream the converging region; **b** contour values of the axial velocity components w at the upstream Poiseuille flow section z_0; **c** contour values of the axial velocity components w at the downstream section z_{max}

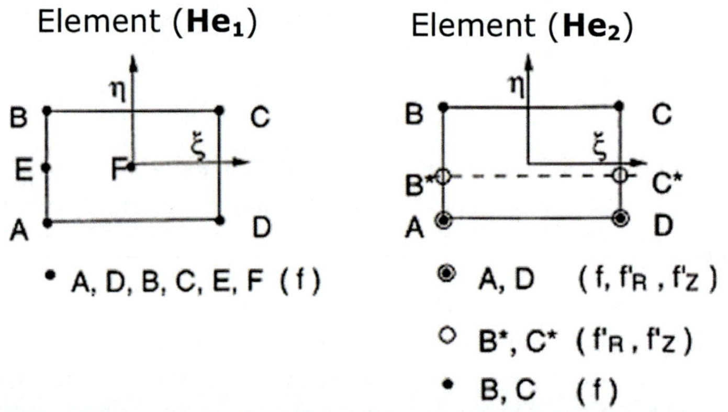

Fig. 5.18 Elements of type (**He**$_1$) and (**He**$_2$)

20° and α = 40°, at different angles φ have revealed a stress peak in the vicinity of the re-entrant corner, in qualitative agreement with literature results.

5.3.2 *Viscoelastic Fluid in the Converging Domain*

The viscoelastic case is retained here for the same geometry, with a similar approach to the Newtonian case, with a mixed formulation. The components T^{ij} of the stress tensor $\underline{T}$ are unknowns, together with the primary variable f, g and p. One still considers the stream tubes separately, from the wall to the central region. The equations involved in the computation are

1. The dynamic Eqs. (5.74)–(5.76), written together with the classic boundary conditions.
2. The boundary condition equation of the form of (5.101).
3. The constitutive equation, which corresponds to a K-BKZ memory-integral codeformational model, written as ([11, 12])

$$\underline{\underline{T}}(t) = \int_{-\infty}^{t} m(t-\tau)h(I_1, I_2)\underline{\underline{C}}_t^{-1}(\tau)\mathrm{d}\tau. \tag{5.122}$$

The procedure for evaluating the kinematics has been provided in Sect. (5.2.4) of this chapter.

1. *Discretization—Unknowns approximation*

In the computations, a stream band is defined as that previously presented in Fig. 5.17a. A parallelepiped is obtained by using variables (R, θ, Z). Sections of the mesh, defined by planes φ = Constant, determine stream bands $\mathrm{B_i}$ which are similar to the example of Fig. 5.17a.

The transformation function is assumed to be quadratic. Though elements of type (**He$_1$**), in Fig. 5.18a are generally used in the flow domain, elements (**He$_2$**) may be retained for computations in the peripheral stream tube, with the memory-integral constitutive equation.

2. *Applications to two-dimensional and three-dimensional geometries*

The relevant equations for an elementary stream tube are solved by the Trust-Region optimization algorithm previously depicted in Chap. 4.

For the three-dimensional geometry, the flow rate parameter used here is the apparent wall shear rate γ_{aw} given by

$$\gamma_{aw} = 4Q/\left(\pi r_0^3\right), \tag{5.123}$$

where Q is the volume rate and r_0 the downstream tube radius. The method used for the calculations in a peripheral stream tube is similar to the procedure given in paragraph 4.5.3.3.4 for the flow in an axisymmetric contraction, by considering only a peripheral stream tube. Singularity effects close to the section of contraction can also be emphasized and are found to be consistent with experimental values [13]. In computations, the modified Hermitian element (**He$_2$**) is found to be more robust than element (**He$_1$**).

3. *Results for the flow domain with a threefold rotational geometry*

The domain is the same as the geometry considered in Fig. 5.6. Here, one shows some results corresponding to truncated cones of angles 20° and 40° which join the upstream circular cylinder. The downstream tube has a triangular basis where the velocity component w at the downstream Poiseuille flow section $z = z_2$ is computed numerically. Flow streamlines are determined by the intersection of surfaces f and g:

Computed streamlines of Fig. 5.19, presented for two different azimuthal angles, are rather similar. However, the computed stresses presented in Fig. 5.20 point out significant differences versus the azimuth angle.

Similar computations can also be made for domains involving a threefold rotational geometry, as for the geometry presented in Figs. 5.21 and 5.22. Upstream and downstream Poiseuille velocity contour values of this example are shown in Fig. 5.16.

Regarding the stability of the assumption for computations in the total flow domain by considering sub-domains (the stream tubes) and the global non-linear boundary condition (5.105), it is still not possible, at the present, to provide a mathematical proof. This is an *open theoretical problem.* However, the main tests performed in two-dimensional and three-dimensional simulations provide satisfactory computed data.

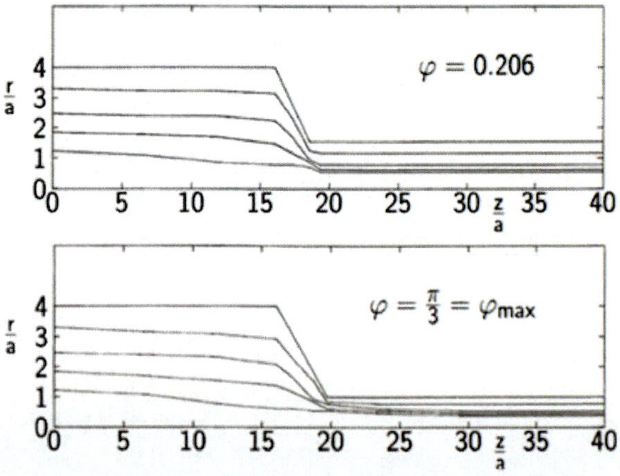

Fig. 5.19 Computed surfaces f as functions of the azimuthal angle (convergent angle = 40°), for a viscoelastic equation*

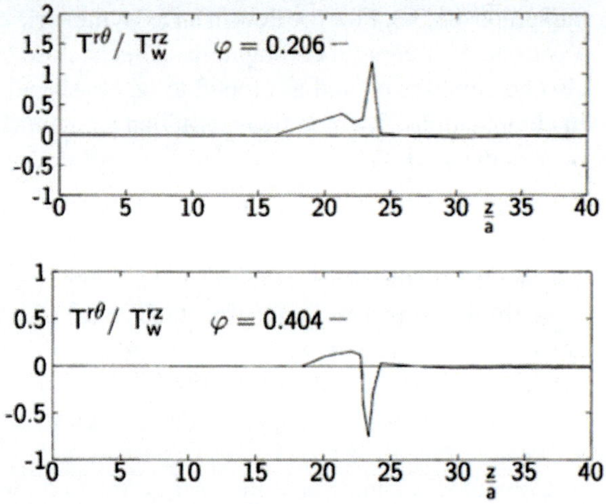

Fig. 5.20 Computed surfaces f as functions of the azimuthal angle (convergent angle = 40°), for the K-BKZ viscoelastic equation*

5.3.3 Swell Problem: Duct of Square Cross-Section

The flow relates to the three-dimensional swell problem for a duct of constant cross-section along the z-axis. The duct length is considered as large enough to assume fully developed flow at sections $z \leq z_1$. The flow domain is defined such that $z_1 \leq z \leq z_2$. Section z_2 starts the solid flow region in the free jet (Fig. 5.23).

*Reprinted from Ref. [10]. Copyright 1997, with permission from Elsevier

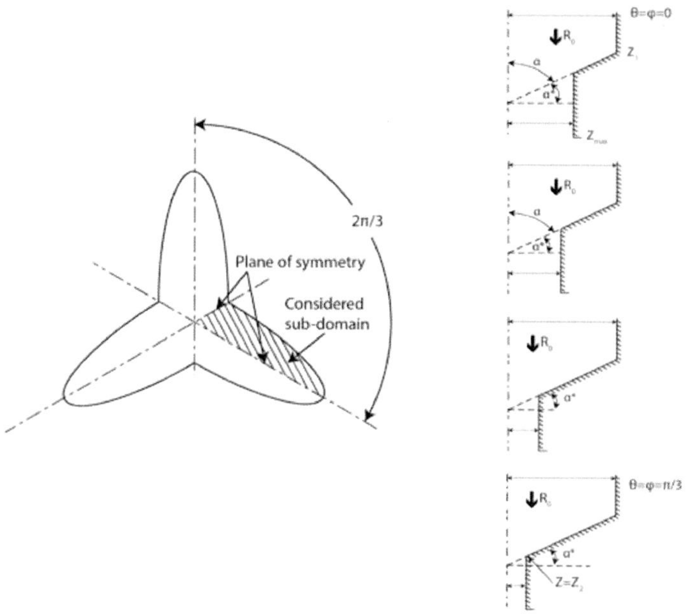

(a) (b)

Fig. 5.21 **a** Downstream section of a duct, **b** convergent profiles of the flow domain

Fig. 5.22 Contour values in the upstream circular domain and corresponding contours at the downstream domain in the computation zone, for a non-Newtonian viscous fluid

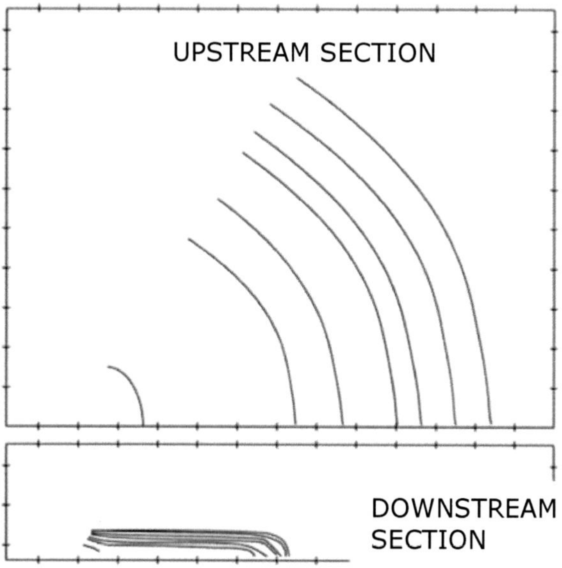

5.3.3.1 Swell Computations with a Peripheral Stream Tube: Newtonian Case

The physical and mapped domains Ω and Ω^* corresponding to the transformation $T: \Omega \rightarrow \Omega^*$ are illustrated by the example of Fig. 5.23, with cylindrical coordinates (r, θ, z). The mapped domain is defined as a straight cylinder whose basis is identical with the upstream cross-section A0 of the physical domain Ω at $z = z_1$. This section limits the fully developed flow region where the kinematics are known. Domains Ω and Ω^* are related by the transformation functions f and g. Figure 5.24 illustrates the respective cross-sections of stream tubes related to contour values of the non-zero components w for the Poiseuille flow of a Newtonian fluid in a duct of rectangular cross-section ($z = z_2$) an those related to the solid flow velocity at $z = z_2$.

1. *Streamline approximation—Discretization of unknowns*

The approximated relationship is adopted in a form close to that already considered in Chap. 4 (Sect. 4.4.2.2, two-dimensional case). The mapped function f is written as a function involving the azimuth angle φ such that

$$f(R, \varphi, Z) = C_1(R, \varphi, Z) + A(R, \varphi)q(Z)[1 - \exp[-(\varphi, Z))B(\varphi, Z)] \quad (5.124)$$

in the tube, for $Z_1 \leq Z \leq Z_0$. In the jet, for $Z_0 \leq Z \leq Z_2$, we assume equation

$$f(R, \varphi, Z) = f(R, \varphi, Z_2) - A(R, \varphi)\exp[-(Z - Z_0)B(\varphi, Z)]. \quad (5.125)$$

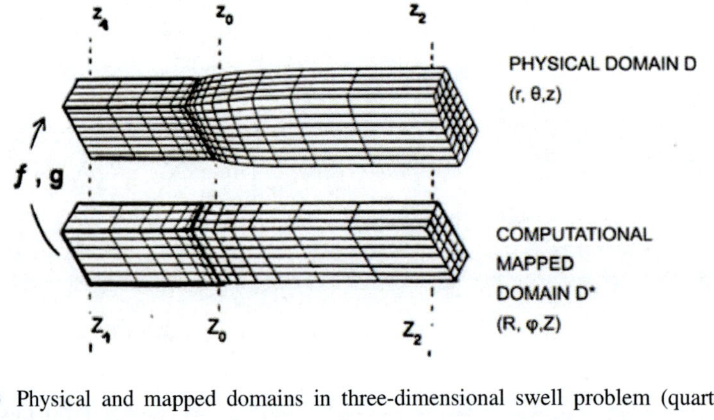

Fig. 5.23 Physical and mapped domains in three-dimensional swell problem (quarter of the domains)

Fig. 5.24 Upstream and downstream cross-sections of stream tubes for the three-dimensional swell problem, with velocity contour values

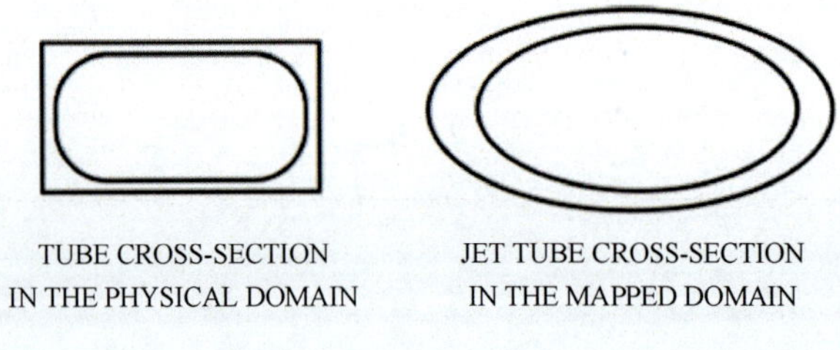

The functions $q(Z)$ and $C_1(R, \varphi, Z)$ have the same meaning as in the two-dimensional case (Chap. 4). The peripheral stream tube involves the wall and the unknown free surface.

Figure 5.25 shows flattened views of stream tube noted B in the physical flow domain Ω and of its mapped domain Ω * for a rectangular duct. The stream tube is limited outside by the boundary surface S_0 involving the wall W_0 and the free surface S_0 and inside by a stream surface f_3 (or S_3) close to the surface S_0. For the mapped stream tube B^*, R_0, φ and Z are used to denote variables related to the outer and internal boundary stream surfaces S_0 and S_3, respectively. These surfaces are closed enough to ensure that the quantities $\| R_0(\varphi, z) - R_3(\varphi, z)\|$ are small. Thus, given a section Z of the flow domain, the mapped peripheral stream tube B^* in domain Ω^*

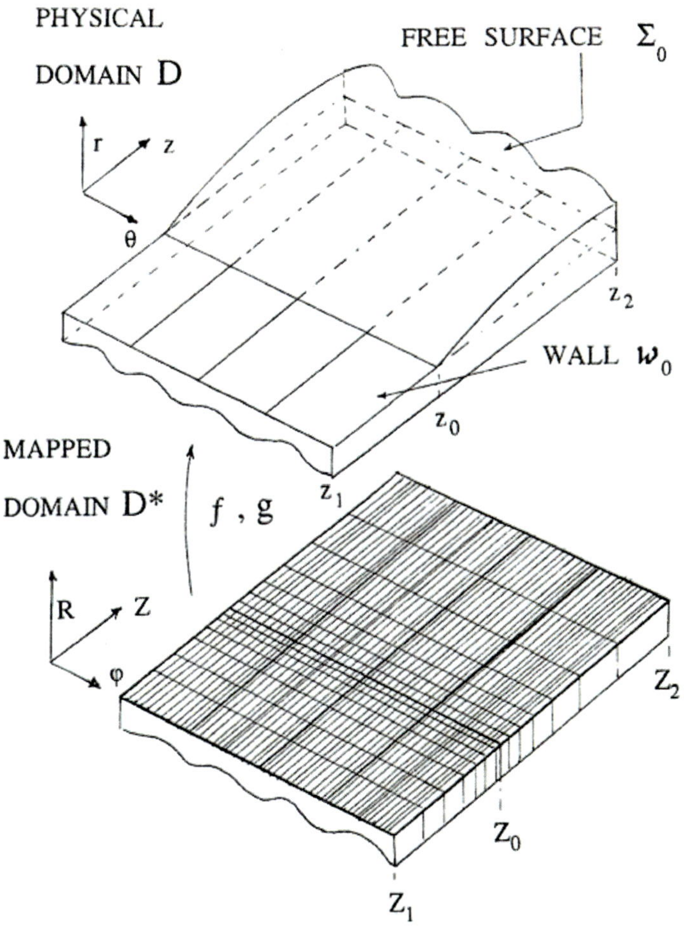

Fig. 5.25 Flattened views of the peripheral stream tube B in the physical domain (Ω and of its transformed computational domain Ω^*. © 1996 by John Wiley & Sons: Ltd: reproduced from Ref. [14] by permission of John Wiley & Sons, Ltd

may be defined by the variables (*R*, φ) such that

$$\boldsymbol{\varphi}_0 \leq \varphi \leq \boldsymbol{\varphi}_1, \quad R_2(\boldsymbol{\varphi}, Z) \leq R \leq R_0(\varphi, Z). \tag{5.126}$$

In Eq. (5.126), φ_0 and φ_I denote the limit values of the azimuth angle φ (maximum difference: $\varphi_I - \varphi_O = 2\pi$ if the duct involves no symmetry). Functions *A* and *B* may be approximated as

$$A(R_2\varphi) = \alpha_0(\varphi) + \alpha_1(\varphi)[R - R_0(\varphi)] + \mathrm{O}(R - R_0(\varphi)]^2), \tag{5.127}$$

$$B(R_2\varphi) = \beta_0(\varphi) + \beta_1(\varphi)[R - R_0(\varphi)] + \mathrm{O}(R - R_0(\varphi)]^2). \tag{5.128}$$

Similarly, the mapping function *g* is expressed in the peripheral stream tube by the equation:

$$g(R, \varphi, Z) = \gamma_d(\varphi, Z) + \gamma_1(\varphi, Z)[R - R_0(\varphi, Z)] + \mathrm{O}\big([R - R_0(\varphi, Z)]^2\big). \tag{5.129}$$

The functions $\alpha_0(\varphi)$, $\alpha_1(\varphi)$, $\beta_0(\varphi)$, $\beta_1(\varphi)$, $\gamma_0(\varphi, Z)$ and $\gamma_1(\varphi, Z)$ of Eqs. (5.127)–(5.129) should be determined numerically.

2. *Governing equations in the peripheral stream tube—Swelling criterion*

- Dynamic equations

Surface tension, inertia and body forces are ignored. There is no traction on the jet at section z_2. Classic dynamic equations are given by Eqs. (5.74)–(5.76) in cylindrical coordinates (*R*, φ, *Z*). Stress components are expressed in terms of the mapping functions *f* and *g*, by using the quantities $\{\alpha_0(\varphi), \alpha_1(\varphi), \beta_0(\varphi), \beta_1(\varphi), \gamma_0(\varphi, Z)\gamma_0(\varphi, Z), \gamma_1(\varphi, Z)\}$, defined by Eqs. (5.127)–(5.129). The equations are written on rectilinear streamlines of the mapped domain Ω^*. Thus, the pressure *p*, assumed to be zero at the solid flow section $Z_2 = z_2$, may be determined by integration using the relation:

$$p(R, \varphi, Z) = \int_{Z_2}^{Z} H(R, \varphi, \zeta)\mathrm{d}\zeta. \tag{5.130}$$

The integrand *H*(*R*, φ, *Z*) is evaluated from the dynamic equations.

The "three-point formula" is adopted for derivatives in terms of *R*, φ, *Z* and *j*. For a function *Y*(*x*) at three points x_1, x_2 and x_3, with respective corresponding values Y_1, Y_2 and Y_3, the x-derivative at x_1 is approximated by

$$\begin{aligned}[\mathrm{d}Y/\mathrm{d}x]_{x_1} &= [Y(x_1) - Y(x_2)]/(x_I - x_2) - [Y(x_2) - Y(x_3)]/(x_2 - x_3) \\ &\quad + [Y(x_1) - Y(x_3)]/(x_1 - x_3).\end{aligned} \tag{5.131}$$

To compute the partial derivatives in the mapped peripheral stream tube, two inner surfaces S_1 and S_2 are used together with the outer surface S_0 (transformed of wall and free surface) and the limiting inner surface S_3.

- *Action of the complementary domain of the peripheral stream tube*

This action, expressed by Eqs. (5.55) and (5.56), is simplified for the flow in a converging duct of threefold rotational symmetry with respect to the z-axis, as considered in Sect. 5.3.1. Thus, for ducts involving symmetries as the swelling flow domain, Eqs. (5.38)–(5.39) reduce to the single scalar equation:

$$\left(\int_{\partial\Omega} \underline{\underline{\sigma}} \cdot \underline{n}\mathrm{d}s\right) \cdot \underline{\varepsilon}_z = \left(\int_{\partial\Omega} (-p\underline{\underline{I}} + \underline{\underline{T}})\underline{n}\mathrm{d}s)\right) \cdot \underline{\varepsilon}_z = 0. \tag{5.132}$$

For a Newtonian fluid, the stress tensor is given by $\underline{\underline{T}} = 2\eta_0\underline{\underline{D}}$, in terms the mapping functions f and g.

The set of governing equations must be considered together with the classical "simple" boundary conditions in terms of the unknowns related to the mapping functions.

3. *Procedure for determining unknown free surface*

The procedure for determining the free jet presents similarities to that already adopted in STM for two-dimensional swell studies with a peripheral stream tube:

- condition on the unknown swell surface Σ_0 (the stress is zero along the free surface):

$$(\mathrm{C}_1) \quad I_1 = \int_{\Sigma_0} \left|(-p\underline{\underline{I}} + \underline{\underline{T}})\right| \mathrm{d}s = \mathbf{0}. \tag{5.133}$$

- action of the complementary domain $\mathrm{B_c}$ of the peripheral stream tube B limited by the upstream and downstream sections, with symmetries for the duct considered. The action of domain $\mathrm{B_c}$ is reduced to the scalar equation:

$$(C_2) \quad I_2 = \int_{\partial B_c} \left|(-p\underline{\underline{I}} + \underline{\underline{T}})\right| \cdot \underline{n}\mathrm{d}s \cdot \underline{\varepsilon}_z = 0. \tag{5.134}$$

Equations (5.133)–(5.134) should be considered with the governing equations in the peripheral stream tube.

4. *Numerical procedure for solving equations*

When one considers governing equations in a peripheral stream tube, it is of interest to recall several advantages of the proposed scheme:

- In contrast to classical methods of flow simulation, the mesh used for swell computations remains unchanged during the iterative process.
- A new free surface does not have to be updated in order to check whether or not condition (5.132) corresponds to a minimum: the unknown swell surface is determined by ***direct computation of the unknowns*** since this equation is involved in the overall set of governing equations. This possibility simplifies the numerical process.

For a sub-domain, STM analysis leads to mathematical constraints which are taken into account together with the relevant equations. The numbers of equations and unknowns are different. The LM algorithm is retained for solving the equations.

5. *Results for a die of square cross-section (Newtonian fluid)*

A peripheral stream tube limited by planes $\theta = g = \varphi_0 = 0$ and $\theta = g = \varphi_1 = \pi/4$ is considered, owing to the symmetry properties of the duct. The iterative procedure permits the direct determination of the optimal free surface, according to the previous considerations. Figure 5.26a presents sections of the computed peripheral stream tube involving the free surfaces S_0 (wall and free surface), S_1 and S_2, for planes z_M = constant. Figure 5.26b shows azimuthal sections of the tube B for planes φ = constant.

The liquid swells or retracts depending on the azimuthal angle φ and are found to be consistent with literature results [14]. The results indicate singularity effects in the vicinity of the exit section and differences in stress levels, for the two sections, as shown in Fig. 5.27.

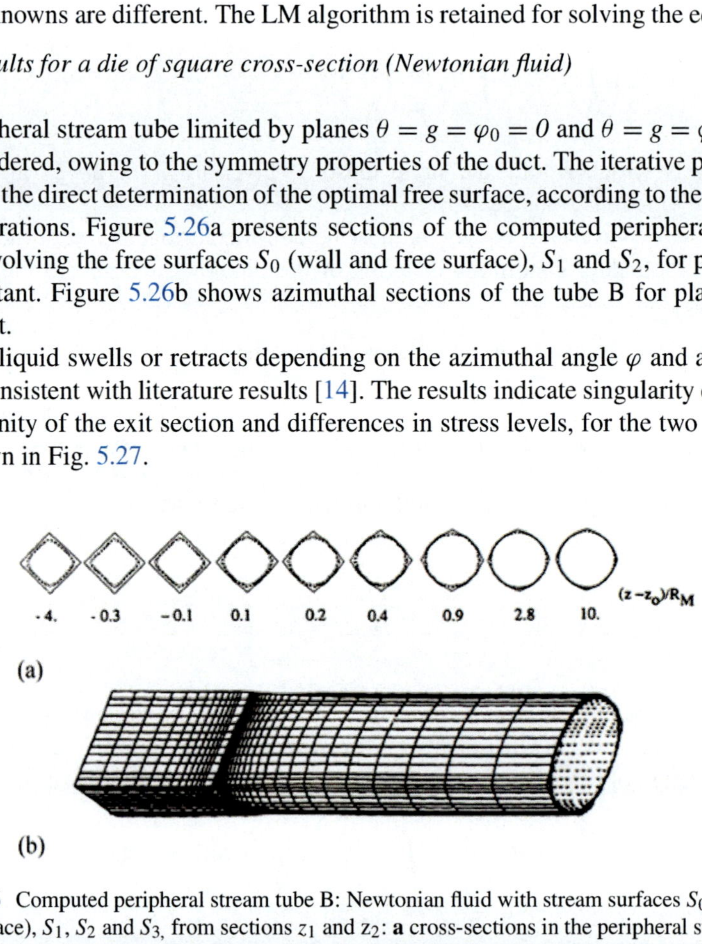

Fig. 5.26 Computed peripheral stream tube B: Newtonian fluid with stream surfaces S_0 (wall and free surface), S_1, S_2 and S_3, from sections z_1 and z_2: **a** cross-sections in the peripheral stream tube (die and jet); **b** three-dimensional view of the flow sub-domain, with discretization points. Ω^*. © 1996 by John Wiley & Sons; Ltd: reproduced from Ref. [14] by permission of John Wiley & Sons, Ltd

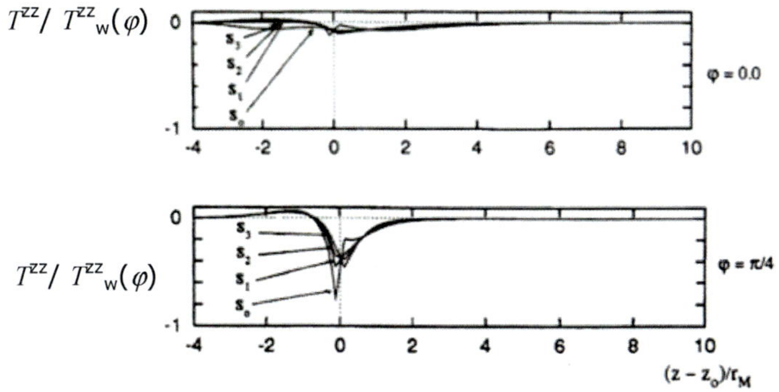

Fig. 5.27 Dimensionless stress $T^{zz}/T^{zz}{}_w(\varphi)$ along streamlines of surfaces S_0, S_1, S_2 and S_3 for azimuthal sections $\theta = g = \varphi$ and $\varphi = \pi/4$ (peripheral stream tube). © 1996 by John Wiley & Sons, Ltd: reproduced from Ref. [14] by permission of John Wiley & sons, Ltd

5.3.3.2 Swell Example for a Viscoelastic Fluid: Duct of Square Cross-Section

A similar STM approach is applied to a viscoelastic constitutive equation, in relation to extrusion experiments of a polyethylene material. The main elements of the problem are summarized in this section.

1. *Constitutive model—Numerical approach*

The experimental data on the material are fit by a memory-integral Wagner constitutive equation [15] expressed by

$$\underline{\underline{T}}(t) = \int_{-\infty}^{t} m(t - t')h(I_1, I_2)\underline{\underline{C}}_t^{-1}(t')\mathrm{d}t'. \tag{5.135}$$

The extrudate swell problem is solved by considering the peripheral stream tube, similarly to the Newtonian case.

2. *Remarks and numerical results*

A peripheral stream tube limited by planes $\theta = g = \varphi_0 = 0$ and planes $\theta = g = \varphi_1 = \pi/4$ is considered. Numerical tests have been performed for different meshes and sizes of the peripheral stream tube.

- *Remarks*

1. A numerical divergence of the algorithm is observed when considering mapped streamlines very close to the wall in domain Ω^*, located at a distance R such that $(R_0(\varphi, Z_1) - R)/R_0(\varphi, Z_1) \leq 0.017$. The azimuth corner $\varphi = 0$ is used as a reference plane (Z_1 denotes the upstream Poiseuille section). This lack

of convergence probably arises from singularity effects corresponding to the junction points between the wall and the free surface.

2. When the size R of the peripheral stream tube B is defined such that $(R_0(\varphi, Z_1)\text{-}R)/R_0(\varphi, Z_1) \leq 0.13$, the computed solutions is found to depend on the initial estimate $\underline{X}_{[0]}$, even at small flow rates.
3. The computed solutions for stream tubes B limited inside by streamlines of abscissa R verifying the condition

$$0.07 \leq (R_0(\varphi, Z_1) - R)/R_0(\varphi, Z_1) \leq 0.13 \tag{5.136}$$

are found to be *stable and consistent for sufficiently refined meshes.* Number of points and planes:

- in the z-direction: 10 points in the tube and 18 points in the jet are adopted for all the flow rates investigated, with refinements in the vicinity of the exit section
- in the φ-direction: 7 azimuth planes are retained.

Using this grid, the number of iterations before convergence is generally found to be of the order of 10.

- *Numerical results*

Comparisons between experimental and STM results, for the free surface at the corner, are shown in Fig. 5.28. Figure 5.30 shows cross-sections of the peripheral stream tube and a view of the computed free surface (Fig. 5.29).

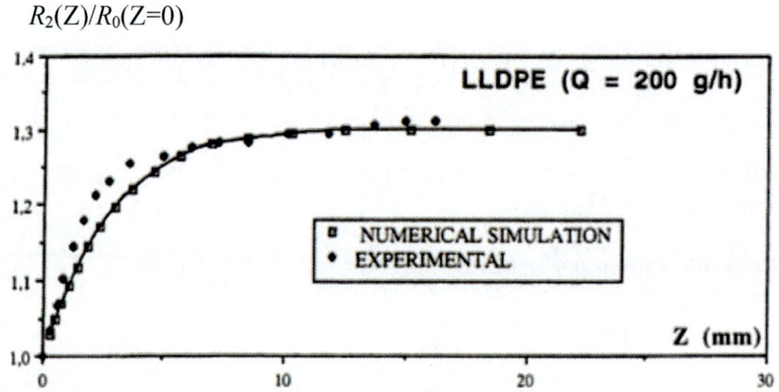

Fig. 5.28 Example of comparisons between the computed jet surface ratio results and the experimental jet data at the free surface at the corner ($\theta = \varphi = 0$)*

*Reprinted from Ref. [14]. Copyright 1999, with permission from Elsevier

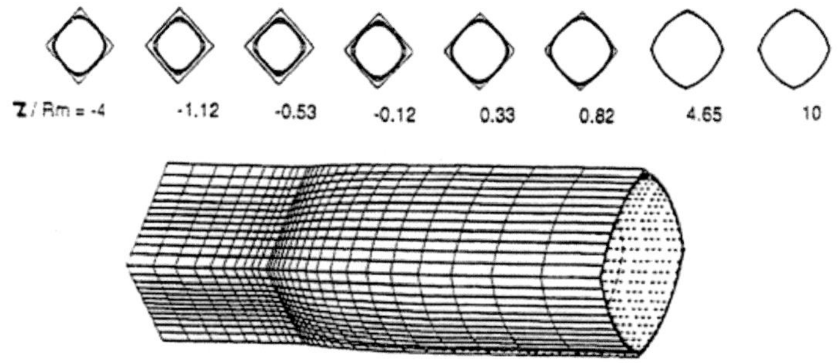

Fig. 5.29 Computed peripheral stream tube; cross-sections in the die and the jet and three-dimensional view of the computed free surface, from sections z_1 to z_2*

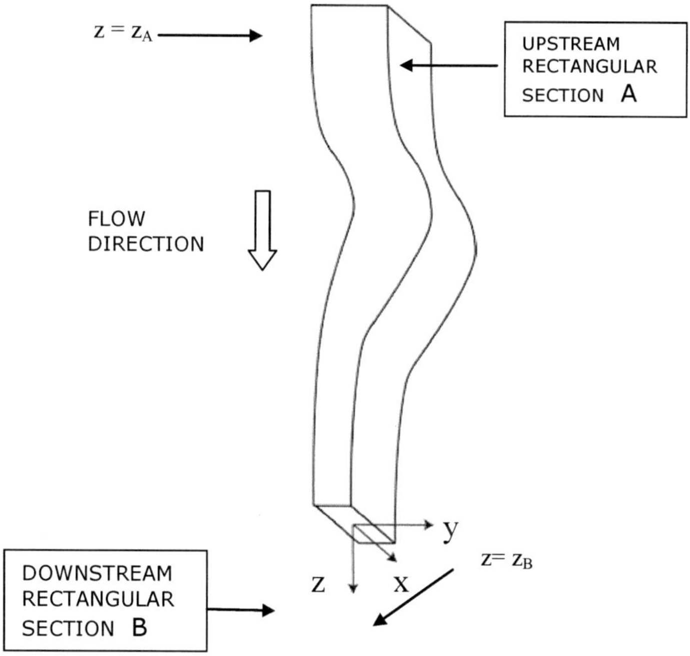

Fig. 5.30 Three-dimensional vertical duct for a three-dimensional flow, where inertial and body forces are ignored. A constant section is considered beyond the downstream section B, thus allowing a fully developed flow

5.4 Concluding Remarks

This chapter has concerned three-dimensional STM studies for confined and free surface flows. Newtonian and non-Newtonian constitutive equations have been considered in the calculations, using a pair of unknown stream functions.

- Computations may lead to satisfactory results when the theoretical model adopted for the numerical experiments is realistic for the material.
- STM applications reveal possibilities of computations to provide robust and efficient numerical models, with a reduced number of unknowns and computer memory.
- The applications concern situations involved in plastic extrusion and fibre operations, which require accurate computational models for predicting rupture zones encountered in industrial processes.
- In the three-dimensional swell flow at the exit of square cross-section, a satisfactory agreement was obtained for the jet shape of linear low-density polyethylene (LLDPE) and STM results, using a memory-integral constitutive equation.
- The examples underline possibilities to adopt different discretization techniques for solving theoretical or engineering problems. Simple elements may be adopted for the primary unknowns. The number of unknowns and CPU time is reduced.
- Analytical forms can be adopted in some cases to facilitate and complete the implementation of the equations.
- In the three-dimensional case, the LM optimization algorithm has proved to be solid and efficient to solve the non-linear STM governing equations.
- In the situations investigated, STM results are found to be in agreement with literature and experimental data. Otherwise, they are generally consistent with the physical behaviour expected from the problem under consideration.

5.5 Example of a Three-Dimensional Problem in STM

Figure 5.30 presents a three-dimensional example of a flow domain proposed for STM calculations, in the steady state.

The respective upstream and downstream sections *A* and *B* of the duct are rectangular. The fully developed flow assumption is retained for two sections:

- section *A*, dimensions X_A and Y_A
- section *B*, dimensions X_B and Y_B, with $X_A > X_B$ and $Y_A > Y_B$.

The sides of the two rectangular sections *A* and *B* are parallel (Fig. 5.25).

Transformation functions *f* and *g* such that $x = f\ (X, Y, Z)$, $y = g\ (X, Y, Z)$, $z = Z$.

It should be of interest to summarize the different global steps of the flow calculations, from the determination of velocities at the upstream and downstream sections towards computation of the flow characteristics, stream surfaces, streamlines and stresses.

One can assume the fluid to be Newtonian and incompressible, of constant viscosity η. At the beginning, a given flow rate Q, in stationary conditions, leads to compute a head loss h at the upstream section.

According to the considered geometry, the whole domain is considered for the computations. The dynamic equations are written with Cartesian variables. The boundary conditions take into account the shape of the four boundary surfaces of the domain. The head loss at the downstream section can be evaluated from the known flow rate at the upstream section. Velocities and their contour values at upstream and downstream sections can be computed by the method proposed in Sect. 5.2.2 or from books and papers [2–7]. A three-dimensional peripheral stream tube and its complementary domain can be considered for the computations, as proposed in Chap. 5.

References

1. Clermont JR (1988) Analysis of incompressible three-dimensional flows using the concept of stream tubes in relation with a transformation of the physical domain. Rheol Acta 27:357–362 [Springer]
2. Roache PJ (1972) Computational fluid dynamics. Hermosa Publishers
3. Normandin M, Clermont J-R, Mahmoud A (2000) Laminar flow calculations in ducts of complex shapes using mapping functions and decomposition of domains. Math Comput Simul 52:21–39. Elsevier
4. Tanner RI (1985) Engineering rheology. Clarendon Press, Oxford
5. Boley BA (1953) Graphical-numerical solution of problems of Saint-Venant torsion and bending. J Appl Mech 20(3):321–326
6. Dodson AG, Townsend P, Walters K (1974) Non-Newtonian flows in pipes of non-circular cross-section. Comput Fluids 2:317–338
7. Ahmeda M, Normandin M, Clermont J-R (1995) Calculation of fully-developed flows of complex fluids in ducts of arbitrary shape, using a mapped circular domain. Commun Num Methods Eng 11:813–820
8. Clermont JR, de la Lande ME (1993) Numerical simulation of three-dimensional duct flows of incompressible fluids by using the stream tube method. Part I: Newtonian equation. Theor Comput Fluid Dyn 4:89–110
9. Bereaux Y, Clermont JR (1997) Numerical simulation of two- and three-dimensional complex flows of viscoelastic fluids using the stream-tube method. Math Comput Simul 44:387–400. Elsevier
10. Clermont JR (1992) Calculation of kinematic histories in two and three-dimensional flows using streamline coordinate functions. Rheol Acta 32:82–93
11. Papanastasiou AC, Scriven LE, Macosco CW (1989) An integral constitutive equation for mixed flows: viscoelastic characterization. J. Rheol 27(4):387–410
12. Luo XL, Mitsoulis E (1990) An efficient algorithm for strain history tracking in finite element computations of non-Newtonian fluids with integral constitutive equations. Int J Num Meth Fluids 11:1015–1031
13. Guillet J, Revenu P, Bereaux Y, Clermont JR (1996) Experimental and numerical study of entry flows of low-density polyethylene melts. Rheol Acta 35:494–507
14. Normandin M, Clermont JR (1996) Three-dimensional extrudate swell: formulation with the stream-tube method and numerical results for a Newtonian fluid. Int J Num Meth Fluids 23:937–952. Wiley
15. Wagner MH (1978) A constitutive analysis of uniaxial elongational flow data of a low-density polyethylene melt. J Non-Newtonian Fluid Mech 4:39–55

Chapter 6
Stream-Tube Method Domain Decomposition Closed Streamlines

6.1 Introduction

This chapter presents STM theoretical elements leading to computations in complex flow domains involving both straight and closed streamlines, using a domain decomposition approach. Local transformation functions in two- or three-dimensional sub-domains may be defined to simulate flows of complex fluids as those requiring evaluation of particle time history. Examples of computations in sub-domains with vortices are given for two-dimensional flows between eccentric cylinders, with Newtonian and more complex models, in relation to journal bearing problems. Different flow conditions are proposed with domain decompositions. The approach, presented here for steady flows of incompressible fluids, provides consistent results when compared with literature data.

6.2 General Transformations: Basic Computational Results with the Stream-Tube Method

6.2.1 *Basic Equations for General Transformations*

Let $x^i\left(x^1 = x, x^2 = y, x^3 = z\right)$ be the Cartesian coordinates related to a basic Euclidean basis $(\boldsymbol{\varepsilon}_i)$: $(\underline{\varepsilon}_1, \underline{\varepsilon}_2, \underline{\varepsilon}_3)$ for a material point M which occupies the position $\underline{X}(x, y, z)$ in $\Omega\left(\Re^3 \supset \Omega\right)$. When considering another coordinate system $\xi^j(\xi^1 = X, \xi^2 = Y, \xi^3 = s)$ for a sub-domain Ω^* of $\Re^3$, a general transformation $x^i(\xi^j)$ from Ω^* to Ω may be defined by the following relations:

$$x = \alpha(X, Y, s);\ y = \beta(X, Y, s);\ Z = \gamma(X, Y, s). \tag{6.1}$$

J.-R. Clermont and A. Ammar, *Stream-Tube Method*,
https://doi.org/10.1007/978-3-030-65470-2_6

The Jacobian $\Delta = \left|\partial(x^i)/\partial(\xi^j)\right|$, assumed to be non-zero, can be expressed by the equation:

$$\Delta = \alpha'_X(\beta' Y \gamma'_m - \beta'_Y \alpha'_Y) - \beta'_X(\alpha'_Y \gamma'_S - \gamma'_Y \alpha'_S) + \gamma'_X(\alpha'_Y \beta'_S - \beta'_Y \alpha'_S), \tag{6.2}$$

where α', β' ..., etc. stand for partial derivatives with respect to the variables X, Y and S. Different partial derivative operators lead us to define quantities denoted by $A_1, A_2, A_3, B_1, B_2, B_3, C_1, C_2$, given by the relations:

$$A_1 = \alpha' Y \beta' s' - \alpha' s' Y;\ A_2 = \beta' Y \gamma' s - \beta'_S \gamma' Y$$
$$A_3 = \gamma' Y \alpha'_S - \gamma'_S \alpha' Y, \tag{6.3}$$

$$B_1 = \alpha'_X \beta'_S - \alpha'_S \beta'_X;\ B_2 = \beta'_X \gamma' s' - \beta'_S \gamma'_X;\ B_3 = \gamma'_X \alpha'_S - \gamma'_S \alpha'_X, \tag{6.4}$$

$$C_1 = \alpha'_X \beta' Y - \alpha' Y'_X;\ C_2 = \beta'_X \gamma'_Y - \beta' Y \gamma' X$$
$$C_3 = \gamma'_X \alpha'_Y - \gamma'_Y \alpha'_X. \tag{6.5}$$

These equations allow to write the Jacobian Δ as

$$\Delta = \alpha'_X A_2 + \alpha'_X A_2 + \gamma'_X A_3. \tag{6.6}$$

Then, the natural basis ($\boldsymbol{e}_i$): ($\underline{e}_1$, $\underline{e}_2$, $\underline{e}_3$) related to the coordinates (X, Y, s) is expressed by

$$\underline{e}_1 = \alpha'_X \underline{\varepsilon}_1 + \beta'_X \underline{\varepsilon}_2 + \gamma'_X \underline{\varepsilon}_3, \tag{6.7}$$

$$\underline{e}_2 = \alpha'_Y \underline{\varepsilon}_1 + \beta' Y \underline{\varepsilon}_2 + \gamma' Y \underline{\varepsilon}, \tag{6.8}$$

$$\underline{e}_3 = \alpha'_s \underline{\varepsilon}_1 + \beta'_s \underline{\varepsilon}_2 + \gamma'_s \underline{\varepsilon}_3. \tag{6.9}$$

Conversely, from Eqs. (6.7) to (6.9), the Cartesian basis vectors may be given in terms of the natural basis by

$$\underline{\varepsilon}_1 = (1/\Delta)\left[A_2 \underline{e}_1 + B_2 \underline{e}_2 + C_2 \underline{e}_3\right], \tag{6.10}$$

$$\underline{\varepsilon}_2 = (1/\Delta)\left[A_3 \underline{e}_1 + B_3 \underline{e}_2 + C_3 \underline{e}_3\right], \tag{6.11}$$

$$\underline{\varepsilon}_3 = (1/\Delta)\left[A_1 \underline{e}_1 + B_1 \underline{e}_2 + C_1 \underline{e}_3\right]. \tag{6.12}$$

It may be shown that the derivative operators $\partial/\partial x$, $\partial/\partial y$ and $\partial/\partial z$ are expressed in terms of $\partial/\partial\xi^j$ by the following equations:

$$\partial/\partial x = (1/\Delta)[A_2\partial/\partial X - B_2\partial/\partial Y + C_2\partial/\partial s], \tag{6.13}$$

$$\partial/\partial y = (1/\Delta)[A_3\partial/\partial X - B_3\partial/\partial Y + C_3\partial/\partial s], \tag{6.14}$$

$$\partial/\partial z = (1/\Delta)[A_1\partial/\partial X - B_1\partial/\partial Y + C_1\partial/\partial s]. \tag{6.15}$$

6.2.2 *Transformations of Sub-domains*

Stream-tube analysis has concerned the main flows of various types of fluids as shown in previous chapters. Pure recirculating flows related to closed streamlines for a two-dimensional journal bearing geometry can be investigated in STM [1]. In both cases, the formulation allows simple mapped computational domains to be defined.

Consider now sub-regions Ω_i that may involve open or closed elementary stream tubes or both. These non-overlapping subdomains are defined such that

$$\Omega = \bigcup_{m=1}^{m=m_0} \Omega_m. \tag{6.16}$$

Starting from an original section z_1, each sub-domain Ω_m is limited by two cross-section planes z_m and z_{m+1} ($m = 1, 2, \ldots, m_o$). These sub-domains may involve open and closed stream tubes denoted by B_0 and B_c, respectively. The regions, a priori unknown in size, correspond to the main flow and vortex regions of the total flow domain Ω.

Consider a sub-domain Ω_m of Ω limited by two cross-section planes at z_m and z_{m+1} (Fig. 6.1) in an axisymmetric flow situation. In Ω_m, a cross-section S_m at $z = \xi m(z_m \leq \xi_m \leq z_{m+1})$ is used as reference section. The mapped domain $\Omega^*{}_m$ is a straight cylinder of basis $S^*{}_m$ identical, in shape, to the reference section $S^*{}_m$. The cylinder consists of mapped straight lines of the physical sub-domain $\Omega^*{}_m$, parallel to the direction of the generants. These transformed streamlines are related to a local variable s in $\Omega^*{}_m$, to be considered as a computational domain. Mapped domains $B^*{}_0$ and $B^*{}_c$ of the respective open and closed stream tubes B_0 and B_c are elementary straight cylinders of $\Omega^*{}_m$. The basis of the cylinder is related to local variables (X, Y).

Flow streamlines of Fig. 6.1 illustrate a two-dimensional analytical case requiring local transformation functions. The upstream, section z_i allows a reference section S_i

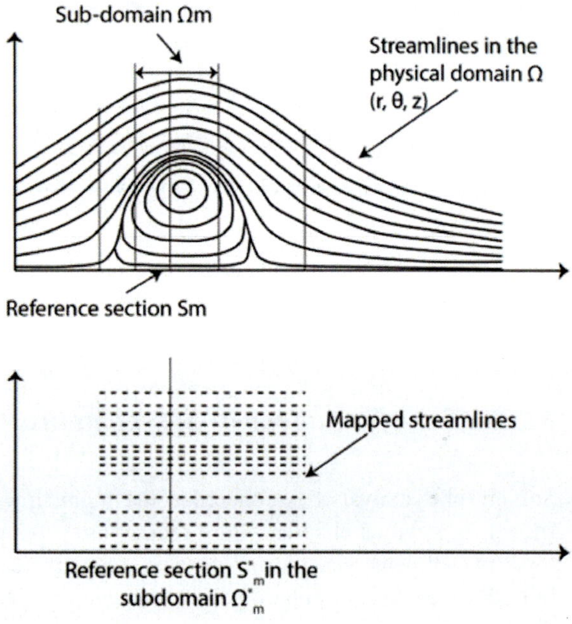

Fig. 6.1 Sub-domain Ω_m and its mapped domain involving rectilinear streamlines for a two-dimensional (axisymmetric) flow

to be defined such that the corresponding reference function Φ_1 is known at section S_i.

Sub-domains $\Omega^*{}_m$, of reference sections S_m, may involve different types of stream tubes.

Let $\Omega^*{}_m$ be elementary sub-domains of rectilinear mapped parallel streamlines related to a reference cross-section $S^*{}_m$ and involving a finite number of mapped stream tubes $B^*{}_i$ (with $\Omega^*_m = \bigcup_{m=1}^{m=m_0} B_{i\,m}$). One investigates elementary transformations T_m defined from a mapped sub-domain $\Omega^*{}_m$, towards a sub-region Ω_m of number *m*. These transformations are defined between two cross-sections z_m and z_{m+1} in the physical domain Ω_m, such that

$$T_m : \Omega_m \to \Omega * m,$$

with

$$M(x, y, z) \to M^*(X, Y, s), \tag{6.17}$$

according to coordinate transformation functions of Eq. (6.1). The variable *s* is related to the curvilinear abscissa χ of a moving particle on its streamline (in a steady flow configuration) defined by

$$\chi = \chi_0 + \int_{t_0}^{t} \left|\underline{V}(\tau)\right| \mathrm{d}\tau. \quad (6.18)$$

χ_0 denotes the abscissa of the reference section $S^*{}_m$ of the mapped sub-domain $\Omega^*{}_m$. $\left|\underline{V}(\tau)\right|$ is the modulus of the velocity vector $\underline{V}$ on the streamlines points, expressed in Cartesian coordinates by the equation:

$$\underline{V} = u(x, y, z)\underline{\varepsilon}_1 + v(x, y, z)\underline{\varepsilon}_2 + w(x, y, z)\underline{\varepsilon}_3. \quad (6.19)$$

As already pointed out in previous chapters for Eq. (6.18), the times t_0 and t are associated with respective positions s_0 and s of a material point.

- *The variable s*

The following variables are now considered:

- s_M ($s_M < +\infty$), the given total length of the streamline in the transformed domain;
- χ_M ($\chi_M < +\infty$), the maximum curvilinear abscissa on every streamline L originated at a point M_0 of the stream tube. One may define the variable s by the following relationship:

$$s = (\chi/\chi_M)s_M. \quad (6.20)$$

To define the variable **s** for a streamline L of zero velocity (presence of a wall where the fluid adheres), the coordinate χ related to a point M is assumed to be the curvilinear distance $\overline{M_0M}$. Such definitions are different from previous specific formulations of STM which do not involve variables directly related to the kinematics.

Figure 6.2 illustrates physical and mapped stream tubes of length s_M corresponding to open and closed streamlines in a three-dimensional flow situation. The case of open stream tubes is shown in Fig. 6.2a. Figure 6.2b depicts a flow sub-domain of Ω_m involving closed stream tubes, related to a reference section, for s

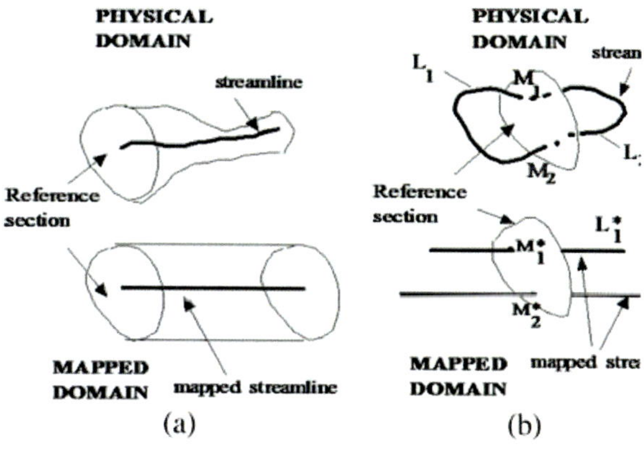

Fig. 6.2 Elementary sub-domains: **a** Physical sub-domain with open streamlines and its transformed domain. **b** Closed streamline L and the corresponding mapped domain

$= s_0$. The sub-region of closed stream tubes, doubly connected, is obtained by the intersection of the stream tube by the reference plane section. On Fig. 6.2b, it may be observed that the closed stream tube involves two separated regions where the main velocity component does not change its sign:

- a closed streamline L can be divided into two curves L_1 and L_2 such that $L = L_1 \cup L_2$, according to the positions of streamline points of zero velocity;
- the curves L_1 and L_2 intersect the reference section at points M_1 and M_2, respectively.

The mapped rectilinear streamlines of L_1 and L_2 are represented in the mapped domain as distinct segments passing through M^*_1 and M^*_2, the respective transformed points of M_1 and M_2. The mapped closed stream tube defines a straight cylinder, of basis included in the reference section of the mapped sub-domain Ω^*_m of Ω_m.

A two-dimensional example of the transformation of a closed streamline L in a mapped sub-domain Ω^*_m with two rectilinear branches related to a reference section is shown in Fig. 6.3.

The elements of Fig. 6.4 may correspond to a two- or a three-dimensional flow situation, with different sub-domains Ω^*_m of reference sections S_m.

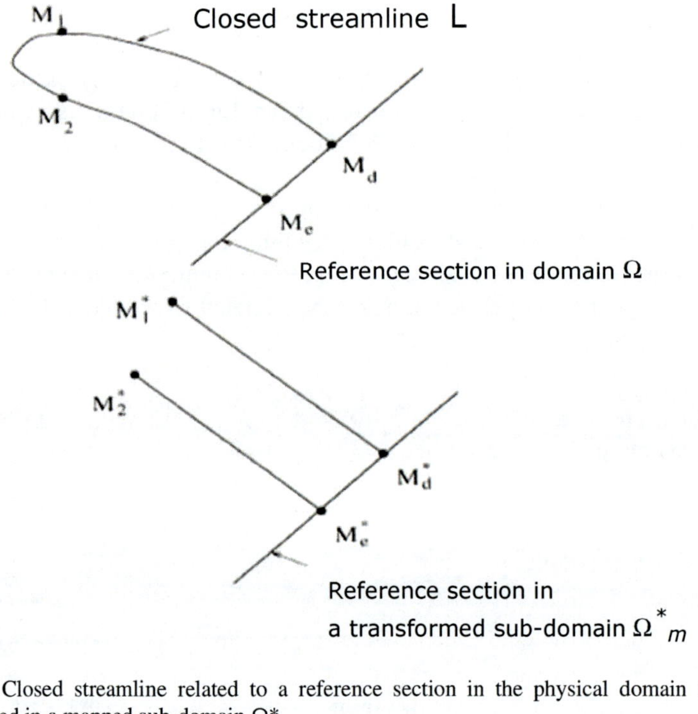

Fig. 6.3 Closed streamline related to a reference section in the physical domain Ω and its transformed in a mapped sub-domain Ω^*_m

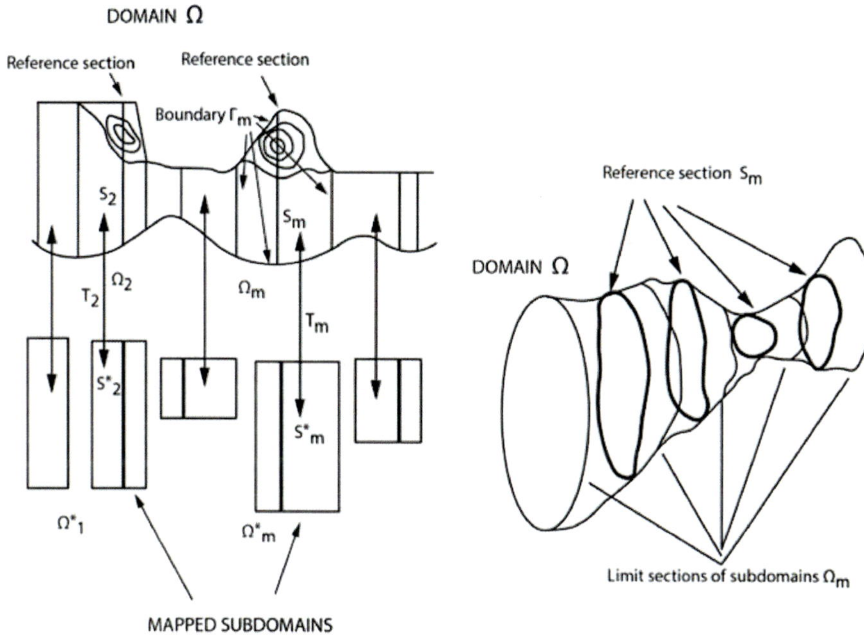

Fig. 6.4 Decomposition of domain Ω into sub-domains Ω_m. Local mappings

6.2.3 *Kinematics: Basic Equations and Unknowns*

Referring to the general definitions for incompressible fluid flows, the velocity vector in terms of a pair a stream functions ψ_1 and ψ_2 is defined by

$$\underline{V} = \nabla\psi_1(x, y, z) \times \nabla\psi_2(x, y, z). \tag{6.21}$$

The use of this equation at a reference section $S(X, Y, s_0)$, where $x = X$, $y = Y$ and $z = s_o$ leads to write the w-velocity component as

$$w(X, Y, s_0) = \frac{\partial\Psi_1(x, y, z)}{\partial X} \cdot \frac{\partial\Psi_2(x, y, z)}{\partial Y} \Big| (X, Y, s_0) - \frac{\partial\Psi_1(x, y, z)}{\partial Y} \cdot \frac{\partial\Psi_2(x, y, z)}{\partial X} \Big| (X, Y, s_0). \tag{6.22}$$

The reference kinematic function ϕ (X, Y) is expressed by

$$\phi(X, Y) = w(X, Y, s_o). \tag{6.23}$$

From Eqs. (6.13) to (6.15), the components (u, v, w) of the velocity vector $\underline{V}$ are expressed as

$$u = \alpha'_s \phi(X, Y)/\Delta; \ \ v = \beta'_s \phi(X, Y)/\Delta; w = \gamma'_s \phi(X, Y)/\Delta, \tag{6.24}$$

where Δ denotes the Jacobian, assumed to be non-singular, written by a relation of the type of Eq. (6.6).

The following relations at the reference section *S* for (*X*, *Y*, s_o) can be assumed as

$$\begin{aligned} \alpha' x(X, Y, s_0) = 1, \quad & \alpha'_Y(X, Y, s_0) = 0 \\ \beta' x(X, Y, s_0) = 0, \quad & \beta'_Y(X, Y, s_o) = 1 \\ \gamma' x(X, Y, s_0) = 0, \quad & \gamma'_Y(X, Y, s_o) = 0. \end{aligned} \tag{6.25}$$

Given a sub-domain Ω_m, the primary unknowns to be considered are the three mapping functions *α, β, γ* and the pressure *p*. The fluid is assumed to adhere to the boundary.

For the moving points, in accordance with the definition of the variable *s,* the set of the governing equations should also involve the following relation (with *w*(*X*, *Ψ*, *σ*) ≠ 0):

$$z = z_0 + \int_{s_0}^{s} w(X, \ Y, \ \xi) \, d\xi \tag{6.26}$$

or, equivalently,

$$\gamma(X, Y, s) = \gamma(X, Y, s_o) + \phi(X, Y) \int_{s_0}^{s} \left[\gamma'_s(X, \ Y, \ \xi)/\Delta(X, \ Y, \ \xi)\right] d\xi. \tag{6.27}$$

In isothermal conditions, the momentum conservation law provides three dynamic equations, written in Cartesian coordinates, by following the approach already defined in STM [2].

The superscripts *i* and *j*, *k*, respectively, associated with components b^i and A^{ik} of a vector $\underline{b}$ and a tensor $\underline{\underline{A}}$, are still related to $x = 1$, $y = 2$ and $z = 3$. The derivatives $\partial/\partial X^i$ are expressed by derivative operators $\partial/\partial \xi^i$ defined by Eqs. (6.13)–(6.15), involving variables (*X*, *Y*, *s)* of the mapped computational domains.

Viscometric shearing flow situations provide simple examples of transformation functions involving global mapping fonctions, as in the Poiseuille case. For pure rotating Couette flow in a plane (*x*, *y*) between two concentric cylinders of radii R_o and R_1 ($R_o < R_1$), global mapping functions may also be expressed by

$$\begin{aligned} x &= \alpha(Y, s) = Y \cos(s/Y) \\ y &= \beta(Y, s) = Y \cos(s/Y) \\ z &= \gamma(X) = X. \end{aligned} \tag{6.28}$$

6.3 Specific Properties: Computational Considerations

6.3.1 *Specific Features of the Analysis*

STM characteristics are recalled here:

(1) The formulation proposed by means of local transformations verifies the incompressibility condition (mass conservation) $\nabla \cdot \underline{V} = 0$. Only momentum conservation equations are written in isothermal cases. These equations $\nabla \cdot \underline{\underline{\sigma}} = \underline{F}$ ($\underline{\underline{\sigma}}$ denotes the stress tensor, $\underline{F}$ the volumic external forces) formally written as

$$E_{mj}(\alpha_i, \beta_i, \gamma_i, p_i) = 0 (j = 1, 2, 3), \text{ for } m = 1, 2, \ldots, m_0 \tag{6.29}$$

should be considered.

(2) Using Eqs. (6.10)–(6.12) and (6.24), it can be shown that the velocity vector $\underline{V}$ in a stream tube is given in the natural basis ($\mathbf{e}_i$): ($\underline{e}_1$, $\underline{e}_2$, $\underline{e}_3$), by the simple relation

$$\underline{V} = [\phi(X, \Psi)/\Delta]\underline{e}_3. \tag{6.30}$$

(3) Simply mapped sub-domains where the streamlines are rectilinear and parallel to the generants of straight cylinders are defined from the transformation (Fig. 6.2).

(4) As in particular flows involving specifically open or closed streamlines, the present formulation avoids particle tracking problems for fluids obeying time-dependent constitutive equations. Time evolution of particles on their pathlines are evaluated by using the w-velocity component expressed by Eq. (6.24), such that

$$t - t_0 = \int_{s_0}^{s} \frac{d\xi}{w(X, Y, \xi)} = \left[\frac{1}{\varphi(X, Y)}\right] \int_{s_0}^{s} \frac{\Delta}{\gamma_S'}\Big|_{(X, Y, \xi)} d\xi. \tag{6.31}$$

In this equation, t_0 is the reference time related to the position of the material point at the reference section (X, Y, s_o) of the sub-domain considered.

(5) The coordinate s is directly related to the curvilinear abscissa of a material point on its pathline. However, this coordinate definition allows consideration of mapped stream tubes of given length.

6.3.2 *Reference Kinematic Functions: Computational Considerations*

It is recalled that, for specific "open stream-tube" problems of previous chapters, the flow is not concerned explicitly by recirculating regions. This leads to consider a single reference kinematic function ϕ. Such function is known and remains unchanged, as also the corresponding reference section, during the iterative numerical computing process. Examples of specific "closed stream-tube" situations [1, 3] requiring a reference section z_o selected arbitrarily to define the streamline transformations are now presented.

The function $\Psi*$ corresponding to the reference function ϕ is unknown and should be determined iteratively in the numerical running process, starting from an initial estimate.

Simple meshes may be defined to approximate the equations and unknowns involved in the STM formulation, according to the geometry of mapped streamlines. The procedure to be achieved here for computing the flow characteristics is essentially related to domain decomposition methods. This requires to write compatibility equations at the interfaces of the sub-domains.

Solution of the problem in the total domain Ω should be obtained by considering local sub-problems stated on the sub-domains Ω_m. This analysis generalizes the STM formulation already depicted. As well-known in computational methods (e.g. [4]), field calculations using sub-domains for different flow conditions have been around for quite a long time.

With domain decomposition techniques, the basic ideas for solving STM problems are summarized as follows (see the example of Fig. 6.3).

The geometrical domain Ω of boundary Γ is divided into M sub-domains $\Omega_m (m = 1, 2, \ldots, M)$. $\Gamma_{m-1} (m = 2, 3, \ldots, M)$ denotes the interface of Ω_{m-1} and Ω_m (Fig. 6.3).

If p_i corresponds to the restriction of the pressure p to sub-domain Ω_i, local transformations α_i (X, Y, s), β_i (X, Y, s) and γ_i (X, Y, s) may be considered to verify

- the kinematic Eq. (6.26)
- the dynamic Eq. (6.29)
- the following boundary condition equations on $\Gamma \cap \Omega_i^*$

$$\begin{aligned} & a_j(\alpha_l, \beta_l, \eta_l, p_i) = 0 \, (j = 1, 2, 3) \\ & b_j(\alpha_l, \beta_l, \eta_l, p_i) = 0 \, (j = 1, 2, 3) \\ & p_i = \pi_I; \end{aligned} \tag{6.32}$$

- the compatibility conditions at the common boundary Γ_i of the sub-domains Ω_i and Ω_{i+1}, such that

$$\begin{aligned} & p(M^*) = p(M_{i+1}^*) \text{ on } \Gamma_i \\ & \underline{V}(M^*) = \underline{V}(M_{i+1}^*) \text{ on } \Gamma_i, \end{aligned} \tag{6.33}$$

where $M_i^* \in \Omega_i$, $M_i^* + 1 \in \Omega_i + 1$,
with $x(M_i^*) = x(M_{i+1}^*)$, $y(M_i^*) = y(M_{i+1}^*)$,
and $z(M_i^*) = Z(M_{i+1}^*)$ on Γ_i.

For every sub-domain Ω_i, one considers a reference section S_i to which a kinematic function ϕ_i corresponds. *This function is generally unknown for the sub-domains*, as pointed out previously.

To solve the problem, the main features of the process are written as follows, in a numerical procedure with the Levenberg–Marquardt algorithm:

(i) Initialization:

Definition of the sub-domains Ω_i $(i = 1, 2, \ldots, M)$ and choice of the reference section S_i.

A geometrical shape of the streamlines is assumed, as also the kinematic function $\phi_{i[0]}$ in every mapped sub-domain Ω_i related to the reference section S_i. The initial guess of the streamlines corresponds to an estimate of the local functions α_i, β_i and γ_i.

(ii) Solving the following set of equations:

- the kinematic Eq. (6.26)
- the dynamic equations are formally written with (6.29) in the total flow domain Ω^*.

6.4 Flows in Ducts

Figure 6.5 presents two flow situations. The geometry shown in Fig. 6.5a, related to the swelling at the exit of a converging domain with a short length tube [5] does not require partitioning of the physical domain. However, an example of duct flows that can be considered by domain decomposition in STM is provided in Fig. 6.5b. This requires known boundary conditions upstream and downstream of the flow domain.

Generally speaking, the boundary updating procedure depends on the problem under consideration. Flow computations in the annulus of two cylinders, of the next section, will provide an illustration of this problem. However, possible vortex regions, notably in complex three-dimensional flows, may lead to insurmountable difficulties to define reference cross-sections and sub-domains.

Unlike flow conditions of Fig. 6.5a, the shape of a two- (or three-) dimensional domain shown in Fig. 6.5b does not a priori allows the STM approach to be considered without domain decomposition. This comes from the possible existence of secondary flows or singularities on the streamlines in the global domain. Such a situation may require adoption of 5 sub-domains, under the assumption of known boundary conditions at the limiting outer upstream and downstream reference flow sections shown in Fig. 6.5b. The general elements for such study are now presented for a flow corresponding to the duct shown in this figure.

1. *Boundary conditions—References sections*

According to the flow domain under consideration, the analysis should be concerned by a domain decomposition method, requiring to write compatibility equations at

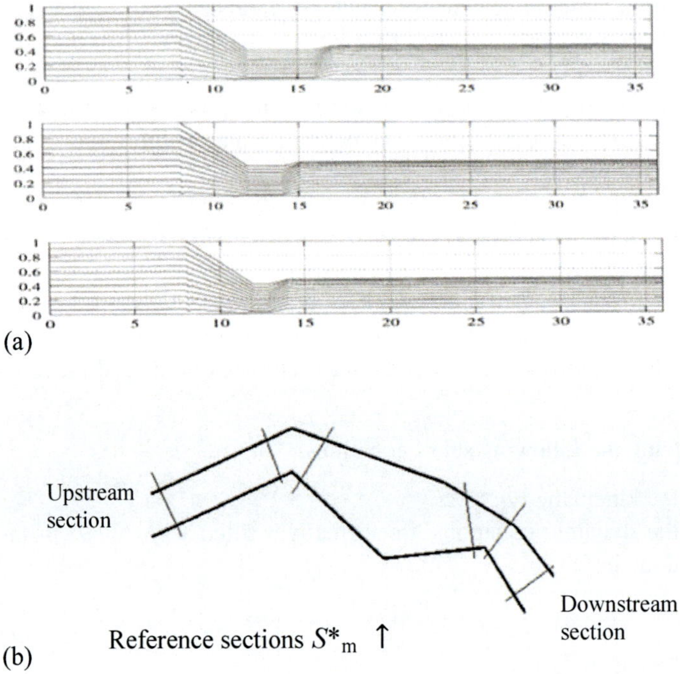

Fig. 6.5 **a** STM computed streamlines and free surface at the exit of cylindrical tubes of small length (Newtonian fluid); **b** Duct geometry requiring domain STM decomposition. The straight lines indicate reference sections for the considered sub-domains

interfaces of sub-domains. No optimization of shape concerning the sub-domains is needed to change with new updated kinematic values at the reference sections of the sub-domains. An initial guess of main flows and vortex zones can be considered unnecessary.

- Reference sections S_m are selected in each sub-domain. The geometry under consideration allows to specify constant common boundaries in the iterative process, though the unknown local transformation functions should be determined for every iteration.
- At the upstream and downstream sections S_1 and $S_{m=6}$ of domain Ω, the kinematic function ϕ_m at the plane section S_m of variables (X, Y) is known. The other kinematic functions should be determined during the iterative process up to convergence of the solving algorithm. We consider local transformations $\alpha_i(X, Y, s)$, $\beta_i(X, Y, s)$ and $\gamma_i(X, Y, s)$ verifying
  - the dynamic Eq. (6.29);
  - the compatibility Eq. (6.33);
  - the constitutive equation for the fluid under consideration
  - the boundary condition equations on the boundaries of the sub-domains

– the compatibility conditions at the common boundary Γ_i of the sub-domains Ω_i and adjacent sub-domains Ω_k, such that

$$p\left(M_i^*\right) = p\left(M_k^*\right) \text{ on } \Gamma_i, \tag{6.34}$$

$$\underline{V}\left(M_i^*\right) = \underline{V}\left(M^*k\right) \text{ on } \Gamma_i, \tag{6.35}$$

where $M_i^* \in \Omega_i$, $M_k^* \in \Omega_k$, with $x\left(M_i^*\right) = x\left(M_k^*\right)$, $y\left(M_i^*\right) = y\left(M_k^*\right)$, $z\left(M_i^*\right) = z\left(M_k^*\right)$ on Γ_i.

For each iteration step $[k]$, the solving procedure is run for the total flow domain Ω. The main features of the algorithm can be written according to the following process, given a numerical method for solving the equations, and based on convergence criteria:

- Definition of the sub-domains Ω_i ($i = 1, 2, \ldots, M = 5$) and choice of the reference sections S_i.
- Initialize the following unknowns:
  - pressure;
  - local mapping functions;
  - kinematics at the reference sections S_i.

This implies that a geometrical shape of the streamlines is assumed, as also the kinematic function $\phi_i[0]$ in each mapped sub-domain Ω_i^*, related to the reference section S_i. The initial guess of the streamlines corresponds to an estimate of the local functions α_i, β_i and γ_i.

2. *Solving the governing equations in the computational sub-domains.*

A constitutive model should be adopted for the material. Writing all the equations involves expressions of compatibility equations at the interfaces of the sub-domains. Those are mathematical constraints for the general governing equations (kinematic and momentum conservation equations). Obviously, the total set of $\mathcal{M}_{\text{eq}}$ discretized equations defines an over-determined system of $\mathcal{M}_{\text{eq}}$ non-linear equations with $\mathcal{N}$ unknowns ($\mathcal{M} > \mathcal{N}$), written as

$$A_j(X_1, X_2, \ldots, X_{\mathcal{N}}) = 0, \quad j = 1, 2, \ldots, \mathcal{M}_{\text{eq}}, \tag{6.36}$$

where $(X_1, X_2, \ldots, X_{\mathcal{N}}) = \underline{\mathbf{X}}$ denotes the vector of the unknowns. Thus, the quadratic objective function $S(\underline{\mathbf{X}})$ can be defined by means of the following matrix product of $\mathcal{A}$ $(\mathbf{X})$ and its transpose ${}^{\mathbf{T}}\mathcal{A}$ $(\underline{\mathbf{X}})$ such that

$$S(\underline{\mathbf{X}}) = {}^{\mathrm{T}}\mathcal{A}(\underline{\mathbf{X}}).\,\mathcal{A}(\underline{\mathbf{X}}) = \sum_{j=1}^{j=M} \mathcal{A}_j^2(X_1, X_2, \ldots, X_N). \tag{6.37}$$

Using the objective function S($\underline{\mathbf{X}}$), the unknowns are determined by means of the Levenberg–Marquardt algorithm related to non-linear least squares problems, as done in [6]. This allows the procedure to converge for an initial estimate fairly far from the solution. The algorithm is proved to be robust and efficient, especially for problems where slight modifications of the unknowns lead to significant changes for the equations (Chap. 4). In the present situation, slight modifications in the position of the streamline points may lead to great changes in the governing equations.

The computed unknowns lead to determine the *pressure, velocities and stresses in the total flow domain.*

6.5 Flows Between Eccentric Cylinders

In this section, the STM approach is applied to computation of two-dimensional flows between eccentric cylinders related to journal bearing problems. Fluids obeying different constitutive equations are considered. The rotating flows are considered in two situations:

(1) there are no circulating flow regions;
(2) the flow domain involve eddies.

6.5.1 *Rotating Flows Without Recirculations: An Example*

The example concerns journal bearing flows. No circulating regions are encountered. The physical geometry is presented in Fig. 6.6, together with the mapped computational domain where the transformed streamlines are concentric circles. The inner cylinder (of radius r_0), rotates with an angular velocity ω and the outer cylinder (of radius r_1) is at rest. A one-to-one transformation is assumed.

1. *Preliminary remarks*

The liquid flows between the two cylinders C_0 and C_1 of radii r_0 and r_1, respectively. *e* denotes the distance between the axes of the cylinders, leading to a relative eccentricity $\in = e/(r_1 - r_0)$. The angular velocity of the shaft is ω_0 (rotating in the trigonometric sense), with fixed conditions of the outer cylinder.

The computational domain $\Omega*$ may be considered as a rectangular domain with variables $R(0 \le R \le R_a)$ and $\Phi(0 \le \Phi \le \Pi)$, according to symmetry conditions, as shown in Fig. 6.7. The two segments $C*_0$ and $C*_1$ are transformed lines of the curves C_0 and C_1 and the reference section corresponds to $\Phi = 0$.

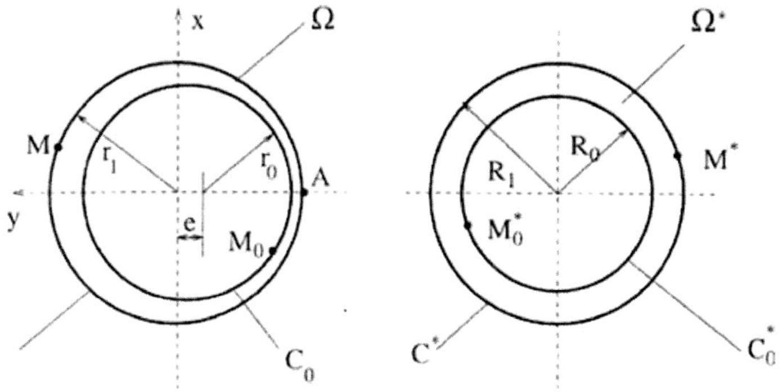

Fig. 6.6 Physical domain Ω and mapped domain Ω* for a one-to-one global mapping function annulus geometry

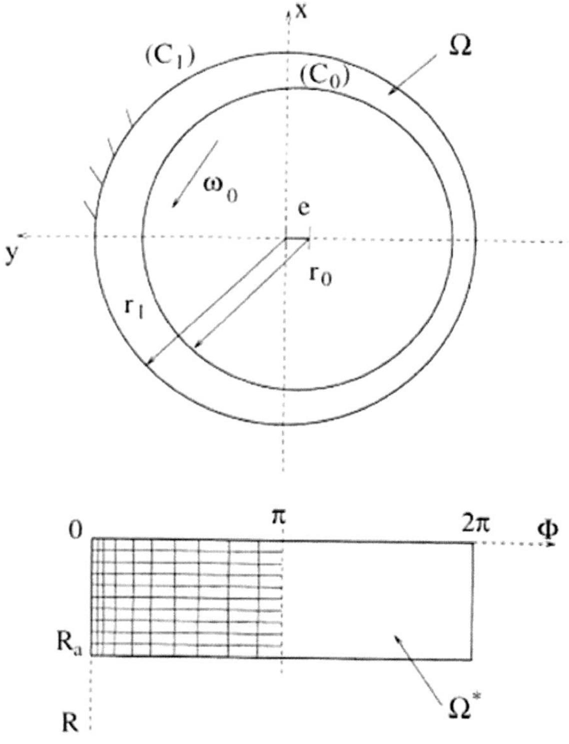

Fig. 6.7 Physical domain of flow between eccentric cylinders. Mapped computational domain Ω* and its rectangular mesh built on the mapped streamlines

2. *Domain transformation—Governing equations*

One considers the following transformation:

$$T : \Omega^* \rightarrow \Omega, \tag{6.38}$$

where $M^*(s, R) \to M(r, \theta)$.

Domain Ω^* is referred to a coordinate system ξ expressed by

$$\xi^1 = s, \quad \xi^2 = R, \tag{6.39}$$

according to the following relations related to local transformations $\alpha(s, R)$ and β (s, R), or λ_a (s, R) and λ_b (s, R) as

$$r = \alpha(s, R) = \lambda_a(s, R) \cdot R; \theta = \beta(s, R) = \theta_b(s, R) \cdot s/R. \tag{6.40}$$

Since no secondary flow regions are considered, a single computational flow domain may be retained, with variables R and Φ. The unknowns are the mapping function λ (R, Φ) and the pressure p.

For points of limiting circular surfaces C (outer surface) and C_0 (inner surface), one may write the following relationships (Fig. 6.7):

$$\begin{aligned}
&\forall M_0^* \in C_0^* \to M_0 \in C_0,\\
&\forall M^* \in C^* \to M \in C_0,\\
&C_0 : (x-a)^2 + (y-b)^2 = (C_0)^2(\lambda_0)^2 = r^2,\\
&\lambda_0 = \frac{e \cdot \cos\Phi + \sqrt{e^2 \cdot \cos^2\Phi - e^2 \cdot + r_0^2}}{R_0},\\
&C : (x-a)^2 + (y-b)^2 = (R_1)^2(\lambda_1)^2 = r_1^2, \lambda_1 = r_1/R_1.
\end{aligned} \tag{6.41}$$

- The pressure p is defined according to an arbitrary value at a point A $(p = p_0)$.
- The boundary conditions along the cylinders are expressed under the no-slip condition at the wall. The periodicity condition for Φ should be taken into account. In this example, the pressure is removed by a derivative of the first and second dynamic equations with respect to y and x, which leads to the unique partial derivative form:

$$\begin{aligned}
\frac{\partial^2 T^{xx}}{\partial x \partial y} + \frac{\partial^2 T^{xy}}{\partial y^2} - \frac{\partial^2 T^{xy}}{\partial x^2} - \frac{\partial^2 T^{xx}}{\partial x \partial y} = \rho\Bigg[&\frac{\partial u \partial u}{\partial y \partial x} + u\frac{\partial^2 u}{\partial x \partial y} + \frac{\partial v \partial u}{\partial y \partial y} + v\frac{\partial^2 u}{\partial y^2}\\
&- \frac{\partial u \partial v}{\partial x \partial x} - \frac{\partial v \partial v}{\partial y \partial x} - u\frac{\partial^2 v}{\partial x^2} - v\frac{\partial^2 v}{\partial x^2}\Bigg].
\end{aligned} \tag{6.42}$$

Then, this equation is written with variables R and Φ of the mapped domain Ω^*.

3. *Constitutive equations for the computations*

The computing examples concern three types of rheological models, with the purpose of comparisons: a Newtonian fluid (comparisons with results by Chawda and Avgousti [7], Dai et al. [8], Okawa et al. [9]; a non-Newtonian purely viscous Carreau fluid (Butler et al. [10]), a viscoelastic Maxwell fluid (Beris et al. [11]).

4. *Comparisons with literature data—Convergence tests for secondary flows*

For the comparisons, the respective Reynolds and Weissenberg numbers Re and We (elastic effects) are considered, using the relations:

$$\mathrm{Re} = \rho\,\frac{\omega_0(r_1 - r_0).r_0}{\eta}, \quad \mathrm{We} = \lambda^0 \cdot \omega_0. \tag{6.43}$$

ω_0 denotes the angular velocity and λ^0 is a characteristic time related to the fluid elasticity. η denotes the dynamic viscosity.

Concerning computations in domain Ω^*, the discretization is performed using finite difference techniques. Unknowns at the mesh points are the mapping function f and the pressure p [5].

Computational STM results concerning comparisons with literature data are provided in Fig. 6.8 (Newtonian fluid) and Fig. 6.9 (*non-Newtonian purely viscous Carreau* and a *viscoelastic Maxwell* fluid. The numerical data are found to be consistent with the literature results presented.

5. *Convergence tests related to secondary flows*

The results are assumed to be related to the appearance of secondary flows: the Jacobian Δ of the transformation is zero when the algorithm diverges. A limit convergence curve, related to flows of a Newtonian fluid between eccentric cylinders, has been obtained from STM calculations. This curve, assumed to be related to the existence of secondary flow regions, is compared to literature data, with different relative eccentricities (Fig. 6.10). The corresponding limit curve is found to be close to other results.

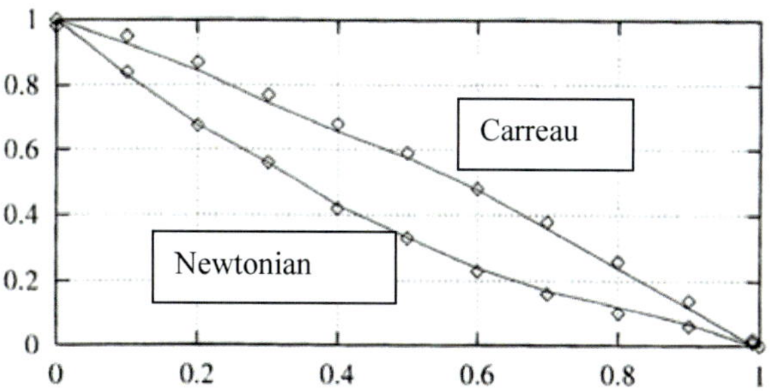

Fig. 6.8 Comparisons of normalized velocity profiles for a relative eccentricity $\epsilon = 0.25$, a rotational velocity $\omega_0 = 5.2\,\mathrm{s}^{-1}$, for $\theta = 0$ and $\theta = \pi$ (—: STM); ✧: Newtonian [7]), ✧: non-Newtonian purely viscous Carreau fluid (—: STM). © 2002 by John Wiley and Sons, Ltd: reproduced from Ref. [6] by permission of John Wiley & sons, Ltd

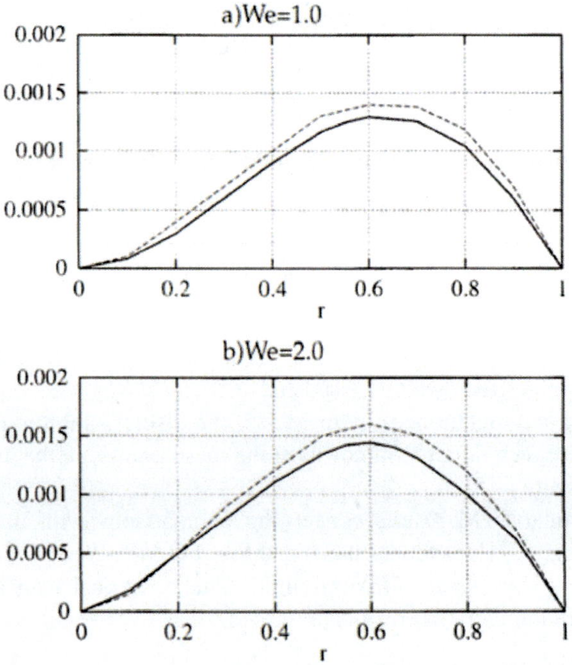

Fig. 6.9 Comparisons of radial normalized velocity profiles v_r for a relative eccentricity $\epsilon = 0.1$, $\theta = \pi/2$. —: STM results, - - -: Beris et al. [11]) with a viscoelastic Maxwell fluid, for **a** We = 1 and **b** We = 2. © 2002 by John Wiley and Sons, Ltd: reproduced from Ref. [6] by permission of John Wiley & sons, Ltd

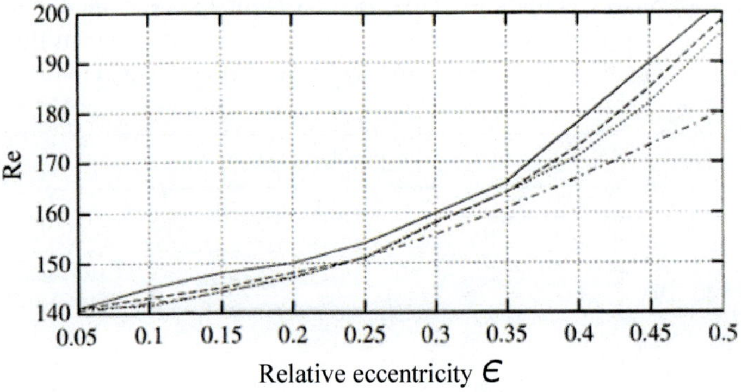

Fig. 6.10 Limit curves corresponding to the appearance of recirculation zones for a Newtonian fluid, where the Reynolds number Re is expressed in terms of the relative eccentricity (—: STM results); - -: Chawda and Avgousti [7], Dai et al. [8], –·–·–: Oikawa et al. [9]). © 2002 by John Wiley and Sons, Ltd: reproduced from Ref. [6] by permission of John Wiley & Sons, Ltd

6.5.2 Two-Dimensional Flows Between Eccentric Cylinders (Journal Bearing Problem) with Recirculating Regions

In this section, flows between eccentric cylinders are assumed to involve recirculation regions related to the eccentricity. A reference section is considered for both domains. Parallel lines to the mapped streamlines are perpendicular to the mapped reference section (Fig. 6.11b).

1. *Transformation and equations*

The situation where the inner cylinder (of radius r_0) rotates with an angular velocity ω and the outer cylinder (of radius r_1) at rest, is still considered.

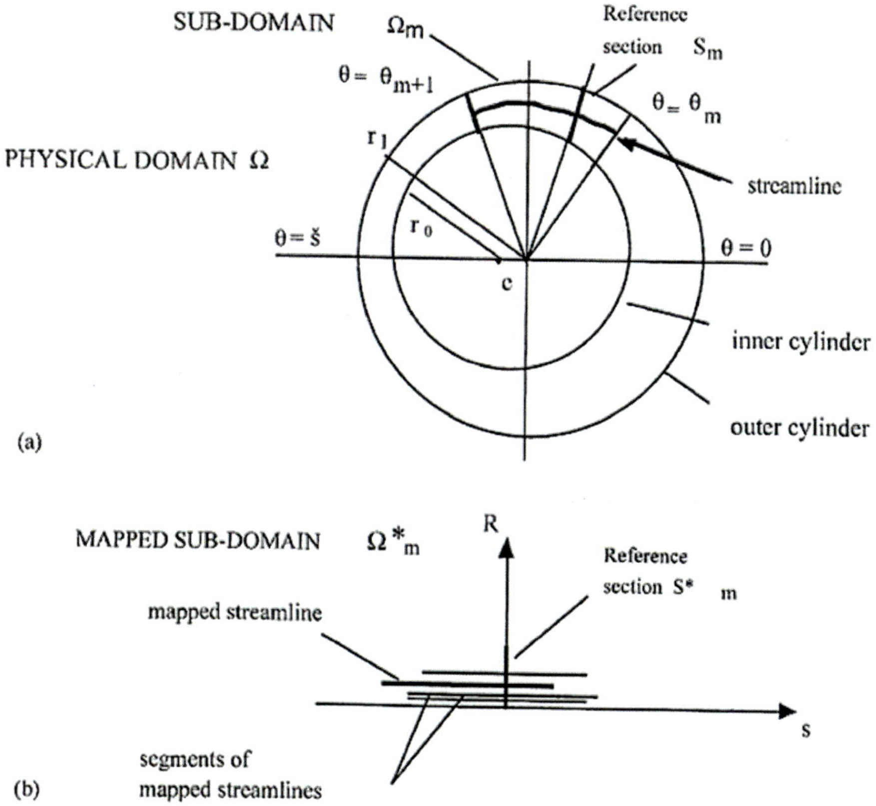

Fig. 6.11 Local transformation for the flow in the eccentric annulus geometry **a** Sub-domain Ω_m of Ω with a reference section S_m and **b** corresponding mapped domain Ω^*_m for the local transformation. The bold curved line between the angles θ_m and θ_{m+1} is a segment of streamline*

*Reprinted from Ref. [12], Copyright 2005, with permission from Elsevier

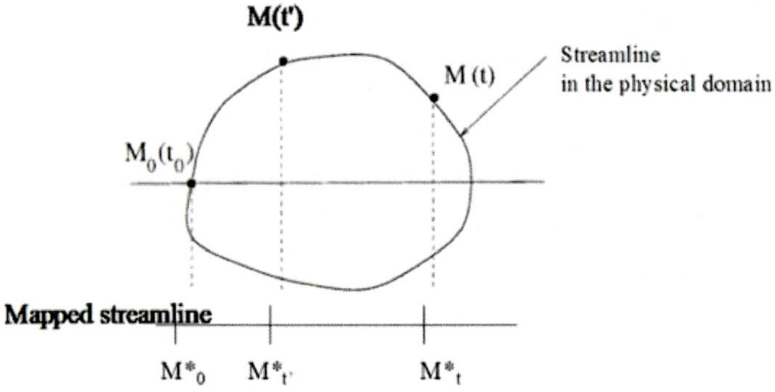

Fig. 6.12 Positions of a material point M on a closed streamline and its position on a mapped streamline

Referring to Newtonian results with such geometry [1, 12] and ignoring inertial effects, two sub-domains Ω_1 and Ω_2 of the half-domain Ω limited by azimuth angles $\theta = 0$ and $\theta = \pi$ are adopted, under a symmetry assumption. It is helpful to show, in Fig. 6.12a–c, different positions of a material point M on a closed streamline, versus time.

The boundary between the two elementary domains Ω_1 and Ω_2 is expected to move during the iterative process. This limit is a priori unknown and should be determined numerically. Under the symmetry assumption of the flow, a half-domain of Ω is retained for the calculations. Approximation of the streamlines, in this half-domain of Ω (Fig. 6.13a), is made at the beginning of the iterative process. The corresponding transformed streamlines are shown in Fig. 6.13b. In the upper region which corresponds to the zone of closed streamlines, two branches of a closed streamline are defined as shown in Fig. 6.3 of this chapter.

Figure 6.13c presents the rectangular grid built on the normalized mapped streamlines. The mesh built on these parallel lines is expected to change during the iteration procedure. Figure 6.14a shows again the physical domain Ω partitioned into sub-domains Ω_1 and Ω_2 and Fig. 6.14b presents the mapped sub-domains $\Omega{*}_1$, $\Omega{*}_2$ and also the boundary conditions. As pointed out previously, the mapped rectilinear streamlines have been normalized at each iteration step.

According to the choice on two sub-domains Ω_1 and Ω_2, for $m = 1$ and 2, let $(x^1, x^2)_m$ be local polar coordinates in sub-domains Ω_m of Ω, defined with respect to the reference flow section $S{*}_m$:

- $\Omega{*}_m$ denotes the transformed domains of a sub-domain Ω_m ($m = 1$ and 2), where the mapped streamlines are parallel to a given direction (Fig. 6.14). Sub-domain Ω_1, of reference section at $\theta = 0$, involves only open streamlines. Closed and open streamlines should be taken into account in sub-domain Ω_2, of reference section at $\theta = \pi$.
- The angle θ_1, a priori unknown, corresponds to the limiting azimuthal section between sub-domains Ω_1 and Ω_2 (Fig. 6.14). A sub-domain $\Omega{*}_m$ is referred to as a coordinate system ξ^j expressed by

$$\xi^1 = s; \quad \xi^2 = R. \tag{6.44}$$

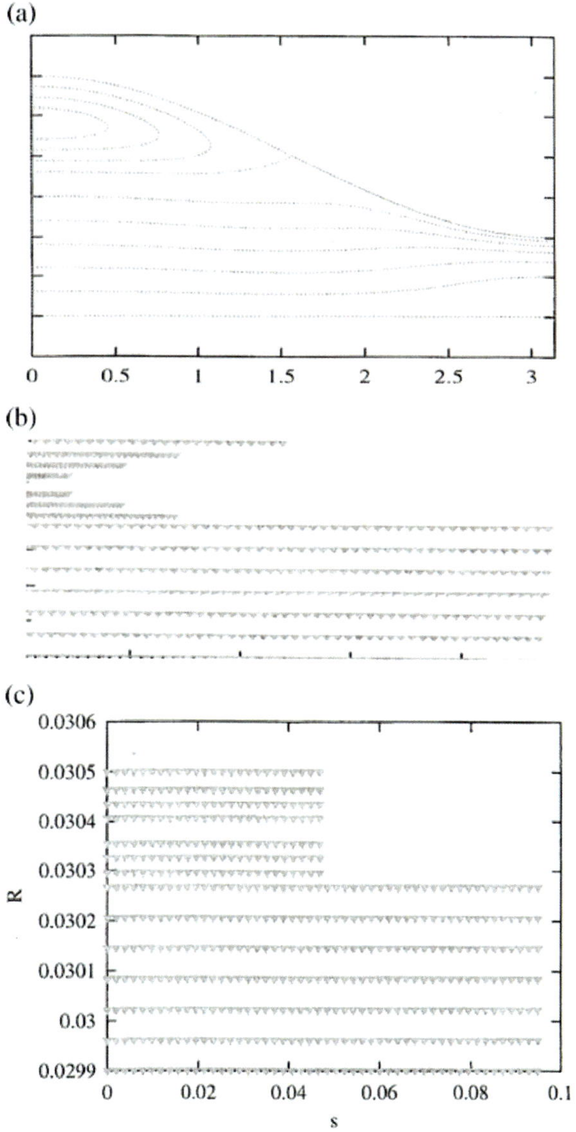

Fig. 6.13 Approximation of the streamlines in a half-domain of Ω at the beginning iterative process and corresponding streamlines, which are dimensionalized during the computational process: **a** approximated streamlines at an iteration step; **b** their corresponding transformed straight streamlines according to Fig. 6.3; **c** a rectangular grid built on the normalized mapped streamlines

- Two reference kinematic functions ϕ_1 and ϕ_2 are considered at sections $\theta = 0$ and π, respectively. The variables are defined according to the following relation, similar to Eq. (6.1):

$$x = \alpha_m(X, Y, s);\ y = \beta_m(X, Y, s);\ z = \gamma_m(X, Y, s). \tag{6.45}$$

The variable R corresponds to the radius r at the reference section of the sub-domains. The two local transformations (for $m = 1$ and 2):

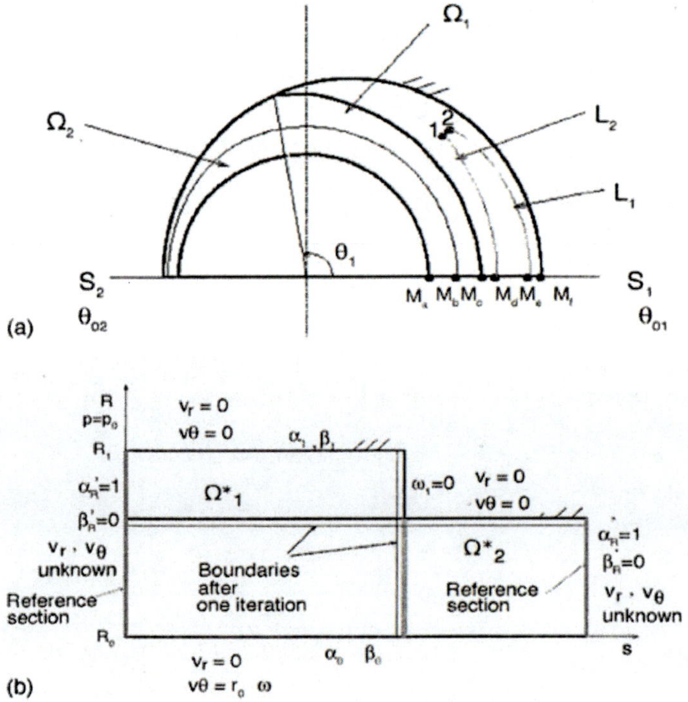

Fig. 6.14 **a** Physical domain Ω partitioned into sub-domains Ω_1 and Ω_2. **b** Mapped sub-domains $\Omega{*}_1$ and $\Omega{*}_2$ and the boundary conditions. The mapped rectilinear streamlines have been normalized The size and positions of the computational sub-domains are modified during the iteration process*

$$T_m : \Omega_m^* \to \Omega_m, \tag{6.46}$$

where $M^*(s, R)$ $M^*(r, \theta)$, involve the unknown local transformations $\alpha(s, R)$ and $\beta(s, R)$, or $\lambda_a(s, R)$ and $\lambda_b(s, R)$ such that

$$r = \alpha(s, R) = \lambda_a(s, R) \cdot R; \theta = \beta(s, R) = \lambda_b(s, R) \cdot s/R. \tag{6.47}$$

- The coordinate R corresponds to the radius r at the reference section $S^*_m(s_0, R)$ of sub-domain Ω^*_m.
- The coordinate s is defined in relation to the curvilinear abscissa χ (Eq. 6.18).

It should be pointed out that the streamline lengths in mapped sub-domains Ω^*_1 and Ω^*_2 have been renormalized for computational purposes.

- The Jacobian defined by

$$\Delta = \alpha_s' \cdot \beta' R - \alpha' R \cdot \beta' s \tag{6.48}$$

---

*Reprinted from Ref. [12], Copyright 2005, with permission from Elsevier

is non-zero. Thus, derivative operators relating the coordinates (r, θ) to the new system of local coordinates (s, R) are expressed by

$$\frac{\partial}{\partial R} = \frac{1}{\Delta\left[\beta'_R \cdot \partial/\partial s - \beta'_s \cdot \partial/\partial R\right]}$$
$$\frac{\partial}{\partial \theta} = \frac{1}{\Delta\left[\alpha'_R \cdot \partial/\partial s - \alpha'_s \cdot \partial/\partial R\right]}. \tag{6.49}$$

Using variables (s, R), the velocity component equations in terms of a stream function $\psi(R, \theta)$, as

$$v_r = \frac{1}{r}\frac{\partial \Psi(r,\ \theta)}{\partial \theta},\ v_\theta = -\frac{\partial \Psi(r,\ \theta)}{\partial r} \tag{6.50}$$

in polar coordinates, are given, in a sub-domain $\Omega*_m$, by the following relations:

$$v_r = -\frac{\alpha'_s(s,\ R)}{\alpha\ (s,\ R) \cdot \Delta}\frac{\mathrm{d}\Psi^*(R)}{\mathrm{d}R}$$
$$v_\theta = -\frac{\beta'_S(s,\ R)}{\Delta}\frac{\mathrm{d}\Psi^*(R)}{\mathrm{d}R}. \tag{6.51}$$

In Eq. (6.51) and for the following relationships, the subscript m indicating the local variables is omitted for the purpose of simplicity. $\Psi^*(R)$ denotes the transformed stream function of Ψ defined from a reference kinematic function at the reference section $S*_m$ of coordinates (s_0, R) such that

$$\frac{\mathrm{d}\Psi * (R)}{\mathrm{d}R} = v_\theta(s = 0,\ R) = H^*(R). \tag{6.52}$$

The relevant unknowns of the problem are expected to be determined iteratively. The function $\Psi^*(R)$ is a priori unknown and must be updated at each step of the numerical procedure.

- Conditions at the reference section $S*_m$:

$$\alpha'_r(s = 0,\ R) = 1; \quad \beta'_s(s = 0,\ R) = 0. \tag{6.53}$$

From Eq. (6.51), the velocity components become

$$v_r = -\frac{\alpha'_s(s,\ R)}{\alpha\ (s,\ R) \cdot \Delta} \cdot H^*(R)$$
$$v_\theta = -\frac{\beta'_s(s,\ R)}{\Delta} H^*(R). \tag{6.54}$$

In the flow domain, the coordinate θ on a streamline of non-zero velocity can be expressed by the following equation related to (6.26):

$$\theta = \theta_0 + H^*(R) \int_{\xi=0}^{\xi=s} \frac{\beta_s'(s,\ R)}{\Delta}\, d\xi. \tag{6.55}$$

2. *Particle tracking for fluids with memory—Multiple sub-domains*

- *Particle tracking*

Concerning the particle tracking for a material point moving on its streamline in a sub-domain $\Omega^*{}_m$, the movement versus time can be readily expressed. Ignoring inertia and body forces, the dynamic equations along two axes (r, θ) of the physical domain Ω, with the local variables (s, R) of the sub-domain Ω_m under consideration, are expressed as

$$\begin{aligned} &\alpha\left[\beta_s' \cdot \partial p/\partial R - \beta_R' \cdot \partial p/\partial s\right] + \beta_s' \cdot \partial\left(\alpha T^{Tr}\right)/\partial R - \beta_R' \cdot \partial\left(\alpha T^{rr}\right)/\partial s \\ &- \beta_R' \cdot \partial\left(\alpha T^{rr}\right)/\partial s - \alpha_s' \cdot \partial\left(\alpha T^{r\theta}\right)/\partial R + \alpha' R \cdot \partial\left(T^{r\theta}\right)/\partial s = 0, \end{aligned} \tag{6.56}$$

$$\begin{aligned} &\alpha_s'.\partial p/\partial R + \alpha_s'.\partial p/\partial s + \beta_s'.\partial\left(\alpha T^{r\theta}\right)/\partial R - \beta_R' \cdot \partial\left(\alpha T^{r\theta}\right)/\partial S \\ &+ \alpha_s'.\partial\left(T^{rr}\right)/\partial R - \alpha' R \cdot \partial\left(T^{rr}\right)/\partial s = 0. \end{aligned} \tag{6.57}$$

- *Remark concerning multiple sub-domains*

For inertial flows between eccentric cylinders, more than two sub-domains of the physical domain Ω are required for the flow analysis. In such cases, the sub-domains should be considered in the total flow domain and, as soon as reference sections are defined, an approach similar to that proposed for two sub-domains can be adopted. Inertial terms are thus written in the dynamic equations.

3. *Constitutive equations*

In this example, inelastic and viscoelastic models are adopted in the flow computations: inelastic models Newtonian equation (constant viscosity); two generalized Newtonian fluids CAR1 and CAR2 (Carreau models, of non-constant viscosity); a viscoelastic K-BKZ fluid,

4. *Numerical procedure and results*

In the present approach, where STM is associated with domain decomposition techniques, the governing equations involve

- dynamic equations;
- boundary condition equations;

- compatibility equations at the interfaces of the sub-domain [1, 3].

The primary unknowns are the pressure and the local mapping functions. Stress components T^{ij}, related to the constitutive equations, are expressed in terms of the transformation functions and their derivatives. Owing to the rectangular mesh defined in the computational sub-domains, finite-difference formulae are adopted to approximate the derivatives. Regular meshes are used to discretize the equations and unknowns.

The functions Φ_m are unknown and evaluated iteratively. The equations in mapped sub-domains Ω_I are solved by the Levenberg–Marquardt optimization algorithm, given an initial estimate from the lubrication theory [11]. In the present application, the reference section remains unchanged during the iterative process. New values of the separating angles between the domains are then obtained. Consequently, in the iterative process, the sizes of the physical and mapped sub-domains Ω_I are modified. New meshes are built in the computational sub-domains and the process is repeated up to verification of the convergence criteria.

- *Results without inertia*

In this case, the calculations involve two sub-domains with the half-domain of the total flow domain, for reasons of symmetry.

- Configuration of flows between eccentric cylinders:

$$r_0 = 5.084\,\text{cm},\ r1 = 6\,\text{cm},\ \varepsilon = 0.5,\ \omega = 150\,\text{rad/s}. \tag{6.58}$$

- Meshes: three regular meshes for the tests M1, M2, M3 in the computational sub-domains (see Fig. 6.14) with 3200, 2450 and 1800 unknowns, respectively.

The evolution in norm of the objective function to 10^{-6} with the three meshes is found to require a number of iterations of 25, 35 and 42, respectively. The differences between the computed solution with the three meshes, for the K-BKZ, the more complex model are not perceptible for the streamline plots.

The computed data concern the use of mesh M2.

Two kinematic examples of results on the computed streamlines are presented in Figs. 6.15 and 6.16, for the Newtonian fluid and the K-BKZ viscoelastic material [12]. The plots are presented in the developed eccentric flow domains.

In Figs. 6.15 and 6.16, computed streamlines reveal significant differences for the same geometric cases investigated. Such results underline the complexity of the fluid behaviour in flows between eccentric cylinders, depending on the rheological properties.

- *Results with inertia*

When taking into account inertial effects in the dynamic equations, the computations are carried out with more than two sub-domains.

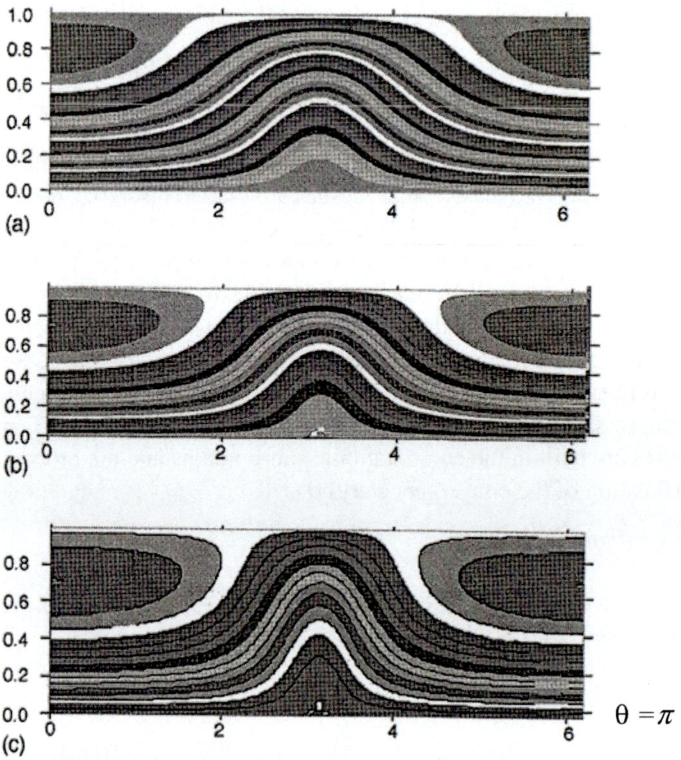

Fig. 6.15 Computed streamlines in the developed flow domain for the Newtonian fluid at different eccentricities: **a** $\epsilon = 0.6$, **b** $\epsilon = 0.7$ and **c** $\epsilon = 0.8$ ($r_0 = 5.084$ cm and $r_1 = 6$ cm)*

Streamlines results for a Newtonian fluid are presented in Fig. 6.17. The configurations have required four sub-domains for the simulations. Computed streamlines shown for three different kinematic conditions illustrate complexity of the flows, from a kinematic point of view. The results also underline STM possibilities to solve complex flow situations with several vortex zones.

6.6 Concluding Remarks

- In this chapter, the decomposition of flow domains into sub-domains by the stream-tube method is defined for calculations in two- and three-dimensional cases.
- It is of interest to predict the appearance of recirculating flow zones between eccentric cylinders, as those detected with the STM approach for an axisymmetric converging geometry (Chap. 4). For a Newtonian fluid, the computations in journal bearings have provided consistent results with literature data.

*Reprinted from Ref. [12], Copyright 2005, with permission from Elsevier

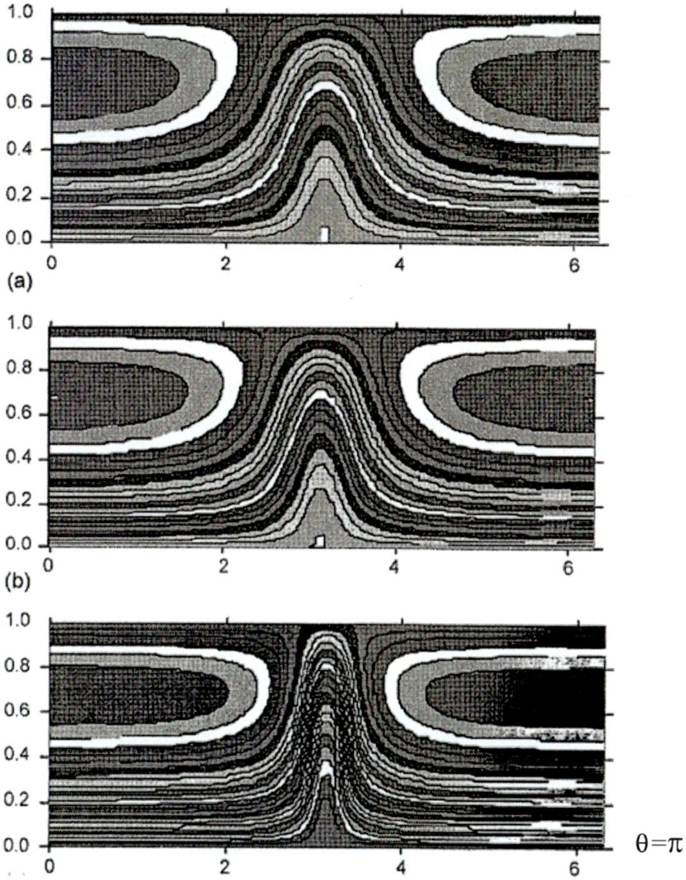

Fig. 6.16 Computed streamlines in the developed flow domain for the K-BKZ fluid at different eccentricities: **a** $\in = 0.6$, **b** $\in = 0.7$ and **c** $\in = 0.8$ ($r_0 = 5.084$ cm and $r_1 = 6$ cm)*

- Non-Newtonian, purely viscous and memory-integral viscoelastic fluids have been considered in flows between eccentric cylinders with recirculating regions, by a domain decomposition method. STM computations reveal differences in the computed streamlines for the three fluids.
- Domain decomposition in STM has led to identify more complicated flow situations arising from inertial conditions.

*Reprinted from Ref. [12], Copyright 2005, with permission from Elsevier

Fig. 6.17 Computed streamlines of a Newtonian fluid for different configurations (**a**), (**b**) and (**c**), under inertial conditions*

References

1. Clermont J-R, Normandin M, Radu D (2000) Some remarks on the concept of stream tubes for numerical simulation of flows of incompressible fluids. Applications. Control Cybern 29(2):535–553
2. Clermont JR, de la Lande ME (1993) Calculation of main flows of a memory integral fluid in an axisymmetric contraction at high Weissenberg numbers. J Non-Newtonian Fluid Mech 46:89–110
3. Clermont J-R, Radu D (1998) Modélisation des écoulements viscoélastiques dans des paliers cylindriques. In: Revue roumaine des sciences techniques. Série de mécanique appliquée, vol 43, no 5, pp 539–549
4. Dinh QV, Glowinski R, Periaux J (1984) Solving elliptic problems by domain decomposition methods with applications. In: Birkoff C, Schoensdat A (eds) Elliptic problem solvers. Academic Press, New York
5. Grecov D (1999) Inst Polytech. Grenoble. PhD thesis
6. Clermont JR, Grecov DR (2002) Numerical study of complex fluids between eccentric cylinders using transformation functions. Int J Num Methods Fluids 40:669–695 [Wiley]
7. Chawda A, Avgousti M (1996) Stability of viscoelastic flow between eccentric cylinders. J Non-Newtonian Fluid Mech 13:97–113
8. Dai R, Dong Q, Szeri A (1992) Flow between eccentric rotating cylinders: bifurcation and stability. Int J Eng Sci 10:1323–1340
9. Oikawa M, Karasudani T, Funakoshi M (1989) Stability of flow between eccentric rotating cylinders. J Phys Soc Jpn 58:2355–2364
10. Berker A, Bouldin MG, Kleis SJ, Van Arsdale WE (1995) Effects of polymer on flow in journal bearing. J Non-Newtonian Fluid Mech 56:333–347
11. Beris A, Armstrong RC, Brown RA (1987) Spectral finite element calculations of the flow of a Maxwell fluid between eccentric cylinders. J Non-Newtonian Fluid Mech 22:129–167
12. Grecov D, Clermont JR (2005) Numerical simulation of non-Newtonian flows between eccentric cylinders by domain decomposition and stream-tube method. J Non-Newtonian Fluid Mech 126:175–185 [Elsevier]

*Reprinted from Ref. [12] Copyright 2005, with permission from Elsevier

Chapter 7
Stream-Tube Method for Unsteady Flows

7.1 Introduction

Stream-Tube Method (STM) concepts are considered in this chapter for studies of isothermal unsteady flows in two- and three-dimensional situations. Notably, the approach provides accurate formulae for evaluating the kinematic histories, required time-dependent constitutive equations. For example, such cases may be encountered in processing rheology and lubrication problems, requiring evaluation of strain or deformation rate tensors. Focusing on incompressible materials, this chapter extends the possibilities of STM to evaluate kinematic quantities in unsteady flows, where streamlines and pathlines are not identical. The approach is considered towards their characteristics at every time t, starting from the rest state or stationary conditions. Fluid materials are moving in a domain Ω of boundary Γ, from an initial time t_0. Domain decompositions of Chap. 6, recalled in this chapter, may also be required in the following formulations. Journal bearing situations are considered in applications, for comparisons of STM results with those from the literature.

7.2 Theoretical Analysis of Unsteady Flows in STM

STM has been considered previously for stationary cases, with global or local transformations functions considered in the governing equations, instead of the velocity as primary unknown. In steady situations, STM avoids problems of numerical diffusion encountered in classical methods, as singularity arising from the center of rotation of a vortex zone. Such complications cannot be improved by convective stabilization techniques and are increased in unsteady cases. Before investigating such cases, it is found of interest to recall some useful information provided in previous chapters.

J.-R. Clermont and A. Ammar, *Stream-Tube Method*,
https://doi.org/10.1007/978-3-030-65470-2_7

7.2.1 Open and Closed Streamlines

Global one-to-one transformations $T_{(t)}$ of a flow domain Ω with open streamlines are related to the use of one or two mapping functions, in two- and three-dimensional cases, respectively. At time t, the corresponding equations, expressed in Cartesian coordinates X^i ($X^1 = x, X^2 = y, X^3 = z$) in domain Ω and its transformed coordinates q^i ($q^1 = X, q^2 = Y, q^3 = Z$) for its mapped domain Ω^* are expressed, in the planar case, by

$$T_{(t)} : \Omega^* \to \Omega : (X, Z) \to (x, z) \tag{7.1}$$

such that

$$x = f(X, Y); \quad z = Z. \tag{7.2}$$

f denotes here the transformation function in Cartesian coordinates for the two-dimensional case.

In the three-dimensional case, it writes

$$T_{(t)} : \Omega^* \to \Omega : (X, Y, Z) \to (x, y, z) \tag{7.3}$$

with

$$x = f(X, Y, Z); \quad y = g(X, Y, Z); \quad z = Z. \tag{7.4}$$

From the above transformations, we define an upstream reference section S_{ref} at $z = z_{ref}$, where the kinematics are known, mapped into a section S_{ref}* of Ω^*, identical in shape to S_{ref}, such that

$$x = f(X, Y, Z_{ref}); \quad y = g(X, Y, Z_{ref}); \quad z_{ref} = X_{ref}. \tag{7.5}$$

The mapped domain Ω^* is a straight cylinder of basis $S^*{}_{ref}$, with transformed streamlines parallel to the mean flow direction.

Similar relations can be obtained with cylindrical coordinates (R, Θ, Z) as

$$r = f(R, \Theta, Z); \quad \theta = g(R, \Theta, Z); \quad z = Z. \tag{7.6}$$

In the cases investigated, the existence of the reference section S_{ref} at $z = z_{ref}$, where the kinematics are known—for boundary conditions—is assumed for computing the velocity field in the flow domain. The incompressibility condition is automatically verified by the formulation, with expressions involving stream functions.

When flow vortices are expected in domain Ω, local mappings are possibly adopted and to be given by the following relations:

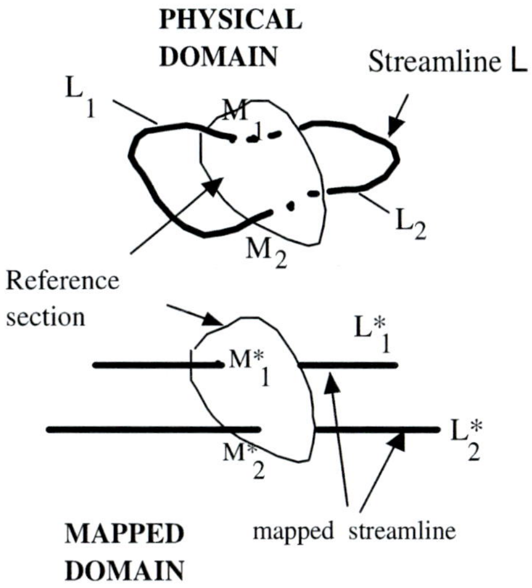

Fig. 7.1 Transformation of a closed streamline L into two branches, related to a mapped reference section

$$T_{m_{(t)}} : \Omega^*{}_m \rightarrow \Omega_m. \tag{7.7}$$

As in Chap. 6, we use domain decomposition such that $\Omega = \bigcup_{m=1}^{m=M} \Omega_m$.

With Cartesian coordinates X^i ($X^1 = x, X^2 = y, X^3 = z$) in domain Ω a coordinate system ξ^j ($\xi^1 = X$, $\xi^2 = Y$, $\xi^3 = s$) is defined for a sub-domain $\Omega^*{}_m$, where the mapped streamlines are parallel segments. The variable s is defined as the length of a segment of streamline in $\Omega^*{}_m$, being zero at a reference section S_m, with two branches of the streamline (Fig. 7.1).

In domain $\Omega^*{}_m$, the existence of a reference section $S^*{}_m$ identical in shape to the reference section S_m of the sub-domain Ω_m of Ω is assumed. Thus, the correspondence between sub-domain Ω_m and a local domain $\Omega^*{}_m$ according to the following relations (Chap. 6) way be written as follows:

$$x = \alpha_m(X, Y, s); \quad y = \beta_m(X, Y, s); \quad z = \gamma_m(X, Y, s). \tag{7.8}$$

For each domain $\Omega^*{}_m$, a kinematic function $\phi^*{}_m$ related to the reference section $S^*{}_m$ is considered, leading to write the velocity components in terms of $\phi^*{}_m$ and variables ξ^j. The Jacobian $\Delta = |\partial(X^i)/\partial(\xi^j)|$, assumed to be non-zero, is given by the equation:

$$\begin{aligned}\Delta = {} & \alpha'_X \left(\beta'_Y \gamma'_S - \beta'_S \gamma'_Y\right) - \beta'_X \left(\alpha'_Y \gamma'_S - \gamma'_Y \alpha'_S\right) \\ & + \gamma'_X \left(\alpha'_Y \beta'_S - \beta'_Y \alpha'_S\right).\end{aligned} \tag{7.9}$$

As already pointed out in Chap. 6, for computing the flow field, the local mapping approach requires to write compatibility equations at the common boundaries of the sub-domains of Ω. Those are written together with the classic conservation equations.

7.2.2 *Domain Transformation of Open Streamlines in Unsteady Flows*

In unsteady conditions, a simply connected bounded domain Ω of boundary Γ is now considered. The assumption of non-existence of secondary flows implies that the pathlines are also open curves, thus leading to define a mean flow direction z for both streamlines and pathlines.

7.2.2.1 Transformation for Open Streamlines and Pathlines

Assumption of a reference upstream section S_{ref}.

- there exists a reference upstream section S_{ref} at $z = Z_{ref(t)}$, where the velocities $\underline{V}_{ref}$ are known.

Using coordinates $(X^1, X^2, X^3) \equiv (x, y, z)$ of the Cartesian basis $(\underline{\varepsilon}_1, \underline{\varepsilon}_2, \underline{\varepsilon}_2)$, three-dimensional velocity vectors are expressed by

$$\underline{V}(x, y, z, t) = u(x, y, z, t)\ \underline{\varepsilon}_1 + v(x, y, z, t)\ \underline{\varepsilon}_2 + w(x, y, z, t)\ \underline{\varepsilon}_3. \tag{7.10}$$

These vectors are tangent to the streamlines at a time t (Fig. 7.2).

Let $\underline{V}_0(x, y, z, t_0)$ be the vector of initial velocity conditions in domain Ω, limited by the upstream section S_{ref}.

Relations (7.6) can be defined for the transformation $T_{s(t)}: \Omega^*(t) \rightarrow \Omega$, at a given time t such that, $\forall t$, the mapped streamlines are parallel and straight.

According to the assumption of open streamlines, the following Jacobian

$$\Delta = \left| \frac{\partial(x, y, z)}{\partial(X, Y, Z)} \right| = f'_X g'_Y - g'_X f'_Y \tag{7.11}$$

of the transformation $T_{s(t)}$ is required to be non-zero. The following derivative operators, involving the Jacobian Δ, are recalled there:

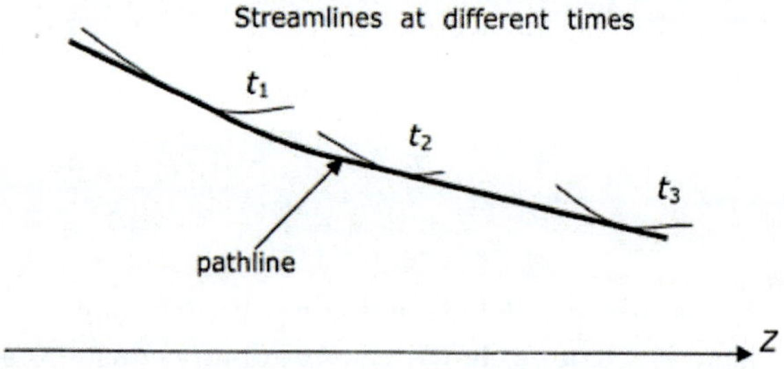

Fig. 7.2 Elements of streamlines and corresponding pathline related to the position of material points at different times t_1, t_2 and t_3

$$\frac{\partial}{\partial x} = \frac{1}{\Delta}\left[g'_Y \frac{\partial}{\partial X} - g'_X \frac{\partial}{\partial Y}\right], \quad \frac{\partial}{\partial y} = \frac{1}{\Delta}\left[-f'_Y \frac{\partial}{\partial X} - f'_X \frac{\partial}{\partial Y}\right],$$
$$\frac{\partial}{\partial z} = \frac{1}{\Delta}\left[\left(f'_Y g'_Z - f'_Z g'_Y\right)\frac{\partial}{\partial X} + \left(f'_Z g'_X - f'_X g'_Z\right)\frac{\partial}{\partial Y}\right] + \frac{\partial}{\partial Z}. \tag{7.12}$$

In the planar case, using Cartesian coordinates (*x*, *z*), the Jacobian $\Delta = \left|\frac{\partial(x,z)}{\partial(X,Z)}\right| = f'_X$ is assumed to be non-zero with the recalled derivative operators:

$$\frac{\partial}{\partial x} = \frac{1}{f'_X}\frac{\partial}{\partial X}; \quad \frac{\partial}{\partial z} = -\frac{f'_z}{f'_X}\frac{\partial}{\partial X} + \frac{\partial}{\partial Z}. \tag{7.13}$$

Trajectories in Flows Without Vortices: Mapping Functions

In this case, the trajectories L_P of the physical domain can be mapped into parallel lines of a domain $\Omega_P{}^*(t)$. Thus, we define a one-to-one transformation $T_{P(t)}$ between domain $\Omega_P{}^*(t)$ and the physical domain Ω using, in $\Omega_P{}^*(t)$, a reference section $S^*{}_{ref}$ identical to the corresponding reference section S_{ref} in Ω as

$$T_{P(t)} : \Omega_P^*(t) \to \Omega, \tag{7.14}$$

with the following relationships, similar to Eq. (7.4):

$$x(t) = \hat{f}(X, Y, Z, t); \quad y(t) = \hat{g}\ (X, Y, Z, t); \quad z = Z(t). \tag{7.15}$$

The mapping functions $\hat{f}$ and $\hat{g}$ are unknown and should be determined at every time *t*. As in the steady case, the assumption of open streamlines and pathlines requires that the Jacobian $\Delta(t) = \left|\frac{\partial(x,y,z)}{\partial(X,Y,Z)}\right|(t)$ of the transformation $T_p(t)$, expressed in the same form as Eq. (7.11), is non-zero. The derivative operators are given versus time by relations (7.12), in terms of the mapping functions $\hat{f}$ and $\hat{g}$.

Referring to the flow direction *z*, the transformed domain $\Omega_P{}^*(t)$ is limited downstream by the curve $\Gamma^*{}_P(t)$, defined by the pathline points at time *t* (Fig. 7.3). For all times $t' \leq t$, the particles move on the rectilinear lines of domain $\Omega^*{}_p(t)$, thus allowing more simplicity for particle tracking. At every time *t*, the position $\underline{M}$ of a material point $\boldsymbol{M}$ of a pathline originating at the position $\underline{M}_{ref} \in S_{ref}$ at a reference time t_0, is given by the vector equation:

$$\underline{M} = \underline{M}_0 + \int_{t_0}^{t} \underline{V}\left(M, t'\right) \mathrm{d}t'. \tag{7.16}$$

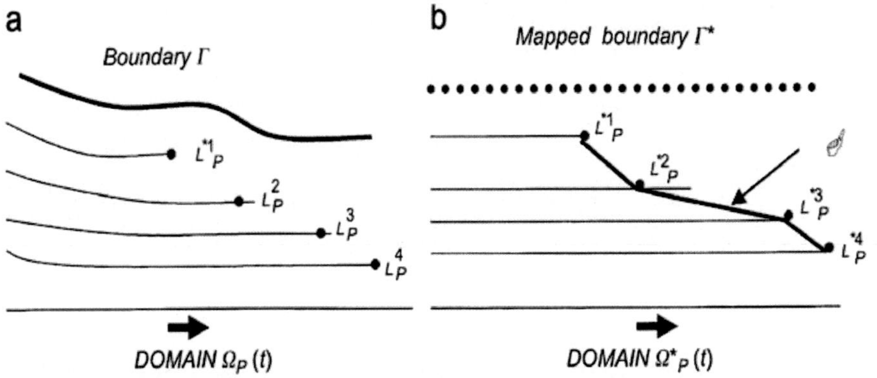

Fig. 7.3 Transformation of trajectories in the case of open streamlines

Material Derivative

In connection to pathline properties, the material point M moves according to the following material derivative:

$$\frac{\mathrm{D}A}{\mathrm{D}t} = \frac{\partial A}{\partial X}\frac{\mathrm{D}X}{\mathrm{D}t} + \frac{\partial A}{\partial Y}\frac{\mathrm{D}Y}{\mathrm{D}t} + \frac{\partial A}{\partial Z}\frac{\mathrm{D}Z}{\mathrm{D}t} = \frac{\partial A}{\partial Z}\frac{\mathrm{D}Z}{\mathrm{D}t}. \tag{7.17}$$

This leads to the following relation for derivatives:

$$\frac{\partial}{\partial Z} = \frac{1}{w(X,\ Y,\ Z,\ t)}\frac{\mathrm{D}}{\mathrm{D}t}. \tag{7.18}$$

7.2.2.2 Kinematics

At any instant, the velocities can be evaluated by STM formulae with mapping functions for streamlines, as detailed in previous chapters. Similarly, to a steady flow with open streamlines, a known velocity profile $w(x, y, z_{ref})$ is assumed at section S_{ref} to which corresponds a kinematic function $\Psi^*(X, Y, t)$. We then get the following relations:

$$u = \frac{\hat{f}'_Z(X, Y, Z, t)\psi^*(X, Y, t)}{\Delta(X, Y, Z, t)}, \tag{7.19}$$

$$v = \frac{\hat{g}'_Z(X, Y, Z, t)\psi^*(X, Y, t)}{\Delta(X, Y, Z, t)}, \tag{7.20}$$

$$w = \frac{\psi^*(X,\ Y, t)}{\Delta(X, Y, Z, t)}. \tag{7.21}$$

Using the mapping functions of the transformation $T_{p(t)}$, one may compute as in steady conditions the natural basis $\underline{b}_i = \frac{\partial M}{\partial X^i}$, leading to write

$$\underline{b}_1 = \hat{f}'_X\,\underline{\varepsilon}_1 + \hat{g}'_X\,\underline{\varepsilon}_2;\, \underline{b}_2 = \hat{f}'_Y\,\underline{\varepsilon}_1 + \hat{g}'_Y\,\underline{\varepsilon}_2;\, \underline{b}_3 = \hat{f}'_Z\,\underline{\varepsilon}_1 + \hat{g}'_Z\underline{\varepsilon}_2 + \underline{\varepsilon}_3. \tag{7.22}$$

Thus, the velocity vector is expressed as

$$\underline{V}(x,\ y,\ z,\ t) = w(X, Y,\ Z, t)\,\underline{b}_3 = \frac{\psi^*(X, Y, t)}{\Delta(X, Y, Z,\ t)}\underline{b}_3(t). \tag{7.23}$$

Problems involving memory-integral rate equations require evaluation of the rate-of-deformation tensor $\underline{\underline{D}}\,(t)$ and of the corotational reference frames $\breve{\underline{e}}_k$ at times $t' \le t$, in order to satisfy the principle of material objectivity [1, 2]. For steady flows, such problems have been considered in the previous Chaps. 4 and 5 for axisymmetric flows (two-dimensional situations). Though significant difficulties arise in three-dimensional problems with corotational models in relation to the determination of the corotational frames [2], computation of the tensor $\underline{\underline{D}}\,(t)$ can be achieved by STM formulae as previously done when purely viscous models are considered in unsteady situations [3, 4].

Codeformational models may be considered by means of concepts already introduced to compute the transformation gradient tensor $\underline{\underline{F}}_t(t')$ $(t' \le t)$. In the following relations, time variables are referred to by subscripts t and t'. With Cartesian coordinates, the transformation gradient is given by the following matrix:

$$\mathbf{F}_t(t') = \left[\frac{\partial x^i_{t'}}{\partial x^j_{t'}}\right]. \tag{7.24}$$

Cauchy and Finger tensors $\underline{\underline{C}}_t(t')$ and $\underline{\underline{B}}(t')$, generally adopted as kinematic quantities for codeformational models are written as $\underline{\underline{C}}_t(t') = \underline{\underline{F}}^T_t(t').\underline{\underline{F}}_t(t')$ and $\underline{\underline{B}}(t') = \underline{\underline{C}}^{-1}_t(t') = \left[\underline{\underline{F}}_t(t')\right]^{-1}.\left[\underline{\underline{F}}^T_t(t')\right]^{-1}$ (Chaps. 3 and 4).

With STM, to solve problems related to codeformational models for unsteady flows with open streamlines, pathline mappings versus time are defined by the transformation $T_{P(t)}$ (Eqs. (7.15)–(7.16)). The previous approach for kinematic histories in steady flow conditions with mapping functions [4] can be adapted to non-stationary flow situations. Similar formulae for strain tensors in steady two- and three-dimensional flows without eddies may be written.

Turning back here to previous elements from the steady analysis, the transformation $T_{P(t)}$ of pathlines is now considered, leading to write, according to Eq. (7.16),

$$X^1_{t'} = X^1_t, \tag{7.25}$$

$$X_{t'}^2 = X_t^2, \tag{7.26}$$

$$X_{t'}^3 = X_t^3 + \int_t^{t'} w(\mathrm{M},\ \xi)\,\mathrm{d}\xi. \tag{7.27}$$

In the natural basis ($\underline{b}_i$), the deformation gradient matrix is given by the simple following form in terms of coordinates q^i ($q^1 = X$, $q^2 = Y$, $q^3 = Z$):

$$[\mathbf{F}]_{b_i} = \begin{bmatrix} 1 & 0 & 0 \\ 0 & 1 & 0 \\ \partial q_{t'}^3/\partial q_t^1 & \partial q_{t'}^3/\partial q_t^2 & \partial q_{t'}^3/\partial q_t^3 \end{bmatrix}, \tag{7.28}$$

where the unknown components are [3]

$$F^{31} = \frac{\partial Z_{t'}}{\partial X_t} =_{t'} \int_{Z_t}^{Z_{t'}} \frac{1}{(w_{\xi'})^2} \frac{\partial w_{\xi'}}{\partial X_\xi} \mathrm{d}Z_{\xi'}, \tag{7.29}$$

$$F^{32} = \frac{\partial Z_{t'}}{\partial Y_t} = w_{t'} \int_{Z_t}^{Z_{t'}} \frac{1}{(w_{\xi'})^2} \frac{\partial w_{\xi'}}{\partial Y_\xi} \mathrm{d}Z_{\xi'}, \tag{7.30}$$

$$F^{33} = \frac{\partial Z_{t'}}{\partial Z_t} = \frac{w_{t'}}{w_t}. \tag{7.31}$$

In these equations, $w_\xi = w(M, \xi)$, $w_{t'} = w\left(M, t'\right)$ and $w_t = w(M,\ t)$ denote the non-zero velocity component of a material point M on its pathline at times ξ, t' and t, respectively. The coordinates (X_ξ, Y_ξ, Z_ξ), ($X_{t'}$, $Y_{t'}$, $Z_{t'}$) and (X_t, Y_t, Z_t) stand for respective positions of the material point M at times ξ , t' and t.

To compute $\underline{\underline{F}}_t(t')$ in the Cartesian basis, the natural basis $\underline{b}_i$ and its reciprocal basis $\underline{\chi}_i$ related to the transformation $T_{P(t)}$ (mapping functions $\hat{f}(t)$ and $\hat{g}(t)$) must be formulated [3]. Thus, components of $\underline{\underline{F}}_t(t')$ in the basis $\underline{\varepsilon}_i$ are obtained from the following matrix equation:

$$[\mathbf{F}]_\varepsilon = {}^T[\mathbf{A}].\,[\mathbf{F}]_b.\left[\overline{\mathbf{A}}\right], \tag{7.32}$$

where matrices $\mathbf{A}$ and $\overline{\mathbf{A}}$ are involved in the expressions related to the natural and reciprocal bases, such that

$$\underline{b}_i = \mathrm{A}_i^k \underline{\varepsilon}_k, \quad \underline{\chi}^j = \overline{\mathrm{A}}_k^j \underline{\varepsilon}_k. \tag{7.33}$$

Thus, the components of the Cauchy and Finger tensors can be computed by using Eqs. (7.31)–(7.32), leading to determine the stress components versus time.

7.2.3 *Domain Transformation for Unsteady Flows with Closed Streamlines*

When vortices are encountered, the fluid flow under consideration can be created:

- from the rest state,
- or by changes brought by non-stationary conditions applied to a steady flow in the physical domain Ω, at an initial time t_0.

For particles moving from the rest state, the hypothesis of a slow flow without vortices can be made at the beginning. As pointed out previously, the assumption of open streamlines and pathlines requires the Jacobians of the domain transformations $T(t)$ and $T_p(t)$ to be non-zero. We recall here that STM numerical results have been obtained for steady flows in ducts and between rotating eccentric cylinders with comparisons with literature data (Chap. 6). A divergence of the algorithms, arising when the Jacobian approaches zero, has been observed for appearance conditions of circulating regions. In a non-stationary process, the onset of vortices can be assumed when, at an instant t^*, the solving process defined for the open streamline approach diverges. Hence, STM techniques for flows with closed streamlines should be necessary. However, the adopted choice is to consider directly the procedure related to flows with closed streamlines. Specifically, the approach concerns more the case of changes of sign for the flow velocities, corresponding to the example of streamlines and pathlines shown in Fig. 7.4. Such considerations are provided for materials obeying rate constitutive equations and codeformational models as well.

7.2.3.1 Rate Constitutive Equations

In principle, rate equations may generally concern the Newtonian and purely viscous fluids as also memory-integral corotational models, usually expressed in terms of the rate-of-deformation tensor $\underline{\underline{D}}(t)$. As already pointed out, insurmountable difficulties practically arise for determining corotational frames in three-dimensional flows. Such

Fig. 7.4 Local view of closed streamlines and a pathline

problems lead to deploy applications only in the two-dimensional case, according to the following approach:

- As in steady conditions, local streamline transformations:
- $T_s^m(t)$: $\Omega*_m \rightarrow \Omega_m$ are defined, using domain decomposition.
- In every domain $\Omega*_m$, the transformed streamlines are parallel segments (Chap. 6) perpendicular to a reference section $S*_m$ identical in shape to a reference section S_m of sub-domain Ω_m (Fig. 7.1). Accordingly, the mapped sub-domain is a straight cylinder of basis $S*_m$.

In the planar case, coordinates q^i ($X^1 = x$, $X^2 = z$) and ξ^j ($\xi^1 = X$, $\xi^2 = s$) for sub-domains Ω_m and $\Omega*_m$, are respectively considered. The variable s is related to the length of a streamline segment, referred to the plane $S*m$, as pointed out in the previous section. The formalism defined for the stationary case can be also applied for every time t, requiring to streamline determination and evaluation of the corresponding stress tensor as a function of $\underline{\underline{D}}\left(t'\right)$ $(t' \leq t)$. Referring to the Cartesian basis $\underline{\varepsilon}_i$, the tensor components of $\underline{\underline{D}}\,(t)$, given by the matrix

$$\mathbf{D}(t) = \begin{bmatrix} \partial u(t)/\partial x & \partial u(t)/\partial z \\ \partial w(t)/\partial x & \partial w(t)/\partial z \end{bmatrix}, \tag{7.34}$$

can be expressed in terms of the variables (X, s) of the mapped domain $\Omega*$, using the basic relations of $T_{m_{(t)}}$.

In the following expressions, the subscript "m" (or superscript "m", in some cases), is generally removed for reasons of brevity, leading to write

$$x = \alpha(X, s);\ z = \beta(X, s). \tag{7.35}$$

The Jacobian $\Delta(t) = \left|\partial\left(q^i\right)/\partial\left(\xi^j\right)\right|$ of $T_s(t)$ is given by

$$\Delta(t) = \alpha'_X\ \beta'_s - \alpha'_s\ \beta'_X, \tag{7.36}$$

and the derivative operators are written as

$$\frac{\partial}{\partial x} = \frac{1}{\Delta(t)}\left[\beta'_s\,\frac{\partial}{\partial X} - \beta'_X\,\frac{\partial}{\partial s}\right], \tag{7.37}$$

$$\frac{\partial}{\partial z} = \frac{1}{\Delta(t)}\left[-\alpha'_s\,\frac{\partial}{\partial X} + \alpha'_X\,\frac{\partial}{\partial s}\right].$$

A reference kinematic function $\phi*$ at section $S*_m$ (where $s = 0$) is defined by

$$\phi^*(X) = w(X, s = 0). \tag{7.38}$$

One wants to express the derivative operators $\partial/\partial x$ and $\partial/\partial z$ in terms of $\partial/\partial X$ and $\partial/\partial s$ [5], with variables (X, s) of the mapped sub-domain Ω^*_m. Thus, the components (u, w) of the velocity vector $\underline{V}(x, z, t) = u(x, z, t)\,\underline{\varepsilon}_1 + w(x, z, t)\,\underline{\varepsilon}\, e'$ are written as

$$u\ (t) = \alpha'_s \varphi^*(X)/\Delta; \quad w(t) = \varphi^*(X)/\Delta. \tag{7.39}$$

Thus, from Eqs. (7.37) and (7.39), elements of the matrix of $\underline{\underline{D}}\,(t)$ (Eq. (7.34)) can be determined as functions of variables (X, s), leading to compute the stress tensor components at any instant t.

7.2.3.2 Codeformational Constitutive Equations

Even though the flow only involves open streamlines at small times $t > t_0$, when the fluid moves from the rest state at an initial time t_0, we focus on the case of closed streamlines and pathlines that may appear in relation to

- the geometry,
- the flow rate,
- the fluid properties.

The relative strain gradient is considered for open or closed pathlines. As in Chap. 6, for domain Ω:

- a finite number of sub-domains Ω_m $(m = 1, \ldots, M)$ where recirculating regions are expected, is defined;
- these sub-domains may also involve open streamlines, to which are associated reference cross-sections S_m.

To these sub-domains, we specify corresponding domains Ω^*_m $(m = 1, \ldots, M)$ where the mapped streamlines are segments parallel to a given direction $\underline{s}$. This leads to define a cylinder of basis S^*m identical to S_m. Kinematic functions $\phi^*_m(S^*_m)$, a priori unknown, are associated to domains Ω^*_m such that

$$\phi^*{}_m\left(P_0 \in S^*{}_m\right) = \underline{v}\ (P_0 \in S_m).\underline{s}. \tag{7.40}$$

As for the previous approach concerning local transformations in the steady case, a new set of variables $\xi^j(X, Y, s)$ is defined, such that

1. (X, Y) are the local variables in the reference section;
2. in this case, s is related to the curvilinear abscissa, that corresponds, differently to the case illustrated in Fig. 7.1, to the *total length of the streamline* limited by two positions P^*_0 and P^* on this streamline (Fig. 7.5), such that $P^*_0 \in S^*_m$ (reference section) implies $s\ (P^*_0) = 0$.

- At every time t, the transformation of streamlines

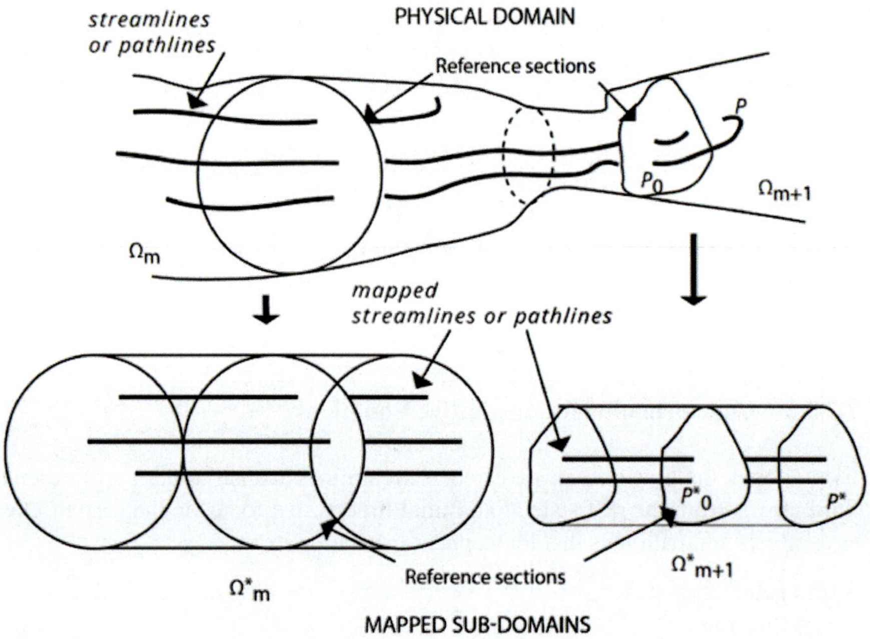

Fig. 7.5 Physical and mapped local sub-domains Ω_m, Ω_{m+1}, $\Omega^*{}_m$ and $\Omega^*{}_{m+1}$ for streamline and pathline transformations

$$T_m^p(t) \, : \, \Omega^*{}_m(t) \to \Omega^*{}_m(t) \tag{7.41}$$

associates a point $P^*(X, Y, s)$ of $\Omega^*{}_m(t)$ to a point $P(x, y, z)$ of domain $\Omega_m(t)$ according to the following equations:

$$x = \alpha_m(X, Y, s, t); \; y = \beta_m(X, Y, s, t); \; z = \gamma_m(X, Y, s, t), \tag{7.42}$$

formally the same as Eq. (7.8), with a Jacobian given by Eq. (7.9), assumed to be non-zero. Denoting by χ the corresponding length of the streamline in the physical sub-domain $\Omega_m(t)$ and omitting the subscript "m" for conciseness, an additional relationship between the three local functions α, β and γ writes

$$\chi = \int_0^s \left[\left(\alpha_s'\right)^2 + \left(\beta_s'\right)^2 + \left(\gamma_s'\right)^2 \right]^{1/2}. \tag{7.43}$$

- A reference kinematic function $\phi'(X, Y, t)$, related to the section $S^*{}_m$ of domain $\Omega^*{}_m$, may be defined by

$$\phi^*(X, Y, t) = w(X, Y, s = 0, t). \tag{7.44}$$

The use of derivative operators $\partial/\partial x$, $\partial/\partial y$ and $\partial/\partial z$ in terms of $\partial/\partial X$, $\partial/\partial Y$ and $\partial/\partial s$ leads to express the velocity components (u, v, w) as

$$u = \alpha_s' \phi^*(X, Y)/\Delta, \quad v = \beta_s' \phi^*(X, Y)/\Delta, \quad w = \gamma_s' \phi^*(X, Y)/\Delta, \tag{7.45}$$

where Δ denotes the Jacobian. The relations involved in (7.46) verify the mass conservation law.

- Looking for the position of material points at time t, local pathline transformations $T_m^p(t)$ are now considered for determining the kinematics variables and the stresses. Domains $\Omega^*{}_m(t)$ and $\Omega_m(t)$ are retained according to the relation:

$$T_m^p(\mathrm{t}) : \Omega^*{}_m(t) \to \Omega_m(t). \tag{7.46}$$

These transformations are defined such that *the trajectories*, in the mapped domain $\Omega^*{}_m$, *are segments parallel to a given direction* $\underline{d}$. For further computational purposes, one can normalize lengths of the mapped pathlines (as done in Chap. 6). Reference sections $S^*{}_m$ and S_m are considered. Thus, the following local relationships are written:

$$x = \hat{\alpha}_m\,(X, Y, \sigma, t),\ y = \hat{\beta}_m(X, Y, \sigma, t),\ z = \hat{\gamma}_m(X, Y, \sigma, t), \tag{7.47}$$

with spatial variables $X^i \equiv (x, y, z)$ and $\xi^k \equiv (X, Y, \sigma)$ for domains $\Omega_m(t)$ and $\Omega^*{}_m(t)$, respectively. The third variable σ of Eq. (7.47) is related to the curvilinear abscissa on a pathline, referred to section $S^*{}_m$

- The local mapping functions require the Jacobian of the transformation to be nonzero. The same reference kinematic function is retained, looking for the position of material points at time t, leading to write, similarly to Eq. (7.47), the velocities in terms of the local pathline transformation functions as follows:

$$u = \hat{\alpha}_s' \phi^*(X, Y)/\Delta, \quad v = \hat{\beta}_s' \phi^*(X, Y)/\Delta, \quad w = \hat{\gamma}_s' \phi^*(X, Y)/\Delta. \tag{7.48}$$

These expressions verify the incompressibility condition. For using memory-integral codeformational models, evaluation of the components of strain tensor $\underline{\underline{F}}_t(t')$ at times $t' \leq t$ is necessary. The same approach as that proposed for open streamlines can be applied, related to coordinates $(X^i) \equiv (x, y, z)$ and $(\xi^j) \equiv (X, Y, \sigma)$ for sub-domains Ω_m and $\Omega^*{}_m$, respectively. To express components of the strain tensor for each sub-domain, the use of coordinates $(\xi^j)t'$ and $(\xi^j)t$ leads to write, similarly to Eqs. (7.25)–(7.27):

$$X_{t'} = X_t; \quad Y_{t'} = Y_t; \quad \sigma_{t'} = \sigma_t + \int_t^{t'} w(M, \xi)\, \mathrm{d}\xi. \tag{7.49}$$

This leads to obtain components of $\underline{\underline{F}}_t(t')$ under the form given by Eq. (7.28). The kinematic terms corresponding to the local sub-domains are computed by relations similar to Eqs. (7.25)–(7.27). Here, the positions of material points in the sub-domains under consideration are taken into account.

7.3 Examples: Flows Between Concentric and Eccentric Cylinders for Newtonian, Anelastic and Viscoelastic Fluids

In the following examples, two-dimensional non-stationary flows between eccentric cylinders are considered under different kinematic conditions, with Newtonian and non-Newtonian fluids.

7.3.1 Flow Characteristics: Rheological Models for the Applications

It of interest to recall in Figs. 7.5 and 7.6 some basic elements already presented in Chap. 6 on flows between eccentric cylinders. This notably concerns decomposition into sub-domains involving only open streamlines (in Ω_1) and closed streamlines (in Ω_2), as also their corresponding transformed domains.

Given a time t in a non-steady flow situation, let $\Omega{*}_m$ be an elementary sub-domain of rectilinear parallel streamlines related to a reference cross-section $S{*}_m$ identical in shape to the reference section S_m of a sub-domain Ω_m of Ω. We now consider elementary transformations T_m defined from a mapped sub-domain $\Omega{*}_m(t)$ towards a sub-region $\Omega_m(t)$ of the physical domain Ω, such that T_m: $\Omega{*}_m(t) \rightarrow \Omega_m(t)$. In this section, we provide again specific information on the flow characteristics given in Chap. 6, related to the characteristics in the eccentric flow geometry.

1. *Specific flow characteristics*

Figure 7.6 shows the half-domain Ω_m of the real domain Ω between the two eccentric cylinders, referred to polar coordinates ($x^1 = r$, $x^2 = \theta$) limited by the azimuthal angles $\theta = 0$ and $\theta = \pi$, under symmetry assumptions. The inner cylinder, of radius r_0, rotates with an angular velocity $\omega(t)$ and the outer cylinder, of radius r_1, is at rest. The parameter e denotes the distance between the axes of the cylinders. The eccentricity is defined as $\varepsilon = \frac{\mathrm{e}}{r_1 - r_0}$, which is zero when concentric cylinders are considered.

Let $(x^1, x^2)_m$ be local polar coordinates in sub-domains Ω_m of Ω defined with respect to the reference flow section S_m. Using a reference section $S{*}_m$ identical in shape to S_m, we denote by $\Omega{*}_m$ the corresponding domain of sub-domain Ω_m, where the mapped streamlines are parallel to a given direction.

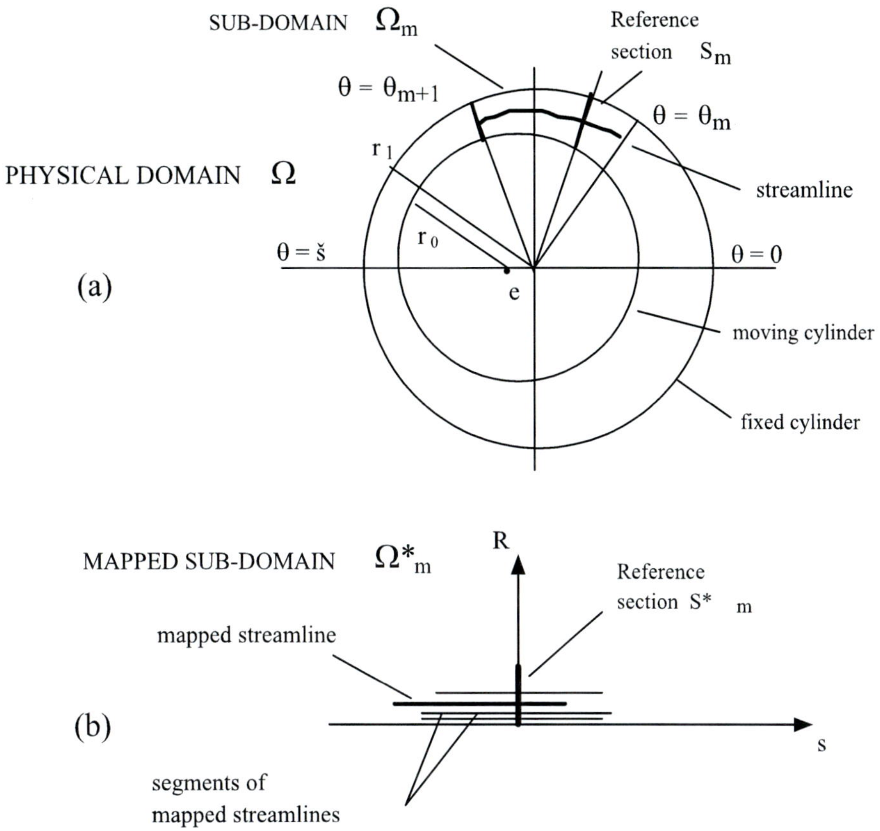

Fig. 7.6 Local transformation functions: **a** sub-domain Ω_m of Ω in the eccentric annulus geometry, involving a reference section S_m; **b** corresponding mapped domain $\Omega^*{}_m$ for the local transformation. Reprinted by permission from Springer Nature Customer Service Centre GmbH from [3], Copyright 2008

- As done in Chap. 6, two elementary sub-domains Ω_1 and Ω_2 such that $\Omega = \Omega_1 \cup \Omega_2$, shown in Fig. 7.7, are considered. The first sub-domain Ω_1 involves only open streamlines. For sub-domain Ω_2, open and closed streamlines are to be taken into account. The angle θ_1 corresponds to the limit Γ_1 between sub-domains Ω_1 and Ω_2 (Fig. 7.7) and is a priori unknown. The reference sections adopted for the respective sub-domains Ω_1 and Ω_2 are the azimuthal sections $\theta = 0$ and $\theta = \pi$.
- Sub-domains $\Omega^*{}_m$ are referred to a coordinate system ξ^j expressed by $\xi^1 = s$; $\xi^2 = R$. For the two sub-domains, two reference kinematic functions ϕ_1 and ϕ_2 are defined at sections $\theta_1 = 0$ and $\theta_2 = \pi$, respectively. The variable s is defined in relation to the curvilinear abscissa χ of a moving particle on its streamline, given by $\chi = \chi_0 + \int_{\tau=t_0}^{t'=t} \left|\underline{V(\tau)}\right| d\tau$, where χ_0 denotes the abscissa corresponding to the reference section $S^*{}_m(s_0, R)$ and $\left|\underline{V(\tau)}\right|$ the modulus of the velocity vector on the streamline [6].

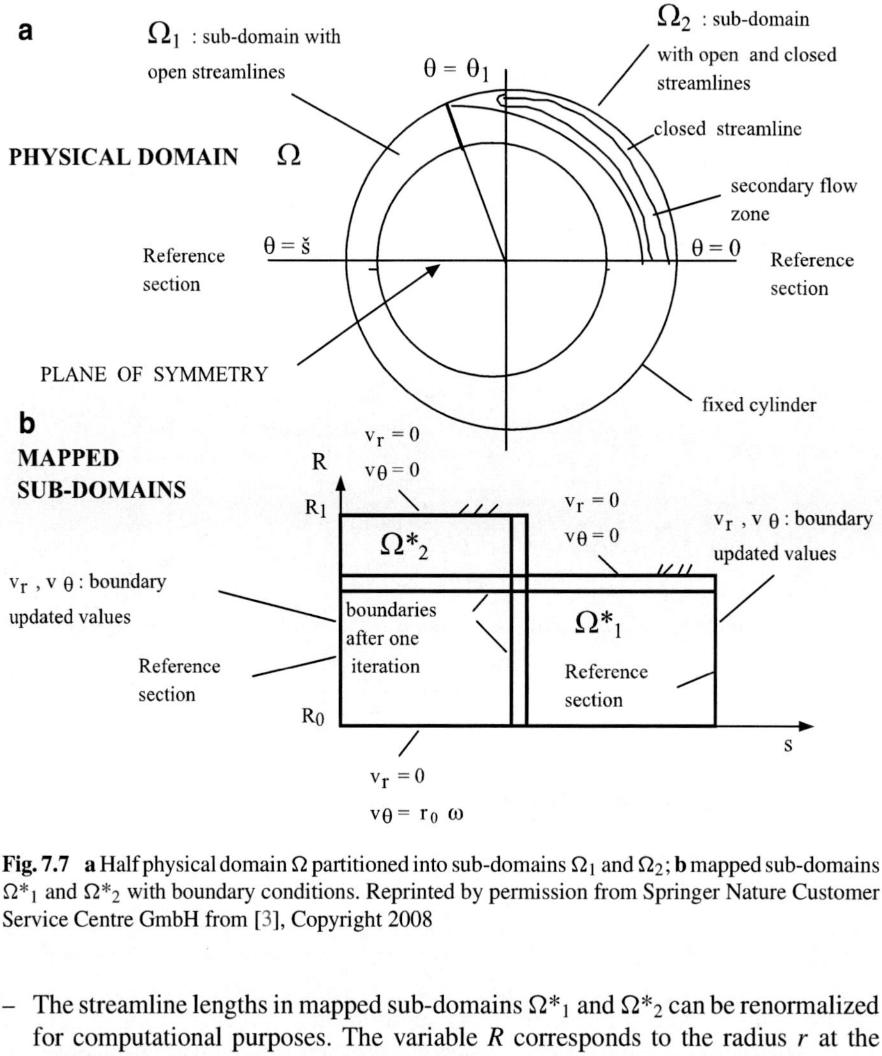

Fig. 7.7 **a** Half physical domain Ω partitioned into sub-domains Ω_1 and Ω_2; **b** mapped sub-domains Ω^*_1 and Ω^*_2 with boundary conditions. Reprinted by permission from Springer Nature Customer Service Centre GmbH from [3], Copyright 2008

- The streamline lengths in mapped sub-domains Ω^*_1 and Ω^*_2 can be renormalized for computational purposes. The variable R corresponds to the radius r at the reference sections of sub-domains Ω^*_m. The two local transformations T_m: $\Omega^*_m \rightarrow \Omega_m$ $(m = 1, 2)$ involve unknown functions $\alpha(s, R)$ and $\beta(s, R)$ (or $\lambda_a(s, R)$ and $\lambda_b(s, R)$) such that

$$r = \alpha(s, R) = \lambda_a(s, R) \cdot R; \quad \theta = \beta(s, R) = \lambda_b(s, R)\frac{s}{R}. \tag{7.50}$$

- The one-to-one assumption for the local transformations T_m implies that the Jacobian $\Delta = \left|\frac{\partial(r,\theta)}{\partial(R,s)}\right| = \alpha'_s\ \beta'_R - \alpha'_R\ \beta'_s\ \alpha'_R$ is non-zero. Derivative operators relating coordinates (r, θ) to local coordinates (s, R) are thus given by

$$\frac{\partial}{\partial r} = \frac{1}{\Delta}\left[\beta'_R \frac{\partial}{\partial s} - \beta'_s \frac{\partial}{\partial R}\right]; \quad \frac{\partial}{\partial \theta} = \frac{1}{\Delta}\left[\alpha'_R \frac{\partial}{\partial s} - \alpha'_s \frac{\partial}{\partial R}\right]. \tag{7.51}$$

- The velocity component equations $v_r = \frac{1}{r}\frac{\partial\Psi}{\partial\theta}$, $v_\theta = -\frac{\partial\Psi}{\partial r}$ in terms of a stream function $\Psi(r, \theta)$, which verify mass conservation, are thus given in terms of variables (*s, R*) by the following relations:

$$v_r = -\frac{\alpha'_s(s, R)}{\alpha\ (s, R).\Delta}\,\frac{\partial\Psi^*(R)}{\partial R};\quad v_\theta = -\frac{\beta'_s(s, R)}{.\Delta}\,\frac{\partial\Psi^*(R)}{\partial R}. \tag{7.52}$$

- For the purpose of simplicity, the subscript "*m*" related to local variables in Eq. (7.52) and the following relationships is omitted. $\Psi^*(R)$ denotes the transformed stream function of Ψ, written by starting from a reference kinematic function at section S^*_m of coordinates (s_0, R) such that $\frac{\partial\Psi^*(R)}{\partial R} = v_\theta(s = 0, R)$. The function $\Psi^*(R)$, a priori unknown, must be updated at each step of the procedure [7]. At the reference section, the following conditions write

$$\alpha'_R(s = 0, R) = 1\ ;\ \beta'_s(s = 0, R) = 0. \tag{7.53}$$

According to the definition of the reference functions at sections $\theta_1 = 0$ and $\theta_2 = \pi$ in the mapped sub-domains Ω^*_m, it may be assumed that the position of streamline segments, of variable length, remain unchanged versus time in the mapped sub-domains. Thus, given a differentiable function $A\ (s, R, t)$, we may write, using variables of the mapped domain Ω^* and the velocity component $w\,(s, R, t)$, the following relation:

$$\frac{\mathrm{D}A\ (s, R, t\ ))}{\mathrm{D}t} = \frac{\partial A(s, R, t)}{\partial t} + w(s, R, t)\frac{\partial A(R, Z, t)}{\partial Z}, \tag{7.54}$$

where $\frac{\mathrm{D}}{\mathrm{D}t}$ denotes the material derivative operator.

2. *Rheological models*

The examples concern the following constitutive equations:

- a Newtonian fluid (symbol "NEWT" for the results) of viscosity η_0 = 0.3 Pa s, where $\underline{\underline{\tau}} = 2\eta_0\underline{\underline{D}}$ and $\underline{\underline{D}}$ the rate-of-deformation tensor,
- a Carreau–Yasuda law [1] for fluids "CAR1" and "CAR2" (respective viscosities at zero shear rate: 6.0 Pa s and 0.4 Pa s at zero shear rate: Mori et al. [6]),
- a viscoelastic differential Maxwell fluid (UCM), of relaxation time $\lambda = 0.5$ s (UCM model [8]).

7.3.2 Dynamic Equations and Solving Procedure

According to the previous considerations, only the dynamic equations associated to initial and boundary conditions must be taken into account, written in the general form:

$$\rho \frac{\mathrm{D}\underline{V}}{\mathrm{D}t} = \underline{\nabla} \cdot \left(-p\underline{\underline{I}} + \underline{\underline{T}} \right), \tag{7.55}$$

where ρ is the fluid density, p the pressure, $\underline{\underline{I}}$ the unit tensor and $\underline{\underline{T}}$ the extra-stress tensor. The material derivative $\frac{\mathrm{D}}{\mathrm{D}t}$ is expressed by Eq. (7.54). Using the respective velocity, pressure and stress components $u,\ w,\ p,\ T^{rr},\ T^{r\theta}$, the dynamic equations involving the inertial terms can be written as [3]:

$$\begin{aligned} \rho\left[\tfrac{\partial u^n}{\partial t} + \tfrac{\partial u^n}{\partial Z} w^n\right] = \tfrac{1}{\Delta^{(n)}} \Big\{ \alpha\Big[\beta'_s \tfrac{\partial p^n}{\partial R} - \beta'_R \tfrac{\partial p^n}{\partial s}\Big] + \beta'_s \frac{\partial\left(\alpha T^{rr(n)}\right)}{\partial R} - \beta'_R \frac{\partial\left(\alpha T^{rr(n)}\right)}{\partial s} \\ -\beta'_R \frac{\partial\left(\alpha T^{rr(n)}\right)}{\partial s} - \alpha'_s \frac{\partial\left(\alpha T^{r\theta(n)}\right)}{\partial R} + \alpha'_R \frac{\partial\left(T^{r\theta(n)}\right)}{\partial s} \Big\}, \end{aligned} \tag{7.56}$$

$$\begin{aligned} \rho\left[\tfrac{\partial w^n}{\partial t} + \tfrac{\partial w^n}{\partial Z} w^n\right] = \tfrac{1}{\Delta^{(n)}} \Big\{ \Big[\alpha'_R \tfrac{\partial p^n}{\partial R} - \alpha'_s \tfrac{\partial p^n}{\partial s}\Big] + \Big[\beta'_s \frac{\partial\left(\alpha T^{rr(n)}\right)}{\partial R} - \beta'_R \frac{\partial\left(\alpha T^{rr(n)}\right)}{\partial s}\Big] \\ -\alpha'_s \frac{\partial\left(T^{rr(n)}\right)}{\partial s} + \alpha'_R \frac{\partial\left(\alpha T^{r\theta(n)}\right)}{\partial s} \Big\}. \end{aligned} \tag{7.57}$$

In Eqs. (7.56)–(7.57), the superscript (n) is related to the computed solution at time $t = t^n$ from the previous step at $t = t^{n-1}$. The superscripts of the tensor components, expressed in terms of the local mapping functions and their derivatives, are still related to variables (r, θ) of the physical sub-domains. We thus obtain, in every sub-domain Ω^*_m, non-linear dynamic equations of the general form:

$$E_1[\alpha^{(n)}, \beta^{(n)}, p^{(n)}]_{(s,R)} = 0,\ E_2[\alpha^{(n)}, \beta^{(n)}, p^{(n)}]_{(s,R)} = 0, \tag{7.58}$$

to be used together with the compatibility equations for the streamlines, notably at the common boundaries of the sub-domains.

Solving procedure

With the pressure and the local mapping functions as primary unknowns, the governing equations involve:

- the dynamic and constitutive equations,
- boundary conditions and compatibility equations for velocity and pressure at the interfaces.

Owing to symmetry properties, the half-domain Ω of the annulus geometry is retained for the computations.

The numerical procedure, close to that detailed in Chap. 6, for steady flow conditions, is reconsidered:

- at each time, reference kinematic functions for the sub-domains are defined;
- depending on the specifications, the common boundaries may be unknown and should be determined numerically as for the unknown local functions.

– Given a time t, at each iteration step, the solving procedure is run for the total flow domain Ω. The main features of the algorithm can be written according to the following process, given a numerical procedure for solving the equations, and based on convergence criteria.

- The iterative approach implies that a geometrical shape of the streamlines is assumed, as also the kinematic function $\Phi_{i[0]}$ in each mapped sub-domain Ω^*_i, related to the reference section S_i. The initial guess of the streamlines corresponds to an estimate of the local functions α_i, β_i.

Owing to the simplicity of the mapped domains, finite difference schemes are retained to discretize the equations and unknowns, as in steady flows (Chap. 6).

The compatibility equations at the interfaces of the sub-domains are mathematical constraints for the general governing equations (kinematic equation for streamline and momentum conservation equations).

- A quadratic objective function S is defined by means of the following matrix product of [**A**] ([X]) and its transpose ^{T}[**A**] ([X]) (Chap. 6).
- For the half-domain $\Omega = \Omega_1 \bigcup \Omega_2$, the interface of Ω_1 and Ω_2 corresponds to the azimuthal angle θ_1, as pointed out previously.

The unknowns are determined by means of the Levenberg–Marquardt algorithm. In the present application, the reference sections remain unchanged during the iterative process.

Compatibility equations

The compatibility equations at the interfaces of the sub-domains are mathematical constraints for the general governing equations (kinematic equation for streamline and momentum conservation equations) (e.g. [7]), such that the total set of the discretized equations defines an over-determined system of M non-linear equations with N unknowns ($M > N$), formally written as

$$A_j(X_1, X_2, \ \ldots, \ X_N) = 0, \ (j = 1, 2, \ldots, M),$$

where $[X] = (X_1, X_2, ..., X_N)$ denotes the vector of the unknowns.

A quadratic objective function S is considered by means of the following matrix product of [**A**] ([X]) and its transpose ^{T}[**A**] ([X]) as

$$\mathrm{S}([\mathbf{X}]) =^{\mathrm{T}} [\mathbf{A}]([\mathbf{X}]).\ [\mathbf{A}]([\mathbf{X}]) = \sum_{l=1}^{j=M} A_j^2(X_1, X_2, ..., X_N). \tag{7.59}$$

For the half-domain $\Omega = \Omega_1 \cup \Omega_2$, the interface of Ω_1 and Ω_2 corresponds to the azimuthal angle θ_1. Using this objective function, the unknowns are determined by means of the Levenberg–Marquardt optimization algorithm.

In the present application, the reference sections remain unchanged during the iterative process.

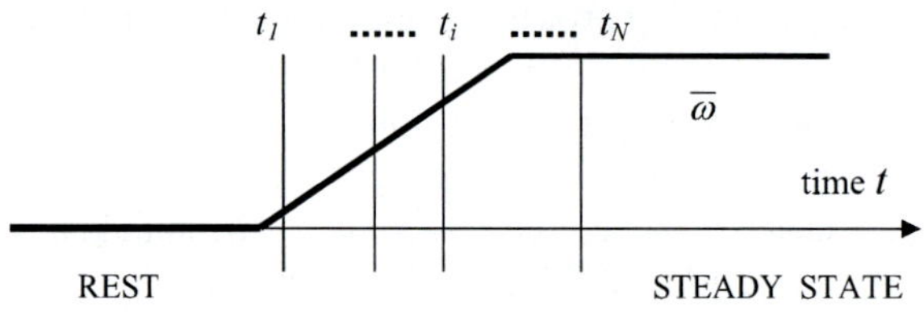

Fig. 7.8 Kinematic conditions for the rotating inner cylinder in a start-up problem

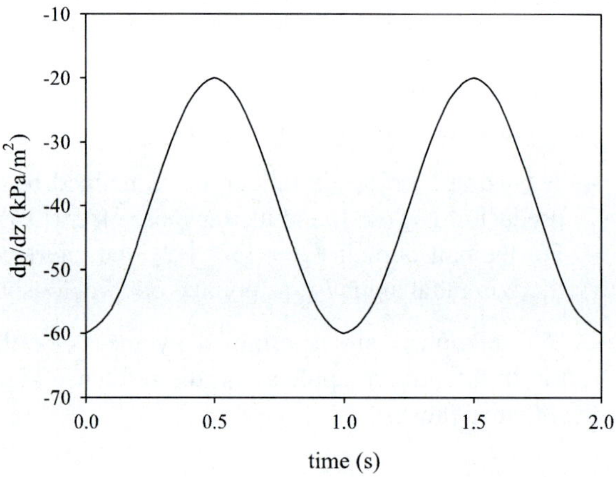

Fig. 7.9 Profile of the pressure gradient versus time for pulsating flow conditions

- *Kinematic conditions*

The computations reported here concern two kinematic flow configurations:

i) The flow created by a linear start-up of the inner cylinder (Fig. 7.8) with an eccentricity $\in = 0.6$, from the rest state to an angular speed $\omega = 50$ rad/s, with the following geometric parameters: $r_0 = 5.05$ cm; $r_1 = 6.0$ cm. For one sub-domain: grid of 35 points in the r-direction and 250 points in the θ-direction.
ii) Pulsating flows between concentric and eccentric cylinders: amplitude of pressure gradient shown in Fig. 7.9 [6]. Number of points for one sub-domain: 100 in the r-direction, 500 in the θ-direction, raised up to 800 for an eccentricity $\in = 0.2$.

7.3.3 Numerical Results

1. *Results for linear start-up flows: Newtonian, CAR1 and CAR2 fluids*

In start-up conditions, STM results are compared with those resulting from the FLUENT code: final rotating rate 5.24 rad/s (Fig. 7.10).

The flow characteristics depend on the rheological parameters (Fig. 7.11a, b): dimensionless radius $R^* = (r - r_1)/(r_0 - r_1)$ versus the angle θ (from 0 to 2π). The final computed times for attaining the sstate flows are different, owing to rheological properties (Fig. 7.12).

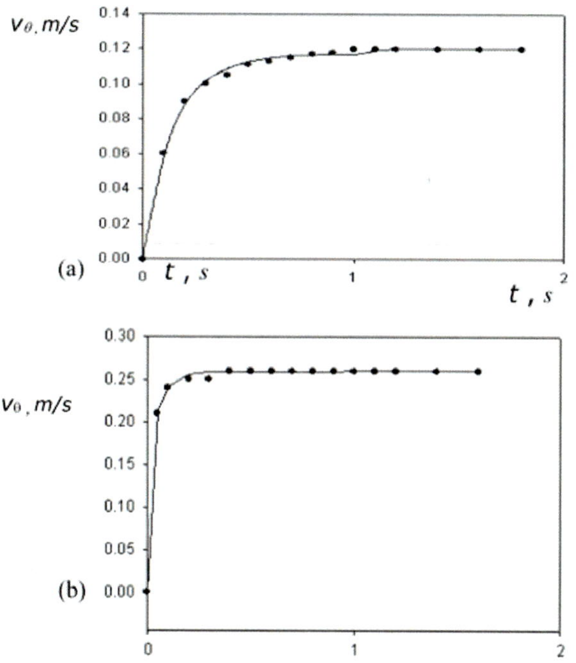

Fig. 7.10 Numerical predictions (Newtonian fluid) of v_θ-velocity component versus time at the dimensionless radius $r^* = 0.206$ for two azimuth sections: (—): FLUENT results; (•) STM data **a** $\theta = 0$; **b** $\theta = \pi$ (eccentric geometry $\in = 0.7$, $r_0 = 6$ cm, $r_1 = 5.08$ cm, linear start-up conditions). Reprinted by permission from Springer Nature Customer Service Centre GmbH from [3], Copyright 2008

2. *Linear start-up flows: UCM fluid*

The computed data with the UCM constitutive equation allow to observe differences in results with anelastic models (Fig. 7.13a, b).

3. *Pulsating fluids in eccentric cylinders*

The pressure gradient is expressed by [6]

$$\frac{\partial p}{\partial z} = -\overline{p}\left[1 + \kappa \ \exp\left(j2\pi\chi_p t\right)\right], \tag{7.60}$$

where $\overline{p}$ denotes the mean pressure gradient, χ_p the pulsating frequency and κ the ratio of the pulsating amplitude to the steady pressure gradient (Fig. 7.10). The same geometry than that of the experiments, defined by $r_0 = 0.30$ cm; $r_1 - 0.35$ cm, is adopted. The flow enhancement I is defined by [6]

$$I = \frac{Q_\chi - Q_s}{Q_s}, \tag{7.61}$$

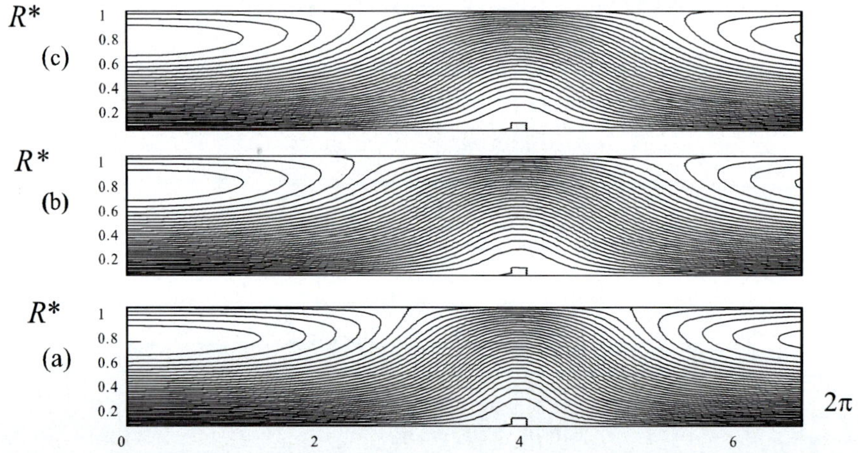

Fig. 7.11 Computed streamlines in the developed domain for a Newtonian fluid at different times: **a** $t_1 = 1.0 \times 10^{-2}$ s; **b** $t_2 = 2.4 \times 10^{-1}$ s; **c** $t_3 = 5.45 \times 10^{-1}$ s (linear start-up conditions). Reprinted by permission from Springer Nature Customer Service Centre GmbH from [3], Copyright 2008

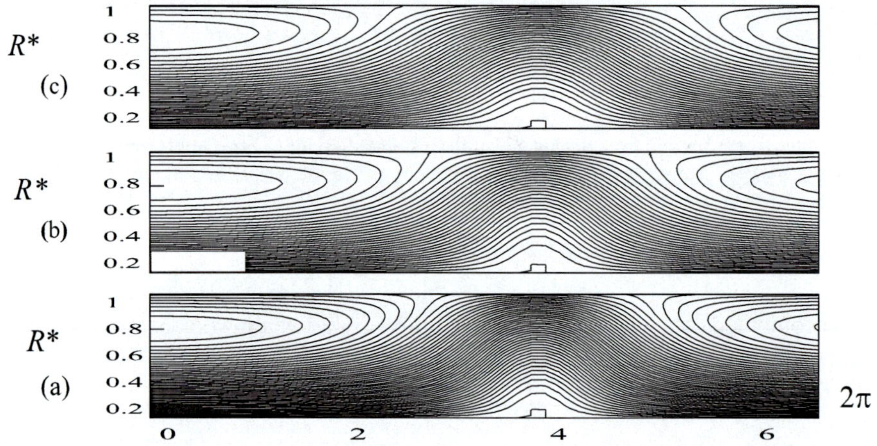

Fig. 7.12 Computed streamlines in the developed domain for the CAR1 fluid in a half (developed) domain at different times **a** $t_1 = 1.0$ s; **b** $t_2 = 3.67$ s; **c** $t_3 = 8.35 \times 10^{-1}$ s (linear start-up conditions). Reprinted by permission from Springer Nature Customer Service Centre GmbH from [3], Copyright 2008

where Q_s and Q_χ are the flow rates corresponding to the steady and pulsating conditions, respectively: frequency retained for the simulations $\chi_p = 1.0$ Hz.

Examples concern two fluids (eccentricity $\in = 0.2$ for the cylinders). Profiles of the dimensionless velocity component v^*_θ in one oscillation cycle are given in Fig. 7.14a, b at different times, for the azimuthal angle $\theta = \pi$.

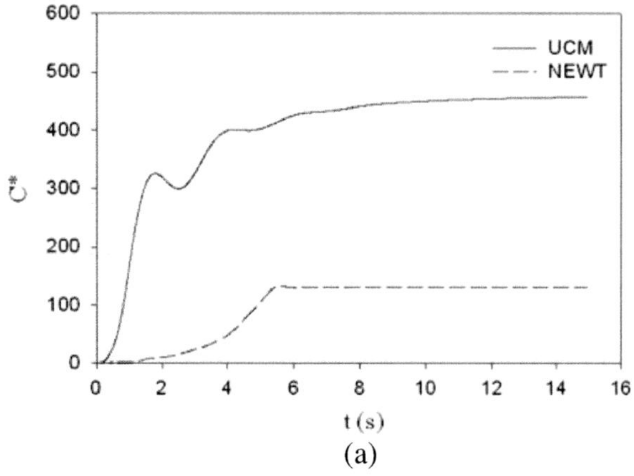

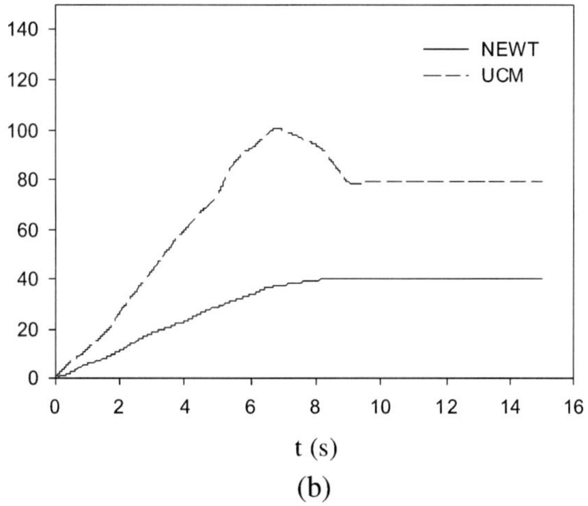

Fig. 7.13 Dimensionless variations versus time for the Newtonian and UCM fluids: **a** torque C^* on the inner cylinder; **b** load F^* on the inner cylinder

7.4 Concluding Remarks

- The geometrical STM approach defined in the steady case is adopted for non-stationary conditions. Transformed rectilinear parallel lines may be considered for the pathlines versus time.
- In non-stationary cases, the STM domain decomposition can be depicted under one of the following conditions:
  - the geometry of the domain allows splitting the total duct into sub-areas involving only open streamlines;
  - the domain may involve closed streamlines, thus allowing a geometric decomposition into sub-domains.
- Open or closed streamlines are adopted in non-stationary cases provided reference sections are defined in the physical and transformed sub-domains.

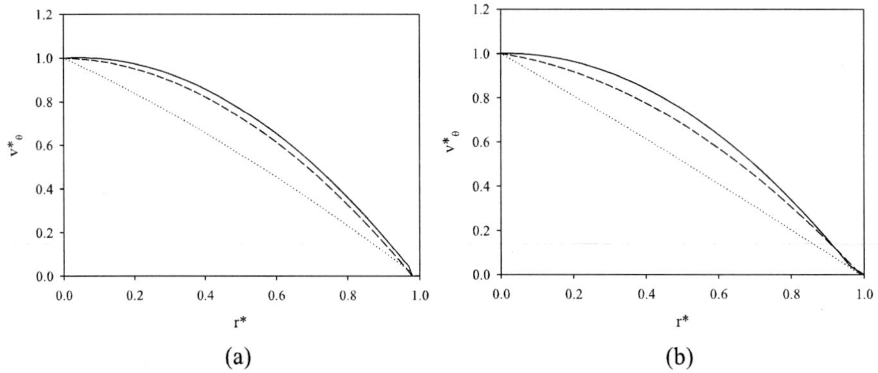

Fig. 7.14 Profiles of the dimensionless velocity component v^*_θ versus a dimensionless radius r^*, for $\theta = \pi$, in one oscillation cycle, at three different times $t_1 = 0.25$ s (……), $t_2 = 0.5$ s (———), $t_3 = 0.75$ s (—): **a** CAR1 fluid; **b** CAR2 fluid. The eccentricity is $\in = 0.2$. Differences are found for the velocities (higher shear rate at the wall of the outer cylinder with CAR1 fluid). Reprinted by permission from Springer Nature Customer Service Centre GmbH from [3], Copyright 2008

- In two-dimensional situations, examples are given for pulsatile and oscillating flows through eccentric cylinders, for different constitutive equations. Still here, satisfactory comparisons of STM results with experimental data should be underlined.
- In mapped computational domains, particle tracking problems are avoided. For two- and three-dimensional cases, these distinguishing features are of particular interest for transient numerical calculations.

References

1. Bird RB, Armstrong RC, Hassager O (1987) Dynamics of polymeric liquids, vol 1. Kinetic theory. Wiley and Sons
2. Clermont J-R (1992) Calculation of kinematic histories in two and three-dimensional flows using streamline coordinate functions. Rheol Acta 32:82–93
3. Grecov D, Clermont J-R (2008) Numerical simulations of non-stationary flows of non-Newtonian fluids between concentric and eccentric cylinders by Stream-Tube Method and domain decomposition. Rheol Acta 47:609–620 [Springer]
4. Clermont J-R, Grecov D (2009) A theoretical analysis of unsteady incompressible flows for time-dependent constitutive equations based on domain transformations. Int J Non-Linear Mech 44:709–715
5. Clermont J-R, Normandin M, Radu D (2000) Some remarks on the concept of stream tubes for numerical simulation of flows of incompressible fluids. Appl Control Cybern 29(2):535–553
6. Mori N, Wakabayashi T, Horikawa A, Namura K (1984) Measurements of pulsating and oscillating flows of non-Newtonian fluids through concentric and eccentric cylinders. Rheol Acta 23:508–513

7. Grecov Radu D, Normandin M, Clermont J-R (2002) A numerical approach for computing flows by local transformations and domain decomposition using an optimization algorithm. Comput Method Appl Mech Eng 191(39–40):4401–4419
8. Crochet MJ, Davies AR, Walters K (1984) Numerical simulation of non-Newtonian flow. Rheology series, vol 1. Elsevier (1984)

Chapter 8
Stream-Tube Method for Thermal Flows and Solid Mechanics

8.1 Introduction

In this chapter, applications of Stream-Tube Method are proposed for non-isothermal flows. It is well known that temperature effects are of importance for industrial situations, notably in complex flows of polymers and manufacturing processes. The examples shown here are found to be interesting and useful enough to be considered now in two-dimensional flow configurations, for different temperature conditions. Examples are proposed for two fluids obeying Newtonian and viscoelastic memory-integral equations.

One purpose of this chapter also concerns the findings of approximate numerical solutions in the field of solid mechanics, with the aid of variational concepts defined with finite elements. Applications are provided for the bending problem in elasticity.

8.2 Thermal Flows in Stream-Tube Method

STM is adopted here for incompressible materials, in non-isothermal conditions. A polymer used in industry is selected in the stick–slip case and two-dimensional confined flows under temperature variations at the boundaries.

8.2.1 Stream-Tube Method and the Thermal Problem

8.2.1.1 STM Equations and Conservation Laws for Non-isothermal Cases in Two-Dimensional Ducts

In two-dimensional non-isothermal cases, a velocity/pressure formulation is firstly considered. Assuming that the internal energy depends only on the temperature (pure

J.-R. Clermont and A. Ammar, *Stream-Tube Method*,
https://doi.org/10.1007/978-3-030-65470-2_8

entropy elasticity) as pointed out by Al-Mubaiyedh et al. [1], the following form of the energy equation writes (Chap. 2):

$$\rho C_p \frac{\mathrm{D}T}{\mathrm{D}t} = k\nabla^2 T + \underline{\underline{\tau}} : \underline{\underline{D}}, \tag{8.1}$$

where T denotes the temperature field, ρ the fluid density. The heat capacity C_P and the thermal conductivity k are constant parameters. $\underline{\underline{D}}$ is the rate-of-deformation tensor. The symbol ":" denotes the scalar product and ∇^2 the Laplacian operator. The conservation laws (mass, momentum conservation and energy) are written in relation to the constitutive equations. To relate the flow characteristics to thermal effects, the Peclet number is adopted by using the equation:

$$Pe = \frac{\rho\, C_p \overline{w}^* R^*}{k}. \tag{8.2}$$

In this equation, R^* and $\overline{w}^*$ denote the respective radius and average velocity, when the domain under consideration is concerned by an upstream Poiseuille flow section.

Some elements provided in Chap. 2 are recalled in the following.

An axisymmetric physical domain Ω is mapped into a computational domain Ω^*, in flows without vortices. Inertia and body forces are ignored. Cylindrical coordinates (r, θ, z) and (R, Θ, Z) are adopted for domains Ω and Ω^*, respectively. An upstream section z_p where the kinematics are known is referred in the physical flow domain. As generally considered in STM, the mapped domain Ω^* is defined such that its upstream section at $Z = Z_p$ is identical to the original one for $z = z_p$. The transformation $\mathcal{T}$: $\Omega^* \to \Omega$ is thus defined by the following equations:

$$r = f(R, \Theta, Z), \quad \theta = \Theta, \quad z = Z. \tag{8.3}$$

The Jacobian Δ of the transformation $\mathcal{T}$, expressed in the axisymmetric case by $\Delta = \left|\frac{\partial(r,z)}{\partial(R,Z)}\right| = f'_R$ is assumed to be non-zero (flows with open streamlines). Thus, the derivative operators, already considered in previous chapters are expressed as

$$\frac{\partial}{\partial r} = \frac{1}{f'_R} \cdot \frac{\partial}{\partial R}, \quad \frac{\partial}{\partial z} = -\frac{f'_R}{f'_R} \cdot \frac{\partial}{\partial R} + \frac{\partial}{\partial Z}. \tag{8.4}$$

The velocity profile w_p at the upstream reference section z_p is assumed to be known. The components of the velocity vector $\underline{V} = u(r, z)\underline{e}_1 + w(r, z)\underline{e}_3$, expressed in the orthonormal basis $(\underline{e}_1, \underline{e}_3)$ of cylindrical coordinates $(r = 1, \theta = 2, z = 3)$, are given by (Chap. 3)

$$w = \frac{w_p}{f \cdot f'_R} (\text{for } R \neq 0),\; u = w \cdot f'_Z \text{ or } u = \frac{w_p \cdot f'_Z}{f \cdot f'_R} (\text{for } R \neq 0). \tag{8.5}$$

Mass conservation is verified by velocity equations (8.5). Numerical calculations, in the non-isothermal case, require the writing of the dynamic equations and energy conservation law (8.1) with boundary conditions. When ignoring inertia and body forces, the vector form of the dynamic equations is written as

$$\underline{0} = \underline{\nabla} \cdot (-p\underline{\underline{I}} + \underline{\underline{\tau}}), \tag{8.6}$$

where p denotes the isotropic part of the stress tensor, $\underline{\underline{I}}$ the unit tensor and $\underline{\underline{\tau}}$ the extra-stress tensor. In STM, by using the derivative operators previously recalled, the following dynamic equations can be written as follows:

$$-\frac{\partial p}{\partial R} + \frac{\partial \tau^{rr}}{\partial R} + \frac{f'_R}{f} \cdot \left(\tau^{rr} - \tau^{\theta\theta}\right) + f'_R \cdot \frac{\partial \tau^{rz}}{\partial Z} - f'_z \cdot \frac{\partial \tau^{rz}}{\partial R} = 0$$
$$f'z \cdot \frac{\partial p}{\partial R} - f'_R \cdot \frac{\partial p}{\partial z} + f'_R \cdot \frac{\partial \tau^{zz}}{\partial z} - f'z \cdot \frac{\partial \tau^{zz}}{\partial R} + \frac{f'R}{f} \cdot \tau^{rz} + \frac{\partial \tau^{rz}}{\partial R} = 0. \tag{8.7}$$

In Eq. (8.7), the superscripts of the tensor components are still related to variables (r, z) of the physical domain. As for the velocities, the components of $\underline{\underline{\tau}}$ are expressed in terms of the mapping function f and its derivatives upon variables (R, Z). In a primary formulation which involves as unknowns the mapping function and the pressure, the set of non-linear equations can be written according to the general form $E_1[f, p]_{(R,Z)} = 0$, $E_2[f, p]_{(R,Z)} = 0$ in the mapped rectangular domain Ω^*. Otherwise, the stress components are considered in a mixed formulation involving the components of $\underline{\underline{\tau}}$ and the pressure.

8.2.2 *Energy Equation with Finite Element Approach*

When investigating two-dimensional situations, the following approach is considered as

(i) the mechanical problem is solved in the context of STM, using discretization procedures generally adopted for the equations;
(ii) the thermal problem is considered with the finite element method.

The energy equation (8.1) is written in the physical domain Ω, given the kinematics and the pressure. Using a Galerkin finite element method, the corresponding form of Eq. (8.1), we obtain, for an element Ω^e, the corresponding form of Eq. (8.1) as

$$\int_{\Omega^e} \rho C_p \underline{V} \cdot \underline{\nabla}(T) \Lambda_T^i \mathrm{d}\Omega^e = \int_{\Omega^e} k\, \nabla^2(T) \Lambda_T^i \mathrm{d}\Omega^e + \int_{\Omega^e} (\underline{\underline{\tau}} : \underline{\underline{D}}) \Lambda_T^i \mathrm{d}\Omega^e, \tag{8.8}$$

where ∇^2 denotes the Laplacian operator and Λ_T^i the weighting function. By integrating by parts the diffusion term, we get the following linear equation (weak form of Eq. (8.2)) in terms of the unknown T:

$$\int_{\Omega^e} \rho C_p \underline{V} \cdot \underline{\nabla}(T) \Lambda_T^i \, \mathrm{d}\Omega^e - \int_{\Gamma^e} \Phi_N \Lambda_T^i \mathrm{d}\Gamma^e + \int_{\Omega^e} k \, \underline{\nabla}(T) \cdot \underline{\nabla}(\Lambda_T^i) \mathrm{d}\Omega^e = \int_{\Omega^e} (\underline{\underline{\tau}} : \underline{\underline{D}}) \Lambda_T^i \mathrm{d}\Omega^e. \tag{8.9}$$

In this equation, Φ_N denotes the heat flux (normal to the surface) through the boundary Γ of the element.

The kinematic quantities are derived from the transformation function f by relations (8.3). To ensure continuity of first-order derivatives for the velocity components of Eq. (8.5), we adopt a cubic approximation of the w-component using Hermite elements (Dhatt and Touzot [2]) similar to those previously considered in [3] for velocity/stress/pressure formulations. In the physical domain Ω, the cubic Hermite element is a trapezoid where two sides of which are segments of streamlines, the other sides being parallel to the r-axis. The element is rectangular in the computational domain Ω^*. Such consideration allows to express the w-velocity component as (Fig. 8.1)

$$w = \sum_{i=1}^{12} \tilde{\mathrm{N}}_i \cdot w_i. \tag{8.10}$$

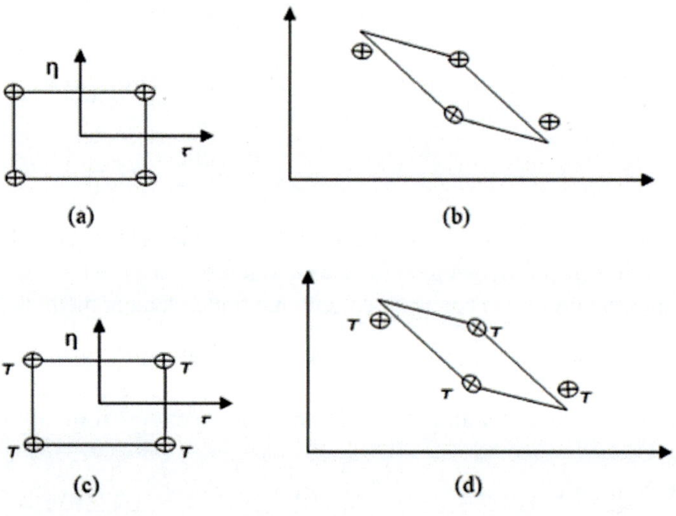

Fig. 8.1 Cubic Hermite element in a planar case for the velocity components **a** in the computational domain; **b** in the physical domain—bilinear element for the temperature T; **c** in the mapped domain; **d** in the physical domain

In Eq. (8.9), $\tilde{N}_i$ denotes the interpolation functions in the real domain with nodal values w_i corresponding to w_i, $\frac{\partial w_i}{\partial r}$, $\frac{\partial w_i}{\partial z}$, determined by means of Eqs. (8.3) in the STM isothermal case. Then, from the same equation, the u-velocity component is evaluated by the relation $u = w \cdot f_Z'$. The unknown temperature is approximated by a linear form, as

$$T = \sum_{i=1}^{4} \tilde{M}_i \cdot T_i, \tag{8.11}$$

with interpolation functions $\tilde{M}_i$ and nodal temperature values T_i. Using Eq. (8.8), a linear system is solved in terms of the unknown temperature. The great sensitivity of the convective term of Eq. (8.9) resulting from significant velocity gradients in the vicinity of the singularity should be underlined. The SUPG (Streamline Upwind/Petrov–Galerkin) scheme (Brooks and Hughes [4]) is adopted. Accordingly, the weighting function Λ_T^i, related to the convective term, is changed into another function Λ_T^{i*} expressed by

$$\Lambda_T^{i*} = \Lambda_T^i + \frac{\beta(\Delta h)}{2|\underline{V}|}\underline{V} \cdot \nabla(\Lambda_T^i). \tag{8.12}$$

In this equation, Δh stands for a mesh segment and β is a parameter given by

$$\beta = \tan(Pe)^{-1} - \frac{1}{Pe}, \tag{8.13}$$

where

$$Pe = \frac{\left|\underline{\underline{V}}\right| \Delta h}{2 \overset{c}{k}} \tag{8.14}$$

is the local Peclet number. $\overset{c}{k}$ denotes the thermal conductivity of the fluid.

The non-isothermal conditions lead to define an iterative decoupled algorithm.

8.2.3 *Two-Dimensional Examples: Ducts with Restriction Zones: Stick–slip: Converging Flows*

In the non-isothermal case, comments on computations considered for axisymmetric and planar duct flows of polymer melts are provided in the following cases:

- tube involving a restriction: "REST"
- tube of constant cross-section with slip at the wall: "DUCT"

- converging domain: "CONV" with two different temperatures at the boundaries

STM equations and temperature-dependent problems may concern many applications on thermal effects, notably those involving singularities at the boundaries. The rheological behaviour corresponds to a memory-integral constitutive equation the data of which depend on the temperature [5]. A Newtonian fluid is also adopted in the calculations.

Viscoelastic polymer processing systems and flows encountered in industry generally involve heating or cooling imposed at the boundaries and heat resulting from viscous dissipation. Heat transfer is relatively small, owing to the low thermal conductivity of polymeric fluids. Large temperature gradients in flow domains may be observed, notably in situations involving singularities in the boundary conditions (e.g. Yesilata et al. [6]). Investigation of such problems aims to ensure a better performance for processes as extrusion or injection molding towards final products.

The study of flows in cylindrical domains with contracting sections is of interest in industry for various materials. Slip of polymer melts, solutions and other materials along the wall of solid surfaces is often dealt with, for example, in chemical engineering processes, owing to its influence on flow characteristics in ducts (e.g. [6–8]). The stick–slip phenomenon has been investigated in Chap. 3, in isothermal conditions.

In the examples presented in this chapter, a Newtonian fluid and a viscoelastic memory-integral Wagner model are adopted for the non-isothermal simulations. Such viscoelastic equation is consistent with experimental data for a linear low-density polyethylene melt (LLDPE) [9]. In the literature, numerical simulations of the transition adherence-to-slip problem have generally concerned inelastic or differential models under isothermal conditions. The Wagner equation, more realistic for numerous applications as the flow examples considered here, is proposed here in a coupled approach involving finite differences and finite elements.

1. *The non-linear constitutive equation—Temperature dependence*

The Wagner model model [10] is related to rheological properties of the LLDPE (Low-Density Polyethylene) melt at 160 °C [5]. The extra-stress tensor $\underline{\underline{\tau}}$ is given, at time t, by the following memory-integral relation:

$$\underline{\underline{\tau}}(t) = \int_{-\infty}^{t} \sum_{p=1}^{7} \frac{a_p}{\lambda_p} \exp\left(-\frac{(t-t')}{\lambda_p}\right) \frac{\underline{\underline{C}}_t^{-1}(t')}{1+\alpha\left[\beta\, tr\underline{\underline{C}}_t^{-1}(t') + (1-\beta) tr\underline{\underline{C}}_t(t') - 3\right]^{b/2}} \mathrm{d}t'. \tag{8.15}$$

Relaxation times λ_p and moduli times a_p are concerned in this equation [9]. The stress evaluation requires particle tracking versus time.

Components of the deformation tensors involved in the Wagner equation may be computed with analytical formulae in the computational domain Ω^*, unlike classic formulations. The relevant formulae have been provided using the natural basis vectors ($\underline{E}_R$, $\underline{E}_\Theta$, $\underline{E}_Z$) related to a material point M, with the curvilinear coordinates (R, Θ, Z), according to the relations:

$$\underline{E}_R = \frac{\partial \underline{M}}{\partial R}, \underline{E}_\Theta = \frac{\partial \underline{M}}{\partial \Theta}, \underline{E}_Z = \frac{\partial \underline{M}}{\partial Z}. \tag{8.16}$$

Stress components of the Wagner model are evaluated by Gauss–Laguerre formulae [11].

In the non-isothermal case, the temperature dependence of the constitutive equations is required. According to the Arrhenius model, based on the theory of reaction rate (e.g. [11]), the rheological parameters may be expressed in terms of a scale factor a_T given by

$$a_T = \exp\left[\frac{E}{R^*}\left(\frac{1}{T} - \frac{1}{T_0}\right)\right], \tag{8.17}$$

where E is the activation energy, R^* the universal gas constant. T and T_0 denote the respective fluid and reference temperatures (in Kelvin). Then, a_T is used as a shift factor according to the William–Landel–Ferry (WLF) time–temperature superposition principle [12]. The following relations for the viscosity function and the relaxation times of the Wagner model are given by

$$\eta[\dot{\gamma}(T), T] = a_T \cdot \eta[\dot{\gamma}(T0), T0], \ \lambda_p(T) = a_T \cdot \lambda_p(T_0). \tag{8.18}$$

Mixed velocity conditions at the wall are concerned for flow domains with stick-slip conditions. The singularity equations arising from the abrupt change in conditions for the velocity vector $\underline{V}$ at the boundary ($\Gamma = \Gamma_1 \bigcup \Gamma_2$), are written again, according to the following conditions:

$$\begin{cases} \forall M \in (\Gamma_1) : \underline{V} = \underline{0} \\ \forall M \in (\Gamma_2) : \underline{V} \neq \underline{0}. \end{cases} \tag{8.19}$$

Components of the deformation tensors involved in the Wagner equation may be computed with analytical formulae in the computational domain Ω^*, unlike classic formulations. The relevant formulae have been provided using the natural basis vectors ($\underline{E}_R$, $\underline{E}_\Theta$, $\underline{E}_Z$) related to a material point M, with the curvilinear coordinates (R, Θ, Z), according to the relations:

$$\underline{E}_R = \frac{\partial \underline{M}}{\partial R}, \underline{E}_\Theta = \frac{\partial \underline{M}}{\partial \Theta}, \underline{E}_Z = \frac{\partial \underline{M}}{\partial Z}. \tag{8.20}$$

Stress components of the Wagner model are evaluated by Gauss–Laguerre formulae [11].

In the non-isothermal case, the temperature dependence of the constitutive equations is required. According to the Arrhenius model, based on the theory of reaction rate (e.g. [11]), the rheological parameters may be expressed in terms of a scale factor a_T given by

$$a_T = \exp\left[\frac{E}{R^*}\left(\frac{1}{T} - \frac{1}{T_0}\right)\right], \tag{8.21}$$

where E is the activation energy, R^* the universal gas constant. T and T_0 denote the respective fluid and reference temperatures (in Kelvin). Then, a_T is used as a shift factor according to the WLF time–temperature superposition principle [12]. The following relations for the viscosity function and the relaxation times of the Wagner model are given by

$$\eta[\dot{\gamma}(T), T] = a_T \cdot \eta[\dot{\gamma}(T0), T0], \lambda_p(T) = a_T \cdot \lambda_p(T_0). \tag{8.22}$$

2. *Boundary conditions—Discretization*

In the non-isothermal problem, with variables of the computational domain Ω^*, the boundary conditions related to geometries of Fig. 8.5 are expressed as follows.

- At the upstream section, the mapping function f, the temperature T and the velocity profiles are known according to the conditions:

$$\frac{\partial f(R,0)}{\partial Z} = 0, \tag{8.23}$$

$$w(R, Z_p) = w_p(R), \tag{8.24}$$

$$T(R,0) = Tp(R). \tag{8.25}$$

- At the wall of the flow domains, f and the wall temperature T_w are known, with the following specifications:

$$f_{\text{wall}} = f(R_0, Z), \quad w(r_{\text{wall}}, Z) = 0, \tag{8.26}$$

for the adherence–slip conditions or

$$\begin{cases} w_{M\in\Gamma_1}(R_0, Z) = 0, \ z \le z_s, \\ \\ \tau^{rz}_{M\in\Gamma_2}(R_0, Z) = 0, \ z > z_s, \end{cases} \tag{8.27}$$

$$T(R_0, Z) = T_p, \quad \text{with} \quad T_p \ge T_0, \tag{8.28}$$

where T_0 denotes the reference temperature.

- Along the axis of symmetry, f is known ($f = 0$) and we get

$$\frac{\partial p(0,Z)}{\partial R} = 0, \tag{8.29}$$

$$\frac{\partial T(0,Z)}{\partial R} = 0. \tag{8.30}$$

- At the limiting downstream section z_2, Neumann conditions are written as

$$\frac{\partial f(R,Z_2)}{\partial Z} = 0, \tag{8.31}$$

$$\frac{\partial \tau^{rz}(R,Z_2)}{\partial Z} = 0, \tag{8.32}$$

$$\frac{\partial w(R,Z_2)}{\partial Z} = 0, \tag{8.33}$$

$$\frac{\partial T(R,Z_2)}{\partial Z} = 0. \tag{8.34}$$

The quantities f, p and T are unknowns. Numerical tests indicate that the assumption of plug flow at the downstream section has ensured a constant temperature field for $z \geq z_2$. The lengths upstream and downstream the transition section (at $z = z_s$) of the computational domain are denoted by L_1 and L_2, respectively (Fig. 8.X). The flow geometries, of upstream radius $R_0 = 0.01$ m require axial lengths large enough for insuring the validity of the imposed boundary conditions, in the non-isothermal case.

3. *Approximations*

The first-order derivatives of the mapping function, the pressure and the stress components of Eq. (8.9) are considered:

- in the R-direction : central-difference formulae;
- in the Z-direction: mixed approach involving summation of both upwind and downstream schemes.

The discretized set of the non-linear equations is solved by the *Newton–Raphson algorithm.*

The elements for velocity and temperatures in physical and mapped domain Ω and Ω^*, respectively, are shown in Fig. 8.1a–d.

The non-isothermal problem leads to define an iterative decoupled algorithm the main features of which may be summarized as follows:

(i) Given the temperature T, solve the STM problem, involving the dynamic equations and the constitutive equation. This leads to evaluate the mapping function f and the pressure p (the isothermal problem);
(ii) Given f and p, compute T by means of the energy equation (the temperature problem), from a velocity/pressure formulation;

(iii) Coupling is achieved by a fixed-point iteration, characterized by specifications on the norm of the dynamic equations and stability for the velocity and the temperature.

8.2.3.1 Domain with a Contracting Section

Such flow domains are concerned in various industrial activities and diseases (i.e. blood problems). In the example considered, the wall geometry (Fig. 8.2) is expressed by the relation:

$$f(R_0, Z)/R_0 = 1 + H\exp[-\varepsilon(Z - Z_s)/R_0], \tag{8.35}$$

where H and ε denote, respectively, the amplitude related to the amplitude of the wall related to the upstream section of the duct and the coefficient of variation of the wall. Conditions at the uspstream and wall domains are of the same type as those already provided previously. For duct flows in this book with a constant wall temperature.

At the downstream section, where fully developed flow conditions are defined, Neumann type of boundary conditions are written, in order to avoid possible oscillations resulting from an inappropriate length of the duct under consideration. We may write the following relationships, where T denotes the temperature (Fig. 8.3):

$$\begin{aligned} &\partial f(R, Z_{\max})/\partial Z = 0,\ \partial\tau r^{rr}, Z_{\max}\big)/\partial Z = 0, \\ &\partial w(R, Z_{\max})/\partial Z = 0,\ \partial T(R, Z_{\max})/\partial Z = 0. \end{aligned} \tag{8.36}$$

We now present some comments related to results obtained in the non-isothermal problem, the temperature influence is taken into account by the Arrhenius equation. To obtain an optimal length of the downstream domain allowing an established thermal regime, a mean flow velocity in the upstream flow zone is set up at 0.01 m/s. In this case, the length of the downstream zone is set at 2470 R_0, the radius of the upstream flow section. The governing equations are the dynamic and the energy equations. We wish to indicate that the Peclet number Pe is 309, which indicates that

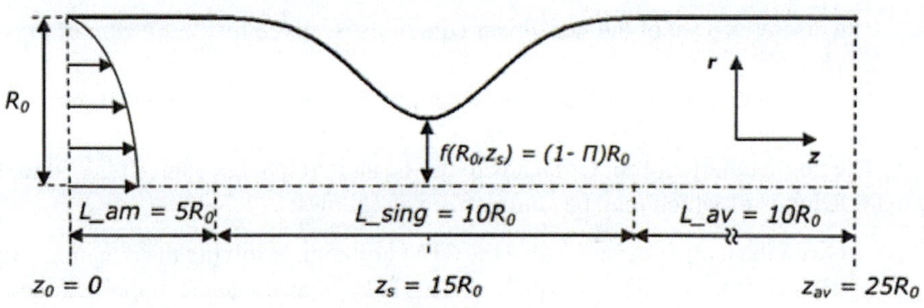

Fig. 8.2 Half-domains CS with a contracting section CS

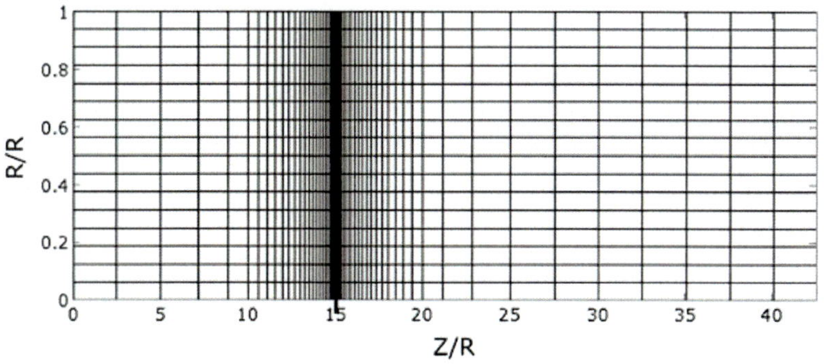

Fig. 8.3 Mesh in the computational domain with smaller elements (versus z) in the vicinity of the contracting section

the heat transfer is dominated by convective effects, with a Reynolds number of the order of 10^{-6}.

Temperature difference variations ΔT_{cp} between the wall and the centreline are expressed in terms of wall temperature *Tp*. The results indicate that the temperature difference Δ_{cp} is reduced when the wall temperature increased. This observation indicates that the wall temperature is inverse versus the term of self-heating in the energy equation, which is reduced with fluid temperature.

Concerning the viscoelastic stress variations for τ^{rr}, $\tau^{\theta\theta}$, τ^{zz} and τ^{rz} in the duct, a decrease of these extra-stress components in absolute value is obtained when the wall temperature increases. Such behaviour should be related to reduction of elasticity and viscosity of the fluid with an increase in the wall temperature.

8.2.3.2 Non-isothermal Stick–Slip Problem

This problem has been previously investigated in the isothermal case (Chap. 4) (Fig. 8.4).

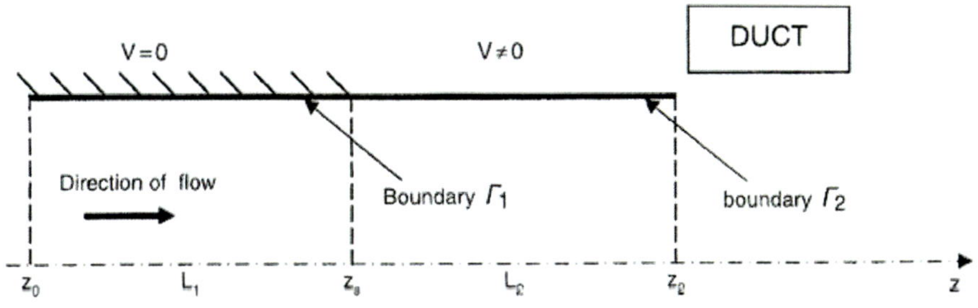

Fig. 8.4 Domain for a transition of adherence to slip flow problem for an axisymmetric geometry (half-domain of a cylindrical tube (DUCT)

For tubes of circular cross-sections (Fig. 8.6), it should be pointed out that the upstream and downstream lengths $L_1 = 15\ R_0$, $L_2 = 43\ R_0$ are adopted in isothermal cases but are required to be $L_1 = 22\ R_0$, $L_2 = 1960\ R_0$ for convergence in non-isothermal conditions; The mesh adopted for the calculations is of the same type than that shown in Fig. 8.3.

The temperature variations on the computed streamlines are determined by solving the governing dynamic and energy equations, using conditions (8.24) corresponding to stick–slip problems. The mesh is refined in the vicinity of the transition zone, similarly to the domain shown in Fig. 8.3. The computed results, not presented here, indicate that The streamline variations, using the memory-integral Wagner model, have indicated that the appearance of streamlines is practically the same in the range of temperatures investigated.

8.2.3.3 Axisymmetric Converging Flow with Two Different Thermal Boundary Conditions

In this case, two thermal boundary conditions are adopted for the upstream and downstream zones (Fig. 8.8): wall temperatures $T_P 1$ at the upstream and converging zones and $T_P 2$ at the downstream domain.

The streamline results, not reported here, are found to be close for the different temperatures considered in the calculations, for the Wagner model as also with Newtonian and purely viscous fluid cases, peaks of temperature profiles have been observed in the vicinity of the re-entrant corner of Fig. 8.5 (converging half-angle: $\alpha = 26.56°$, Wagner model).

The increase in temperature of the moving material in a converging domain comes from the self-heating activity. It is important in zones close to the wall, notably those involving geometrical singularities.

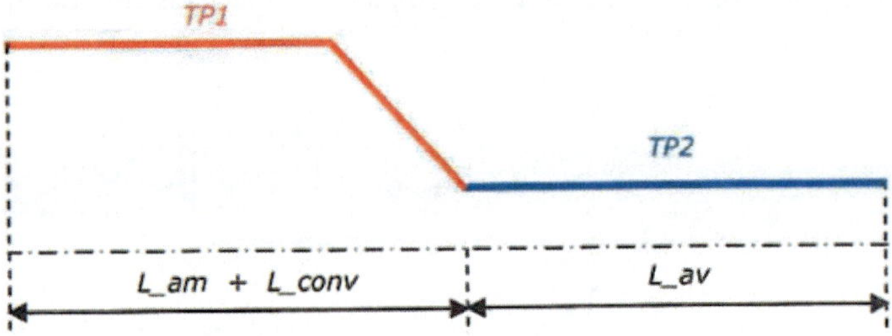

Fig. 8.5 Half-domain for axisymmetric flows in a converging geometry

8.3 Comments on the STM Flow Results

- In relation to applications to different flow considerations: adherence-to-slip problem in a tube, converging and restricting geometries, a viscoelastic model is adopted for non-isothermal STM calculations, notably in the vicinity of the singularity zones.
- In non-isothermal conditions, STM enables to adopt a complex realistic constitutive equation with simple discretization schemes. Solving the governing equations in rectangular computational domains helps to avoid numerical complications.
- The non-isothermal STM approach has led to apply the method for the crystallization process of polymers [13].

8.4 Stream-Tube Method for Solid Mechanics Problems

A two-dimensional situation is considered here. k denotes the mapping function (Fig. 8.6) and we adopt Cartesian coordinates (Z, X) and (z, x) for the reference and physical domains, respectively. One may write

$$\begin{cases} x = k(X, Z) \\ z = Z. \end{cases} \tag{8.37}$$

We also assume that, at an upstream section, we have

$$x_{\max} = k(0, X_{\max}) = X_{\max}. \tag{8.38}$$

Solving the solid problem with STM consists of determining the "good mapping function" that minimizes the stored energy for a loaded elastic body. This problem has been previously solved numerically for fluids. Here, we wish to solve a non-linear matrix system for an incompressible solid problem where the unknowns are the nodal values of the transformation function k and possibly its derivatives [14].

In the planar case, we consider the mapping function k of Eq. (8.33) with the following notations:

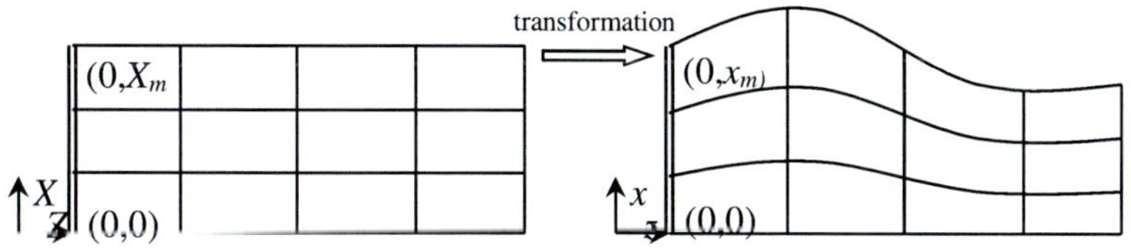

Fig. 8.6 Domain transformation

$$k,_X = \frac{\partial k}{\partial X} \qquad k,_Z = \frac{\partial k}{\partial Z}$$
$$k,_{XX} = \frac{\partial^2 k}{\partial X^2} \qquad k,_{ZZ} = \frac{\partial^2 k}{\partial Z^2} \qquad k,_{XZ} = \frac{\partial^2 k}{\partial X \partial Z}. \tag{8.39}$$

Accordingly, we may express the derivative operators as follows:

$$\frac{\partial}{\partial x} = \frac{1}{k,_X}\frac{\partial}{\partial X}$$
$$\frac{\partial}{\partial z} = \frac{\partial}{\partial Z} - \frac{1}{k,_Z}\frac{\partial}{\partial X}. \tag{8.40}$$

We can also define the boundary conditions at a reference section as

$$W_r(X) = V_{\max}(1 - X^2/X^2_{\max})$$
$$U_r = 0. \tag{8.41}$$

Let $\underline{v}$ be the displacement vector of components u and w in the physical domain. The incompressibility condition can be expressed by

$$w = \frac{W_r}{k,_X} \quad \text{and} \quad u = k,_Z\, w. \tag{8.42}$$

To perform calculations based on energetic concepts, we need to express the derivatives of the displacement components u and w. Thus, using Eqs. (8.39) and (8.42), we get

$$\frac{\partial w}{\partial X} = -\frac{k,_{XX}}{k,_X^2} W_r + \frac{\partial W_r}{\partial X}\frac{1}{k,_X}$$
$$\frac{\partial w}{\partial Z} = -\frac{k,_{XZ}}{k,_X^2} W_r$$
$$\frac{\partial u}{\partial X} = k,_{XZ}\, w + k,_Z\, \frac{\partial w}{\partial X}$$
$$\frac{\partial u}{\partial Z} = k,_{ZZ}\, w + k,_Z\, \frac{\partial w}{\partial Z}. \tag{8.43}$$

Equation (8.40) allow us, by using Eq. (8.42), to express derivatives in the physical domain as follows:

$$\frac{\partial w}{\partial x} = \frac{1}{k,_X}\frac{\partial w}{\partial X}$$
$$\frac{\partial w}{\partial z} = \frac{\partial w}{\partial Z} - k,_Z\, \frac{\partial w}{\partial X}$$
$$\frac{\partial u}{\partial x} = \frac{1}{k,_X}\frac{\partial u}{\partial X}$$

$$\frac{\partial u}{\partial z} = \frac{\partial u}{\partial Z} - k,_Z \frac{\partial u}{\partial X}. \tag{8.44}$$

The mass conservation writes

$$\mathrm{div}(\underline{v}) = \frac{\partial u}{\partial x} + \frac{\partial w}{\partial z} = 0. \tag{8.45}$$

To interpolate the mapping function k for the computations, we retained rectangular Hermite elements, providing accurate approximations for k and its first partial derivatives. Such elements, easily defined on a mesh built on rectilinear streamlines of the reference domain, allow the displacement field to be continuous and derivable at the nodal points and inside the elements.

8.4.1 Formulation Based on Energetic Concepts

We denote the solid stress tensor
$\underline{\underline{\sigma}} = -p\underline{\underline{1}} + 2\mu\underline{\underline{E}}$,

where
μ: is the elasticity shearing modulus,
$\underline{\underline{E}}$: is the strain tensor ($\underline{\underline{E}} = (\underline{\mathrm{grad}v} + \underline{\mathrm{grad}v^{\mathrm{T}}})/2$).
p: is the pressure,
$\underline{\underline{1}}$: is the identity tensor.
Ignoring inertia, the equilibrium equations yield

$$\underline{\mathrm{div}\,\underline{\underline{\sigma}}} = \underline{0}. \tag{8.46}$$

This equation can be multiplied by a kinematically admissible test vector field $\underline{v}^*$ that vanishes where velocity boundary conditions are specified and verifying the incompressibility condition. Integrating by parts the variational formulation, we obtain

$$\int_{\Omega} \underline{\mathrm{div}\,\underline{\underline{\sigma}}}.\underline{v}^*\,\mathrm{d}\Omega = 0 \quad \Leftrightarrow \quad -\int_{\Omega} \underline{\underline{\sigma}} : \underline{\underline{E}}^*\,\mathrm{d}\Omega + \int_{\Gamma} \left(\underline{\underline{\sigma}}_{imp}\underline{n}\right) \underline{v}^*\,\mathrm{d}\Gamma = 0, \tag{8.47}$$

with

$$\underline{\underline{E}}^* = \left(\underline{\underline{\mathrm{grad}v^*}} + \underline{\underline{\mathrm{grad}v^{*\mathrm{T}}}}\right)/2. \tag{8.48}$$

Γ is the boundary of Ω with unit normal vector $\underline{n}$.

A similar form can be obtained by considering the test fields (v^*) as differential elements, leading to

$$-\int_{\Omega} \underline{\underline{\sigma}} : \underline{\underline{dE}}\, \mathrm{d}\Omega + \int_{\Gamma} \left(\underline{\underline{\sigma}}_{imp} \underline{n}\right) \underline{dv}\, \mathrm{d}\Gamma = 0. \tag{8.49}$$

By integration with respect to these differential elements, Eq. (8.43) becomes equivalent to the following statement:

$$\underbrace{-\int_{\Omega} (p\underline{\underline{1}} : \underline{\underline{E}} + \mu \underline{\underline{E}} : \underline{\underline{E}})\, \mathrm{d}\Omega}_{I_{\mathrm{int}}} + \underbrace{\int_{\Gamma} \left(\underline{\underline{\sigma}}_{imp} \underline{n}\right) \underline{v}\, \mathrm{d}\Gamma}_{I_{est}} \text{ is minimum.} \tag{8.50}$$

In this relation, I_{int} and I_{ext} denote, respectively, the internal and the external energy to be minimized. In the following, problems with imposed kinematics (displacement) will be considered.

Since the incompressibility Eq. (8.45) is automatically verified by the mapping function k, we get $\mathrm{tr}(E) = 0$. Consequently, the term $p\underline{\underline{1}} : \underline{\underline{E}}$ vanishes. Thus, in contrast to classic hybrid displacement–pressure finite element formulations, the pressure is not an unknown of the problem.

Our problem is now reduced to the minimization of the quantity

$$I_{\mathrm{int}} = \int_{\Omega} (\mu \underline{\underline{E}} : \underline{\underline{E}})\, \mathrm{d}\Omega. \tag{8.51}$$

In the case of the elastic two-dimensional case, it writes

$$I_{\mathrm{int}} = \int_{\Omega} \mu \left(\left(\frac{\partial u}{\partial x}\right)^2 + \left(\frac{\partial w}{\partial z}\right)^2 + 2\left(\left(\frac{\partial u}{\partial z} + \frac{\partial w}{\partial x}\right)/2\right)^2 \right) \mathrm{d}\Omega. \tag{8.52}$$

Owing to Eqs. (8.42) and (8.43), this energy is a non-linear function of the transformation function and its derivatives. For the rectangular elements of vertices noted 1, 2, 3, 4, we may obtain a (16×16) system such that

$$\underline{\underline{M}}^e(\underline{K}^e)\underline{K}^e = \underline{F}^e(\underline{K}^e), \tag{8.53}$$

and the vector of degrees-of-freedom is

$$\underline{K}^e = \begin{bmatrix} k1 \\ k,_X 1 \\ k,_Z 1 \\ k,_{XZ} 1 \\ \vdots \\ k,_{XZ} 4 \end{bmatrix}. \tag{8.54}$$

The different quantities involved in the system (8.52) are evaluated numerically according to the following equations:

$$M_{ij}^{e} = \frac{\partial^2 I_{\text{int}}}{\partial K_i \partial K_j}, \qquad F_i^e = \frac{\partial I_{\text{int}}}{\partial K_i}. \tag{8.55}$$

Then, by assembling the elementary matrices, we obtain the global system which should be associated to the displacement boundary conditions.

The system is non-linear. It will be solved with Newton–Raphson method. We denote by $\underline{\mathbf{K}}_0$ the initial estimate. For each iteration "n", we denote by δ^n the solution of $\underline{\underline{\mathbf{M}}}(\underline{\mathbf{K}}^n)\,\delta^n = \underline{\mathbf{F}}(\underline{\mathbf{K}}^n)$.

The iterative update of the solution $\underline{\mathbf{K}}$ is done for the next iteration by $\underline{\mathbf{K}}^{n+1} = \underline{\mathbf{K}}^n + \alpha\delta^n$, where α is a step size parameter of values between 0 and 1. It should be close to 0 when the physical domain is very different (in shape) from the reference domain. To optimize the choice of α, we perform an adaptative step size procedure by selecting the value minimizing the quantity:

$$\underline{\underline{\mathbf{M}}}\left(\underline{\mathbf{K}}^n + \alpha\delta^n\right)\alpha\delta^n - \underline{\mathbf{F}}\left(\underline{\mathbf{K}}^n + \alpha\delta^n\right). \tag{8.56}$$

The algorithm can be summarized as follows:

Choice of initial estimate $\underline{\boldsymbol{K}}_0$
Repeat until convergence
For all elements

For all integration points

End

$$F_i^{e,g} = \frac{I_{\text{int}}(K_i + \varepsilon, K_j, \ldots) - I_{\text{int}}(K_i, K_j, \ldots)}{\varepsilon} \tag{8.57}$$

$$M_{ij}^e = M_{ij}^e + M_{ij}^{e,g}. \tag{8.58}$$

Forming global system by assembling elementary matrices
End
Find δ^n solution of $\underline{\underline{\mathbf{M}}}(\underline{\mathbf{K}}^n)\,\delta^n = \underline{\mathbf{F}}(\underline{\mathbf{K}}^n)$
Find optimal parameter (α)
Update $\underline{\mathbf{K}}^{n+1} = \underline{\mathbf{K}}^n + \alpha\delta^n$,
checking convergence by calculation of the norm $\|\underline{\mathbf{F}}(\underline{\mathbf{K}}^{n+1})\|$.

Remark

In case of the application of the STM for elastic incompressible solid, we must not confuse transformed domain and deformed domain (see Fig. 8.7).

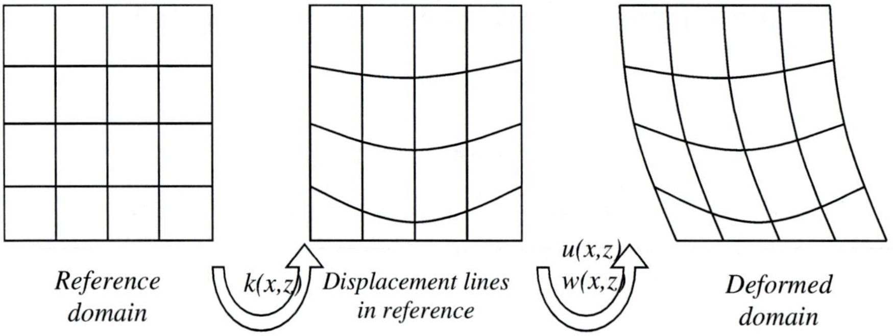

Fig. 8.7 Definition of domains

Fig. 8.8 Solid beam under deformation

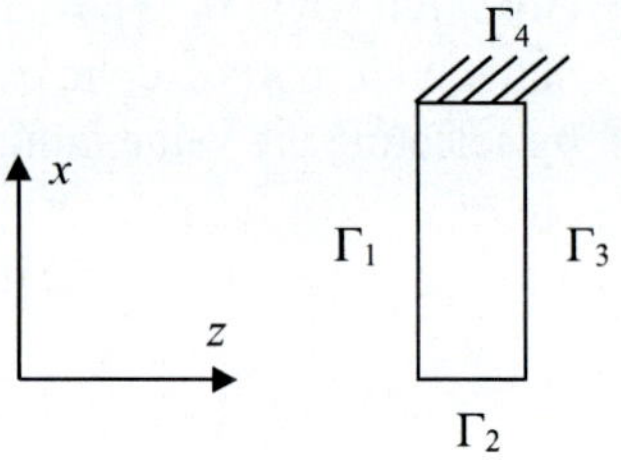

8.4.2 An Example of Results

We consider the following case, shown in Fig. 8.8 [14].

Boundary conditions are considered for the upper and lower plates of the beam. The displacement profiles, not imposed at any section, are determined from the transformation function after convergence of the solving algorithm.

Γ_1 and Γ_3 are the boundaries at $Z_{\min}$ and $Z_{\max}$. Γ_2 and Γ_4 are the boundaries at $X_{\min}$ and $X_{\max}$, respectively.

No stress conditions are applied on Γ_1 and Γ_3,

On Γ_4, we impose a zero displacement.

On Γ_2, we impose a zero displacement in the x-direction.

Consequently, in terms of transformation function, we have

- On Γ_4: we impose k, as
  $K,_X = 1, k,_Z = 0, k,_{XZ} = 0$.
- On Γ_2: we ensure that
  $k, k,_Z = 0, k,_{XZ} = 0$.

In this case, for the purposes of comparisons, we present in Fig. 8.9 our STM results (column 1) and those obtained by means of finite element calculations (linear elasticity, with a Poisson coefficient of 0.4999) (column 2). The computed data, that also concern the displacement lines in the reference domain and contour values in the x- and z-directions, are found to be identical for the two methods.

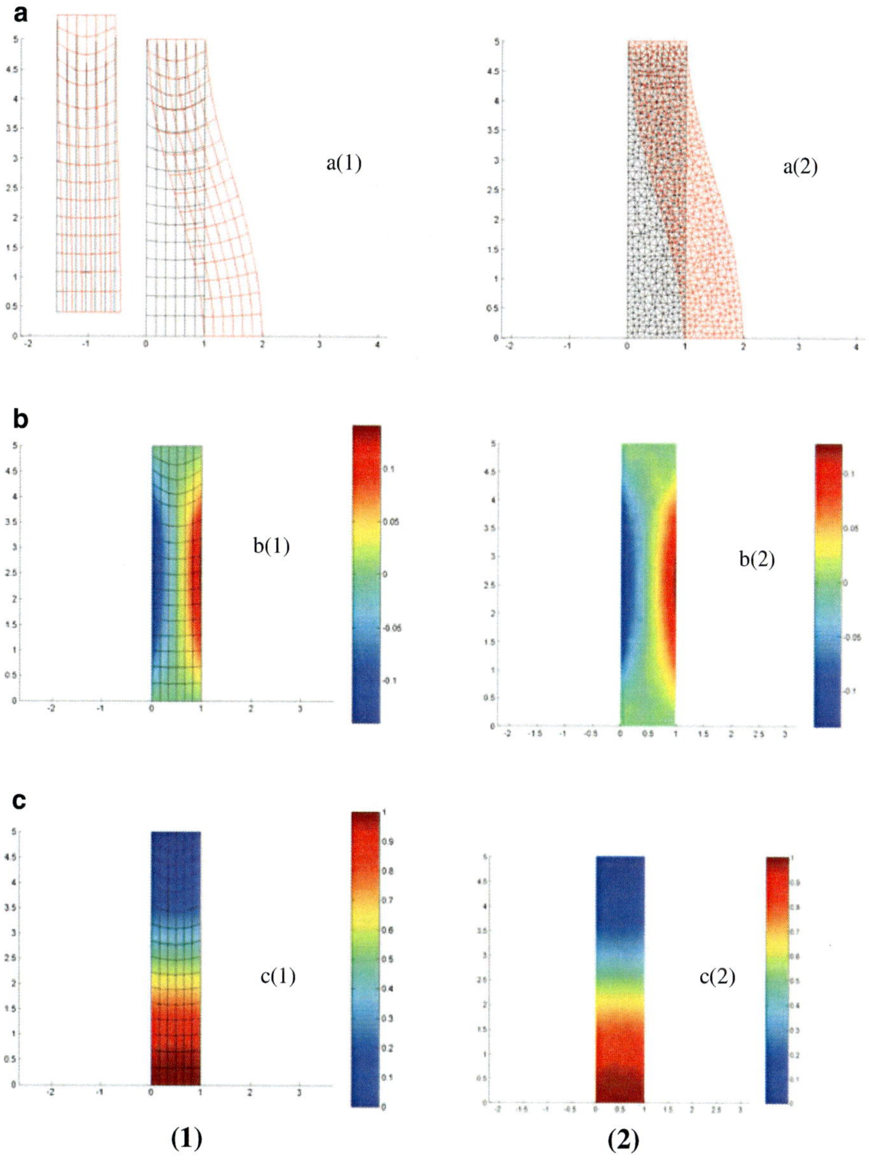

Fig. 8.9 Comparison of STM (1) and FE (2) calculations (2): **a** displacement lines and deformed shape; **b** x-axis displacement isovalues; **c** z-axis displacement isovalues

A similar study is presented in Fig. 8.10, with free displacement on the surface Γ_2.

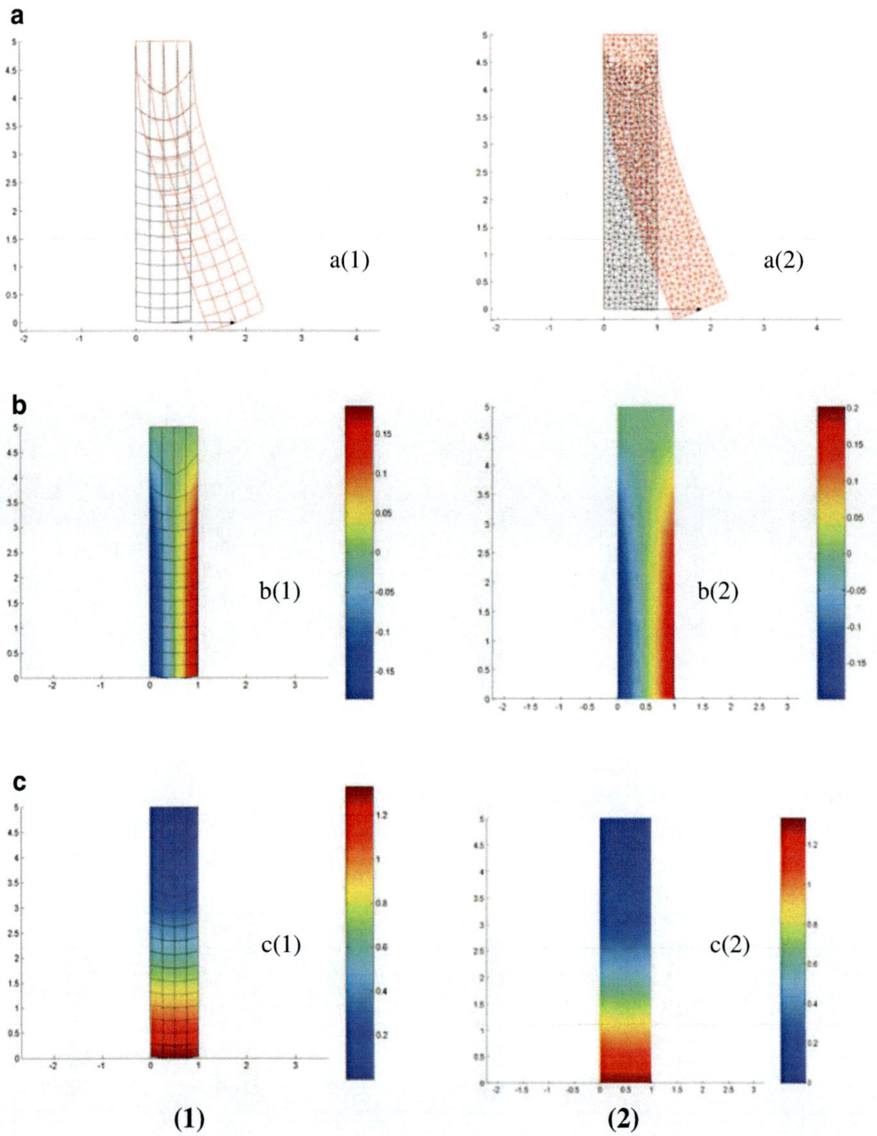

Fig. 8.10 Comparison of STM (1) and FE (2) calculations: **a** displacement lines and deformed shape; **b** x-axis displacement isovalues; **c** z-axis displacement isovalues

8.5 Concluding Remarks

- The STM approach can be applied to fluid and solid mechanics problems, under temperature conditions.

- STM allows the solving of non-isothermal flow problems with adherence-to-slip conditions at the boundaries, for a complex viscoelastic constitutive equation. Calculations have led to provide information on flow characteristics related to different two-dimensional geometries, with thermal boundary conditions for polymeric fluids.
- A variational approach associated to the STM has been proposed to compute movements of solids under load, providing consistent results.

References

1. Al-Mubaiyedh UA, Sreshkumar R, Khomami B (2000) Energetic effects on the stability of viscoelastic Dean Flow. J Non-Newt Fluid Mech 95:277–293
2. Dhatt G, Touzot G (1984) Une présentation de la méthode des éléments finis. Maloine Editeurs, Paris
3. Marchal JM, Crochet MJ (1986) Hermitian finite elements for calculating viscoelastic flow. J Non-Newt Fluid Mech 79:87–2076
4. Brook AN, Hughes TJR (1982) Streamline upwind Petrov-Galerkin formulations for convection-dominated flows with particular emphasis on the incompressible Navier-Stokes equations. Comput Method Appl Mech Eng 32:199–259
5. Chine A, Ammar A, Clermont J-R (2015) Simulations of two-dimensional steady isothermal and non-isothermal steady flows with slip for a viscoelastic memory-integral fluid. Eng Comput 32(8):2318–2342
6. Yesilata B, Oztekin A, Neti S (2000) Non-isothermal viscoelastic flow through an axisymmetric sudden contraction. J Non-Newt Fluid Mech 89:133–164
7. Mooney M (1931) Explicit formulas for slip and fluidity. J Rheol 2:210–222
8. El Kissi N, Piau JM (1994) Adhesion of linear low density polyethylene for flow regimes with sharkskin. J Rheol 38(5):1447–1463
9. Carrot C, Guillet J, Fulchiron R (2001) Converging flow analysis, entrance pressure drops and vortex sizes: measurements and calculated values. Polym Eng Sci 41(1, 2):2095–2107
10. Wagner MH (1976) Analysis of time-dependent non-linear stress-growth data for shear and elongational flow of a low-density polyethylene melt. Rheol Acta 15:136–142
11. Chine A (2007) Simulations numériques d'écoulements de fluides anélastiques et viscoélastiques dans des conduites bidimensionnelles avec prise en compte de la thermodépendance et de la cristallisation. PhD thesis, INPG Grenoble
12. Ferry D (1980) Viscoelastic properties of polymers. Wiley, New-York
13. Chine A, Ammar A, Normandin M, Clermont JR (2007) A numerical approach for polymer flow predictions in two-dimensional geometries, involving crystallization processes. Int J Form Process 10(4):447–470
14. Ammar A, Clermont JR (2005) A finite-element approach in stream-tube method for solving fluid and solid mechanics problems. Mech Res Commun 32(1):65–73

Chapter 9
Micro–Macro Simulations and Stream-Tube Method

9.1 Introduction

Atomistic modelling is the most detailed level of description that can be applied today in rheological studies, using techniques of non-equilibrium molecular dynamics. Such calculations require enormous computer resources, and then they are currently limited to flow geometries of molecular dimensions. Consideration of macroscopic flows found in processing applications calls for less detailed mesoscopic models, such as those of kinetic theory.

Models of kinetic theory provide a coarse-grained description of molecular configurations wherein atomistic processes are ignored. They are meant to display in a more or less accurate fashion the important features that govern the flow-induced evolution of configurations. Over the last few years, different models related to dilute polymers have been evaluated in simple flows by means of stochastic simulation or Brownian dynamics methods.

Kinetic theory models can be very complicated mathematical objects. It is usually not easy to compute their rheological response in rheometric flows, and their use in numerical simulations of complex flows has long been thought impossible. The traditional approach has been to derive from a particular kinetic theory model a macroscopic constitutive equation that relates the viscoelastic stress to the deformation history. One then solves the constitutive model together with the conservation laws using a suitable numerical method, to predict velocity and stress fields in complex flows. The majority of constitutive equations used in continuum numerical simulations are indeed derived (or at least very much inspired) from kinetic theory.

Indeed, derivation of a constitutive equation from a model of kinetic theory usually involves closure approximations of a purely mathematical nature such as decoupling or pre-averaging. It is now widely accepted that closure approximations have a significant impact on rheological predictions for dilute polymer, solutions, or fibre suspensions.

In this context, micro–macro methods of computational rheology that couple the coarse-grained molecular scale of kinetic theory to the macroscopic scale of

J.-R. Clermont and A. Ammar, *Stream-Tube Method*,
https://doi.org/10.1007/978-3-030-65470-2_9

continuum mechanics have an important role to play. On the one hand, this approach is much more demanding in computer resources than more conventional continuum simulations that integrate a constitutive equation to evaluate the viscoelastic contribution of the stress tensor. On the other hand, micro–macro techniques allow the direct use of kinetic theory models and thus avoid the introduction of closure approximations.

Since the early 1990s, the field has developed considerably following the introduction of the CONNFFESSIT method by Ottinger and Laso [25]. Being relatively new, micro–macro techniques have been implemented only for models of kinetic theory with few configurational degrees of freedom, such as non-linear dumbbell models of dilute polymer solutions and single-segment tube models of linear entangled polymers.

Kinetic theory provides two basic building blocks: the diffusion or Fokker–Planck equation that governs the evolution of the distribution function (giving the probability distribution of configurations) and an expression relating the viscoelastic stress to the distribution function. The Fokker–Planck equation has the general form:

$$\frac{d\Psi}{dt} + \frac{\partial}{\partial \underline{X}}\left(\underline{A}\Psi\right) = \frac{1}{2}\frac{\partial}{\partial \underline{X}}\frac{\partial}{\partial \underline{X}} : \left(\underline{\underline{B}}\Psi\right), \tag{9.1}$$

where $\frac{d\Psi}{dt}$ is the material derivative, vector $\underline{X}$ defines the coarse-grained configuration and has a dimension N. Factor $\underline{A}$ is a N-dimensional vector that defines the drift or deterministic component of the molecular model. Finally, $\underline{\underline{B}}$ is a symmetric, positive definite $N \times N$ matrix that embodies the diffusive or stochastic component of molecular model. In general, both $\underline{A}$ and $\underline{\underline{B}}$ (and in consequence the distribution function ψ) depend on the physical coordinates $\underline{x}$, on the configuration coordinates $\underline{X}$ and on the time t.

The second building block of a kinetic theory model is an expression relating the distribution function and the stress. It takes the following form:

$$\underline{\underline{\tau}}_p = \int_C \underline{\underline{g}}\left(\underline{X}\right)\Psi d\underline{X}, \tag{9.2}$$

where C represents the configuration space and $\underline{\underline{g}}$ is a model-dependent tensorial function of configuration. In a complex flow, the velocity field is a priori unknown and stress fields are coupled through the conservation laws. In the isothermal and incompressible case, the conservation of mass and momentum balance are then expressed (neglecting inertia and body forces) by

$$\begin{cases} \nabla \cdot \underline{v} = 0 \\ \nabla \cdot \left(-p\underline{\underline{I}} + \underline{\underline{\tau}}_p + \eta_s \underline{\underline{D}}\right) = 0\ , \end{cases} \tag{9.3}$$

where p is the fluid pressure and $\eta_s \underline{\underline{D}}$ is a purely viscous component ($\underline{\underline{D}}$ being the strain rate tensor and η_s the solvant viscosity). The set of coupled Eqs. (9.1)–(9.3), supplemented with suitable initial and boundary conditions in both physical and configuration spaces, is the generic multiscale formulation. Three basic approaches have been adopted for exploiting the generic multiscale model:

(i) *The continuum approach* wherein a constitutive equation of continuum mechanics that relates the viscoelastic stress to the deformation history is derived from, and replaces altogether, the kinetic theory models (1) and (2). The derivation process usually involves closure approximations. The resulting constitutive model takes the form of a differential, integral or integrodifferential equation. It yields molecular information in terms of averaged quantities, such as the second moment of the distribution: $\int_C \underline{X} \otimes \underline{X} d\underline{X}$
(ii) *The Fokker–Planck approach* wherein one solves the generic problem (9.1)–(9.3) as such, in both configuration and physical spaces. The distribution function is thus computed explicitly as a solution of the Fokker–Planck Eq. (9.1). The viscoelastic stress is computed from relation (9.2).
(iii) *The Stochastic approach* which draws on the mathematical equivalence between the Fokker–Planck Eq. (9.1) and the following stochastic differential equation:

$$\underline{X} = \underline{A} dt + \underline{\underline{b}} d\underline{W}, \tag{9.4}$$

where $\underline{\underline{B}} = \underline{\underline{b}}\underline{\underline{b}}^T$ and $\underline{W}$ is a Wiener stochastic process of dimension N.

In a complex flow, the stochastic differential Eq. (9.4) applies along individual flow trajectories, the time derivative is thus a material derivation. Instead of solving the deterministic Fokker–Planck Eq. (9.1), one solves the associated stochastic differential Eq. (9.4) for a large ensemble of realizations of the stochastic process $\underline{X}$ by means of a suitable numerical technique. The distribution function is not computed explicitly, and the viscoelastic stress of Eq. (9.2) is readily obtained as an ensemble average.

For more details concerning the micro–macro approach, the reader can refer to the review paper [17] and the references therein.

The continuum approach has been adopted throughout the development of computational rheology. First simulations were obtained in the early 1980s. Two decades later, macroscopic numerical techniques based upon the continuum approach remains under active development.

The control of the statistical noise is a major issue in stochastic micro–macro simulations based on the stochastic approach. Moreover, to reconstruct the distribution one needs to operate with an extremely large number of particles, however, in general, only the moments of such distribution are required, which can be computed using a much more reduced population of particles. These difficulties do not arise at all in the Fokker–Planck approach, however, the Fokker–Planck Eq. (9.1) must be solved for the distribution function in both physical and configuration spaces. This necessitates a suitable discretization procedure for all relevant variables, namely position $\underline{x}$, configuration $\underline{X}$ and time t. Until now, the dimensionality of the problem could be daunting and consideration of molecular models with many configurational degrees of freedom did not appear feasible. This probably explains why relatively few studies based on the Fokker–Planck approach have appeared in the literature until very recently at least. In papers [8] and [20], the resolution of the Fokker–Planck equation involving a moderate number of dimensions is considered. An alternative deterministic particle approach has been proposed in [9].

Sometimes the treated model results highly multidimensional and, in this case, the possibility of describing functions from their values at the nodes of a mesh or a grid in the domain of interest results simply prohibitory. Some attempts exist concerning the treatment of multidimensional problems. The interested reader can refer to [7] for a review on sparse grids methods involving sparse tensor product spaces, but despite its optimality, the interpolation is defined in the whole multidimensional domain, and, consequently, only problems defined in spaces of the dimension of the order of tens can be treated. In conclusion, the problematic lied to models defined in multidimensional spaces remains still open, and new efforts must be paid to reach significant improvements in the next years.

Some of the most used kinetic theory models defined from a Fokker–Planck equation have two important particularities: they can be expressed in a separated form (this is the case of multi-bead models) and, in general, realistic molecular models involve springs with finite extension which implies that the distribution function vanishes on the boundary of the domain where the springs conformation is defined. In this case, the separated representation and the definition of tensor product approximation spaces, run perfectly and allow to circumvent the difficulties related to the multidimensional character of kinetic theory models, as proved in [2]. This technique consists of the use of a separated representation of the molecular conformation distribution, that introduced in the variational formulation of the Fokker–Planck equation leads to an iteration procedure that involves at each step the resolution of a small-size non-linear problem. The resolution of those non-linear problems can be performed by using a standard Newton strategy, alternating directions resolution or more sophisticated strategies. Thus, the number of degrees of freedom involved in the discretization of the Fokker–Planck equation can be reduced from $(n_n)^N$ (required in the usual grid/mesh-based strategies) to $(n_n) \times N$ (n_n being the number of nodes involved in the discretization of each conformation coordinate and N the dimension of the conformation space).

In [2], we considered the steady-state solution of some classes of multidimensional partial differential equations, and in particular, those governing the molecular configuration distribution in kinetic theory models of complex fluids. In the conclusions of that chapter was claimed that the resolution of multidimensional transient Fokker–Planck equations could be considered within an incremental time discretization. However, being the time no more than other coordinates, one could expect a coupled space–time resolution. The main difficulties related to a such approach lies in the fact that the initial condition being non-zero the procedure proposed in [2] cannot be applied in a direct manner. Another difficulty lies in the fact that space approximations are built using standard piece-wise functions, and when this kind of approximation is retained also to construct the time interpolation the known inconsistency related to centered differences of time derivatives is encountered. In [3], we propose alternatives to circumvent both difficulties, and then to allow an efficient treatment of multidimensional transient kinetic theory models.

Until now, only simple rheometric flows were addressed. A first attempt in addressing more complex flows was proposed in [27]. The present chapter focuses on the solution of complex fluid models defined in complex geometries. For this purpose, we are defining a complex fluid model, a bit more complex than the ones addressed in our former works. The hypothetical fluid consists of a fibre suspension in which the fibre can aggregate. Thus two populations coexist, the one related to the active fibres (making a parallel with the concepts used in associative polymers models [15, 16, 21]) that participate to the different fibres aggregates, and the other composed of free fibres immersed in the fluid with no direct connection with the other fibres or aggregates. Now, high shear rates are expected destroying the aggregates (therefore increasing the fraction of free tubes) and are reestablished as soon as the shear rate becomes lower enough. Obviously, a such model involves two populations, and then two coupled Fokker–Planck equations. Thus, the results here obtained using a hypothetical model (perhaps unrealistic) could be transferred for addressing more realistic models, as the ones related to associative polymers just referred to.

After describing the model equation related to such hypothetical fluid, we will consider its discretization within the separated representation (finite sums decomposition) framework. The last step will consist of extending this separated representation to cover all the model coordinates (space, configuration and time). This extension will be checked on a contraction flow for the just proposed fluid model.

9.2 A Representative Micro–Macro Model of a Complex Fluid Flow

9.2.1 Macroscopic Equations

The governing equations to be solved concern the mass and momentum conservation laws for incompressible, isothermal steady-state viscoelastic flow. Inertia and body forces are ignored. We may write

$$\nabla \cdot \underline{v} = 0, \tag{9.5}$$

$$\nabla p - \nabla \cdot \underline{\underline{\tau}}_p - 2\eta_s \nabla \cdot \underline{\underline{D}} = 0, \tag{9.6}$$

where $\underline{v} = (u, v, w)$ denotes the velocity field of the fluid, p is the pressure, $\underline{\underline{\tau}}_p$ is the extra-stress tensor due to the presence of fibres and η_s is the viscosity of the solvent. The extra-stress tensor $\underline{\underline{\tau}}_p$ can be obtained by means of a macroscopic, integral or differential, constitutive equation or using a molecular model involving the resolution of a Fokker–Planck equation. In this work, we consider the microscopic model of fibre network.

9.2.2 Microscopic Equations of a Hypothetical Fibre Network Model

A network model introduced firstly by Tanaka and Edwards [28] and [29], is composed of segments and nodes (junctions) (Fig. 9.1). The segment is a microstructure joining two successive nodes and the nodes are the points where the interactions are localized. The internal structure has a transient topology and its dynamics is controlled by the rate of attachment (creation) and detachment (destruction) stochastic and reversible process and depends on the applied deformation.

Network models [15, 16, 21, 26, 30] incorporate common basic features and physical phenomena. Thus, in our case, we consider that a fibre can be active or pendant (dangling). However, the transition from one entity to the other is possible due to flow-induced aggregation/disaggregation. We assume that the total number of active and pendant fibres is constant. The flow is considered homogeneous at the scale of fibres independently of the population that it belongs to.

The previous assumptions, seem implying the necessity of considering a set of two Fokker–Planck equations, one for each fibre populations governing the evolution of the respective fibre orientation distributions $\Psi\left(\underline{x}, \underline{p}, t\right)$ and $\Phi\left(\underline{x}\underline{p}, t\right)$ for the active and pendant populations, respectively. They describe the fraction of active (respectively, pendant) fibres that at point $\underline{x}$ and time t are oriented in the direction

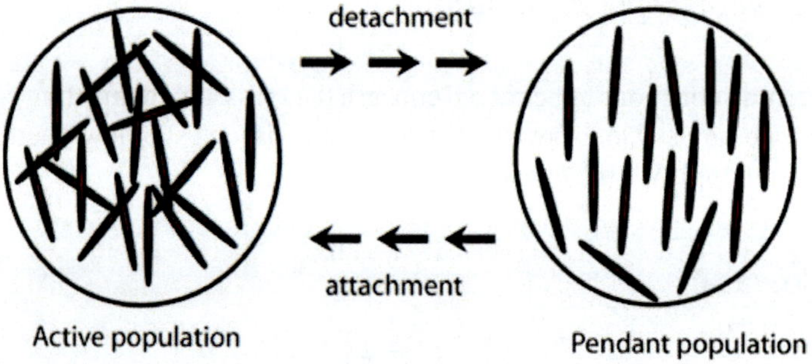

Fig. 9.1 Fibre network model and kinematic phenomena*

*Reprinted from Ref. [34] Copyright 2010, with permission from Elsevier

defined by the unit vector $\underline{p}$. Thus, the orientation distribution function evolution related to the active fibres population $\Psi\left(\underline{x}, \underline{p}, t\right)$ writes

$$\frac{d\Psi}{dt} = -\frac{\partial}{\partial \underline{p}}\left(\frac{d\underline{p}}{dt}\Psi\right) + D_{r1}\frac{\partial^2 \Psi}{\partial \underline{p}^2} + V_d\Psi - V_c\Phi, \tag{9.7}$$

where D_{r1} is the diffusion coefficient associated with the active fibres population, V_d and V_c represent the kinetic terms characterizing, respectively, destruction and aggregation velocities, $\frac{d\underline{p}}{dt}$ denotes the flow-induced fibre rotary velocity that is assumed given by Jeffery's equation:

$$\frac{d\underline{p}}{dt} = \underline{\underline{\Omega}}\,\underline{p} + \lambda\underline{\underline{D}}\,\underline{p} - \lambda\left(\underline{p}^T\underline{\underline{D}}\,\underline{p}\right)\underline{p}, \tag{9.8}$$

where $\underline{\underline{\Omega}}$ and $\underline{\underline{D}}$ are, respectively, the vorticity and the strain rate tensors associated with the fluid flow and λ is a scalar depending on the fibre aspect ratio $r*$:

$$\lambda = \frac{r^{*2} - 1}{r^{*2} + 1}, \tag{9.9}$$

with $r* = \frac{L}{d}$, L and d referring to the length and the diameter of the fibres, respectively.

Equation (9.7) is coupled with the evolution equation for the distribution function of pendant fibres, $\Phi\left(\underline{x}, \underline{p}, t\right)$:

$$\frac{d\Phi}{dt} = -\frac{\partial}{\partial \underline{p}}\left(\frac{d\underline{p}}{dt}\Phi\right) + D_{r2}\frac{\partial^2 \Phi}{\partial \underline{p}^2} - V_d\Psi + V_c\Phi, \tag{9.10}$$

where D_{r2} denotes the diffusion coefficient related to such fibre population.

The conservation equation involving both population distributions writes

$$\int_{\Omega_{\underline{p}}} \left(\Psi\left(\underline{x}, t, \underline{p}\right) + \Phi\left(\underline{x}, t, \underline{p}\right)\right) d\Omega_{\underline{p}} = 1, \quad \forall \underline{x}, \forall t, \tag{9.11}$$

$\Omega_{\underline{p}}$ being the fibre configuration space where the fibre orientation is defined, and that corresponds in the three-dimensional case to the surface of the unit sphere (sphere of unit radius). Equation (9.11) also implies

$$\frac{\partial}{\partial t}\left\{\int_{\Omega_{\underline{p}}} \left(\Psi\left(\underline{x}, t, \underline{p}\right) + \Phi\left(\underline{x}, t, \underline{p}\right)\right) d\Omega_{\underline{p}}\right\} = 0, \quad \forall \underline{x}, \forall t. \tag{9.12}$$

By introducing the notations

$$\begin{cases} \int\limits_{\Omega_{\underline{p}}} \Psi\left(\underline{x}, t, \underline{p}\right). d\Omega_{\underline{p}} = \Theta\left(\underline{x}, t\right) \\ \int\limits_{\Omega_{\underline{p}}} \Phi\left(\underline{x}, t, \underline{p}\right). d\Omega_{\underline{p}} = \Upsilon\left(\underline{x}, t\right) \end{cases}, \tag{9.13}$$

Equation (9.11) becomes

$$\Theta\left(\underline{x}, t\right) + \Upsilon\left(\underline{x}, t\right) = 1, \quad \forall \underline{x}, \forall t. \tag{9.14}$$

We can notice that by adding Eqs. (9.7), (9.9) and integrating over the conformation space $\Omega_{\underline{p}}$, we obtain the fibre conservation balance (9.12).

The integration of the equation Eq. (9.7) in the configuration space $\Omega_{\underline{p}}$ writes

$$\frac{d}{dt} \int\limits_{\Omega_{\underline{p}}} \Psi d\Omega_{\underline{p}} = V_d \int\limits_{\Omega_{\underline{p}}} \Psi d\Omega_{\underline{p}} - V_c \int\limits_{\Omega_{\underline{p}}} \Phi d\Omega_{\underline{p}} \tag{9.15}$$

that using the notation previously introduced results

$$\frac{d\Theta}{dt} = V_d \Theta - V_c (1 - \Theta). \tag{9.16}$$

The expression of the extra-stress tensor $\underline{\underline{\tau}}_p$ related to the fibres presence could be derived by generalizing the usual constitutive equation applied for short fibre suspensions:

$$\begin{aligned} \underline{\underline{\tau}}_p \left(\underline{x}, t\right) = 2\eta_s \Bigg(& N_{p1} \int\limits_{\Omega_{\underline{p}}} \left(\underline{p} \otimes \underline{p} \otimes \underline{p} \otimes \underline{p}\right) \Psi\left(\underline{x}, \underline{p}, t\right) d\underline{p} \\ & + N_{p2} \int\limits_{\Omega_{\underline{p}}} \left(\underline{p} \otimes \underline{p} \otimes \underline{p} \otimes \underline{p}\right) \Phi\left(\underline{x}, \underline{p}, t\right) d\underline{p} \Bigg) : \underline{\underline{D}}, \end{aligned} \tag{9.17}$$

where ":" denotes the tensorial product twice contracted, i.e. the dyadic product.

To solve the associated problem, we consider a decoupled procedure between both the microscopic and the macroscopic scales based on a fixed point strategy (in the steady-state regime) or on an explicit time integration (in the transient regime).

Both scales dialog through the extra-stress tensor, calculated at the microscopic scale from both populations distribution functions according to Eq. (9.17). The extra-stress tensor is then used for updating the suspension flow kinematics (at the macroscopic scale) by solving the resulting anisotropic Stokes problem. The resulting velocity field is then injected to update the orientations distributions by solving (at the microscopic scale) the two coupled Fokker–Planck equations, in which the material derivative (advective terms) makes use of the macroscopic velocity field. The procedure continues until reaching convergence, in steady analysis, or the maximum simulation time (in the transient case).

In what follows we are describing both solvers, the one that serves to solve the two coupled Fokker–Planck equations, and then the flow solver that allows updating the flow kinematics.

9.3 Microscopic Scale: A Separated Representation Solver

In some of our former works (referred in the introduction section), a separated representation technique has been proposed, making use of finite sums decompositions, that allowed efficient solutions of complex fluid models in simple flows. These models only incorporated a single Fokker–Planck equation eventually defined in high dimensional spaces. The interested reader is invited to refer to [2] and [3] for additional information on the basic concepts and the implementation details of that novel technique.

The two Fokker–Planck Eqs. (9.7) and (9.9) have been strongly coupled. They can be written in a compact form by introducing the vector $\underline{\Xi}\left(\underline{p}, t, \underline{x}\right)$:

$$\underline{\Xi}\left(\underline{p}, t, \underline{x}\right) = \begin{pmatrix} \Psi\left(\underline{p}, t, \underline{x}\right) \\ \Phi\left(\underline{p}, t, \underline{x}\right) \end{pmatrix}, \tag{9.18}$$

where we assume that both initial orientation distributions are known at the initial time $t = 0$ defining the vector $\underline{\Xi}_0$ (where for the sake of clarity, the dependence of $\underline{\Xi}_0$ on the $\underline{p}$ and $\underline{x}$ has been voluntary omitted).

Owing to the linearity of the resulting kinetic theory model one could apply a variable transformation in order to define a new couple of unknowns fields subjected to a homogeneous initial condition:

$$\begin{pmatrix} \Psi\left(\underline{p}, t, \underline{x}\right) \\ \Phi\left(\underline{p}, t, \underline{x}\right) \end{pmatrix} = \begin{pmatrix} \Psi\left(\underline{p}, t = 0, \underline{x}\right) \\ \Phi\left(\underline{p}, t = 0, \underline{x}\right) \end{pmatrix} + \begin{pmatrix} \psi\left(\underline{p}, t, \underline{x}\right) \\ \phi\left(\underline{p}, t, \underline{x}\right) \end{pmatrix} = \underline{\Xi}_0\left(\underline{p}, \underline{x}\right) + \underline{\Xi}\left(\underline{p}, t, \underline{x}\right). \tag{9.19}$$

The resulting system can be written in the compact form:

$$\frac{d\Xi(\underline{p},t,\underline{x})}{dt}+\underline{\underline{e}}_0(\underline{p},\underline{x})\Xi(\underline{p},t,\underline{x})+\underline{\underline{e}}_1(\underline{p},\underline{x})\frac{(\partial\,\Xi(\underline{p},t,\underline{x}))}{\partial\underline{p}}+\underline{\underline{e}}_2\frac{\partial^2\Xi(\underline{p},t,\underline{x})}{\partial\underline{p}^2}$$
$$-\underline{\underline{e}}\ {}_0(\underline{p},\underline{x})\Xi_0(\underline{p},\underline{x})-\underline{\underline{e}}_1(\underline{p},\underline{x})\frac{\partial\,\Xi_0(\underline{p},\underline{x})}{\partial\underline{p}}$$
$$-\underline{\underline{e}}2\frac{\partial^2\Xi_0(\underline{p},\underline{x})}{\partial\underline{p}^2}, \tag{9.20}$$

where

$$\underline{\underline{e}}_0\left(\underline{p},\underline{x}\right)=\begin{pmatrix}\frac{\partial\dot{\underline{p}}}{\partial\underline{p}}\left(\underline{p},\underline{x}\right)-V_d\ V_c & \\ V_d & \frac{\partial\dot{\underline{p}}}{\partial\underline{p}}\left(\underline{p},\underline{x}\right)-V_c\end{pmatrix}, \tag{9.21}$$

$$\underline{\underline{e}}_1\left(\underline{p}\right)=\dot{\underline{p}}\begin{pmatrix}1\,0\\0\,1\end{pmatrix}, \tag{9.22}$$

and

$$\underline{\underline{e}}_2=\begin{pmatrix}-D_{r1}\ 0 & \\ 0 & -D_{r2}\end{pmatrix}. \tag{9.23}$$

Now, we consider the variational formulation related to Eq. (9.20) with respect to the conformation coordinates $\underline{p}$, where for the sake of clarity the dependence of all variables on the different coordinates, has been omitted:

$$\int_{\Omega_{\underline{p}}}\underline{\Xi}^*\frac{d\underline{\Xi}}{dt}d\Omega_{\underline{p}}+\int_{\Omega_p}\underline{\Xi}^*\underline{\underline{e}}\ 0\underline{\Xi}d\Omega_{\underline{p}}+\int_{\Omega_{\underline{p}}}\underline{\Xi}^*\underline{\underline{e}}\ 1\frac{\partial\underline{\Xi}}{\partial\underline{p}}d\Omega_{\underline{p}}$$
$$+\int_{\Omega_p}\underline{\Xi}^*\underline{\underline{e}}\ 2\frac{\partial^2\underline{\Xi}}{\partial\underline{p}^2}d\Omega_{\underline{p}}$$
$$=-\int_{\Omega\underline{p}}\underline{\Xi}^*\underline{\underline{e}}\ 0(\underline{p})\underline{\Xi}\ 0d\Omega_{\underline{p}}-\int_{\Omega_{\underline{p}}}\underline{\Xi}^*\underline{\underline{e}}\ 1\frac{\partial\underline{\Xi}_0}{\partial\underline{p}^2}d\Omega_{\underline{p}}$$
$$-\int_{\Omega_{\underline{p}}}\underline{\Xi}^*\underline{\underline{e}}\ 2\frac{\partial^2\underline{\Xi}_0}{\partial\underline{p}^2}d\Omega_{\underline{p}}, \tag{9.24}$$

where $\underline{\Xi}^*$ represents the usual test functions.

Integrating by parts the last term of the first member, we obtain

$$\int_{\Omega_{\underline{p}}}\underline{\Xi}^*\frac{d\underline{\Xi}}{dt}d\Omega_{\underline{p}}+\int_{\Omega_{\underline{p}}}\underline{\Xi}^*\underline{\underline{e}}\ 0\underline{\Xi}d\Omega_{\underline{p}}+\int_{\Omega_{\underline{p}}}\underline{\Xi}^*\underline{\underline{e}}\ 1\frac{\partial\underline{\Xi}}{\partial\underline{p}}d\Omega_{\underline{p}}$$

$$+\int_{\Omega_{\underline{p}}} \frac{\Gamma h_{\underline{p}}}{2} \frac{\partial \underline{\Xi}}{\partial \underline{p}} * \underline{\underline{e}}\ 1 \frac{\partial \underline{\Xi}}{\partial \underline{p}} d\Omega_{\underline{p}} - \int_{\Omega_{\underline{p}}} \frac{\partial \underline{\Xi}}{\partial \underline{p}} * \underline{\underline{e}}\ 2 \frac{\partial \underline{\Xi}}{\partial \underline{p}} d\Omega_{\underline{p}}$$

$$= \int_{\Omega_{\underline{p}}} \underline{\Xi} * \frac{d\underline{\Xi}}{dt} d\Omega_{\underline{p}} + \int_{\Omega_{\underline{p}}} \underline{\Xi} * \underline{\underline{e}}\ 0 \underline{\Xi} d\Omega_{\underline{p}} + \int_{\Omega_{\underline{p}}} \underline{\Xi} * \underline{\underline{e}}\ 1 \frac{\partial \underline{\Xi}}{\partial \underline{p}} d\Omega_{\underline{p}}$$

$$- \int_{\Omega_{\underline{p}}} \frac{\partial \underline{\Xi}}{\partial q} * \underline{\underline{e}}\ 2 \frac{\partial \underline{\Xi}}{\partial \underline{p}} d\Omega_{\underline{p}}$$

$$= - \int_{\Omega_p} \underline{\Xi} * \underline{\underline{e}}\ 0 \underline{\Xi}\ 0 d\Omega_{\underline{p}} - \int_{\Omega_{\underline{p}}} \underline{\Xi} * \underline{\underline{e}}\ 1 \frac{\partial \underline{\Xi}_0}{\partial \underline{p}} d\Omega_{\underline{p}}$$

$$- \int_{\Omega_{\underline{p}}} \underline{\Xi} * \underline{\underline{e}}\ 2 \frac{\partial^2 \underline{\Xi}_0}{\partial \underline{p}^2} d\Omega_{\underline{p}}. \tag{9.25}$$

Owing to the advective character of Eq. (9.25) in the conformation space, a SUPG stabilization is applied on the advective term (the one involving first derivatives with respect to the conformation coordinates):

$$\int_{\Omega_{\underline{p}}} \underline{\Xi} * \frac{d\underline{\Xi}}{dt} d\Omega_{\underline{p}} + \int_{\Omega_{\underline{p}}} \underline{\Xi} * \underline{\underline{e}}\ 0 \underline{\Xi} d\Omega_{\underline{p}} + \int_{\Omega_{\underline{p}}} \underline{\Xi} * \underline{\underline{e}}\ 1 \frac{\partial \underline{\Xi}}{\partial \underline{p}} d\Omega_{\underline{p}}$$

$$+ \int_{\Omega_{\underline{p}}} \frac{\Gamma h_{\underline{p}}}{2} \frac{\partial \underline{\Xi}}{\partial \underline{p}} * \underline{\underline{e}}\ 1 \frac{\partial \underline{\Xi}}{\partial \underline{D}} d\Omega_{\underline{p}} - \int_{\Omega_{\underline{p}}} \frac{\partial \underline{\Xi}}{\partial \underline{p}} * \underline{\underline{e}}\ 2 \frac{\partial \underline{\Xi}}{\partial \underline{p}} d\Omega_{\underline{p}}$$

$$= - \int_{\Omega_p} \underline{\Xi} * \underline{\underline{e}}\ 0 \underline{\Xi}\ 0 d\Omega_{\underline{p}} - \int_{\Omega \underline{p}} \underline{\Xi} * \underline{\underline{e}}\ 1 \frac{\partial \underline{\Xi}}{\partial \underline{p}} 0 d\Omega_{\underline{p}}$$

$$- \int_{\Omega_{\underline{p}}} \underline{\Xi} * \underline{\underline{e}}\ 2 \frac{\partial^2 \underline{\Xi}_0}{\partial \underline{p}^2} d\Omega_{\underline{p}}, \tag{9.26}$$

where Γ is calculated according to

$$\Gamma = \max\left(\left((\coth(P_{e1}) - \frac{1}{P_{e1}}\right), \left((\coth(P_{e2}) - \frac{1}{P_{e2}}\right)\right). \tag{9.27}$$

The Peclet numbers P_{e1} and P_{e2} are given by

$$\begin{cases} P_{e1} = \frac{\|\underline{\underline{e}}_1(\underline{p}, \underline{x})\| h_{\underline{p}}}{2D_{r1}} \\ P_{e2} = \frac{\|\underline{\underline{e}}_1(\underline{p}, \underline{x})\| h_{\underline{p}}}{2D_{r2}} \end{cases}, \tag{9.28}$$

where $h_{\underline{p}}$ represents the characteristic size of the conformation space discretization.

In what follows, we are considering the steady-state solution of a complex flow. For this purpose, we are coupling the separated representation solver just described with the stream-tube method (presented in previous chapters of this book but summarized in the next section) for computing the flow kinematics. The Fokker–Planck equation will be integrated along the flow streamlines. For this purpose, we adopt the method of characteristics. In this context, a point $\underline{x}$ located on a streamline can be represented as $\underline{x}(t)$. In what follows, we consider that both distribution functions: $\psi\left(\underline{p}, t\right)$ and $\phi\left(\underline{p}, t\right)$. Thus, the solution representation in the conformation space is searched in the separated form:

$$\underline{\Xi} = \begin{pmatrix} \psi\left(\underline{p}, t\right) \\ \phi\left(\underline{p}, t\right) \end{pmatrix} = \sum_{j=1}^{n} \begin{pmatrix} \alpha_j E_j\left(\underline{p}\right) F_j(t) \\ \beta_j G_j\left(\underline{p}\right) H_j(t) \end{pmatrix}, \tag{9.29}$$

where again, for the sake of clarity, the dependence of variables on the physical space $\underline{x}$ is omitted.

This solution representation is built-up within an iteration scheme. Thus, if at iteration n the solution involves the known functional couples $E_j\left(\underline{p}\right) F_j(t)$ and $G_j\left(\underline{p}\right) H_j(t)$ (see Eq. (9.29)), then we perform a projection, an enrichment step and a convergence check:

1. *Projection:* By introducing the representation (9.29) into the stabilized variational formulation one could compute (after discretization) the approximation coefficients α_i and β_i, $i \in [1, \cdots, n]$.
2. *Enrichment:* With the just computed approximation (9.29) where both the functional couples and the approximation coefficients are known, we look for the new functional couples: $R\left(\underline{p}\right) S(t)$ and $V\left(\underline{p}\right) W(t)$ by introducing the enriched approximation

$$\underline{\Xi} = \begin{pmatrix} \psi\left(\underline{p}, t\right) \\ \phi\left(\underline{p}, t\right) \end{pmatrix} = \sum_{j=1}^{n} \begin{pmatrix} \alpha_j E_j\left(\underline{p}\right) F_j(t) \\ \beta_j G_j\left(\underline{p}\right) H_j(t) \end{pmatrix} + \begin{pmatrix} R\left(\underline{p}\right) S(t) \\ V\left(\underline{p}\right) W(t) \end{pmatrix} \tag{9.30}$$

again into the variational formulation. The enrichment stage results (after discretization and numerical integration) in a non-linear problem that is solved using an appropriate iteration procedure. This question is addressed later. After convergence of the non-linear solver, we obtain the expression of both functional couples that after normalization, i.e.

$$\begin{cases} E_{n+1}\left(\underline{p}\right) = \frac{R(p)}{\|R(p)\|} \\ F_{n+1}(t) = \frac{S(t)}{\|S(t)\|} \\ G_{n+1}\left(\underline{p}\right) = \frac{V(p)}{\|V(\underline{p})\|} \\ H_{n+1}(t) = \frac{W(t)}{\|W(t)\|} \end{cases} \tag{9.31}$$

leads to the updated separated representation:

$$\underline{\Xi} = \begin{pmatrix} \psi\left(\underline{p}, t\right) \\ \phi\left(\underline{p}, t\right) \end{pmatrix} = \sum_{j=1}^{n+1} \begin{pmatrix} \alpha_j E_j\left(\underline{p}\right) F_j(t) \\ \beta_j G_j\left(\underline{p}\right) H_j(t) \end{pmatrix}. \tag{9.32}$$

Checking the convergence:

This iteration scheme needs the definition of a stopping criterion. Many possibilities exist, being the simplest ones:

- The iteration process stops as soon as $\| R(\underline{p}) S(t) \|$ and $\| V(\underline{p}) W(t) \|$ become small enough. These norms can be evaluated very fast, but even when both become small enough nothing proves that the solution is reached. Only in some cases, we proved that these norms decrease monotonically.
- The other possibility consists of evaluating the residual of the strong formulation of the problem, Eq. (9.20). When the residual becomes small enough the model solution is attained. However, the computational cost of this residual is higher than the one needed for computing the norms of the enrichment functions just introduced.

In what follows we use the stopping criterion based on the residual of the strong formulation. The iteration at which the convergence is attained defined the number of sums involved in the finite sums decomposition.

The solution procedure needs firstly a discrete representation of all the fields involved in the variational formulation. The associated notation is grouped in the first appendix. These discrete representations allow us to transform the variational formulation (9.26) into a matrix form as described in the second appendix for the projection step and in the third appendix for the approximation basis enrichment step.

Here, we would like only to address the main benefits that such separated representation introduces:

- The projection step (at iteration n) needs the solution of a linear system consisting of 2n unknowns (the coefficients α_i and β_i in the expression (9.29). This system is fully populated, but it has a very reduced size (in the order of tens in all the simulations performed until now).

- The enriching step involves the solution of a non-linear problem because now the searched terms are the product of functions. We use an alternating directions scheme (see [3]) in which the functions defined in the conformation space $\underline{p}$ are searched assuming known the ones related to the time coordinate (that were computed at the previous iteration). Thus, from the just-updated function defined in the conformation space $\underline{p}$, the time function can be updated. The iteration procedure continues until reaching convergence (other more sophisticated procedures—e.g. Newton–Raphson—exist but its application does not increase significantly the solver efficiency). The solution of the non-linear enrichment problem is attained in few iterations, that will be noted by P. Thus, if we are using $N_{\underline{p}}$ nodes in the conformation space and N_t time-nodes, and even if we are computing the whole space–time evolution, we must solve linear systems of size $2N_{\underline{p}}$ (related to the discretization of the conformation domain $\Omega_{\underline{p}}$—the unit sphere surface) and one-dimensional backward discrete models related to the time evolution.
- This procedure, that seems to be especially appropriate for solving multidimensional models because here the complexity scales linearly with the dimension of the space (instead of the exponential scaling of mesh-based discretization strategies), allows even in the case just described important CPU time savings. It allows also a simple treatment of non-linear kinetic theory models, where at iteration n, the non-linear terms could be evaluated at the solution defined at the previous iteration.

9.3.1 *Addressing Complex Flows*

The previous analysis is also useful for the treatment of transient models with uniform solutions in the physical space $\Omega_{\underline{x}}$. Such flows are usually encountered in rheometric devices. However, in complex kinematic conditions, the evolution of the different fields with respect to the physical coordinates $\underline{x}$ cannot be neglected. Thus, we must address the discretization of the material derivative $\frac{d\Xi}{dt}$ involved in the weak form of the Fokker–Planck Eq. (9.26).

In this more general situation, one could define a general separated representation consisting of a finite sum decomposition in which each term of the sum involves the product of three functions:

- a function of the physical space $\underline{x}$,
- a function of the conformation space $\underline{p}$.
- and finally a function of time t.

This kind of representation needs for an efficient stabilization of the advection term that applies in the physical space $\Omega_{\underline{x}}$. This route needs further analysis and developments. Thus, in what follows we are focusing on steady complex flows that, as we are proving, can profit from all the analysis previously introduced.

For a steady-state flow, one could integrate the Fokker–Planck equation along the characteristic lines (that coincide in this case with the flow streamlines) from the initial condition assumed to be known at the inflow boundary. Thus, one could expect that the solution at each point P only depends on curvilinear coordinate s associated with that point s_P. We assume that the streamline that passes through point P has its origin s = 0 at the point in which the streamline intersects the domain inflow boundary. The stabilization by upwinding used for the treatment of the time discretization is automatically applied to the discretization on the curvilinear coordinate.

With these ideas in mind, one could write, for the orientation distribution field $\underline{\Xi}$ defined in $\Omega_{\underline{x}} \times \Omega_{\underline{p}}$ the following separated representation:

$$\underline{\Xi}_{st} = \begin{pmatrix} \psi_{st}\left(\underline{p}, s\right) \\ \phi_{st}\left(\underline{p}, s\right) \end{pmatrix} = \sum_{j=1}^{n} \begin{pmatrix} \alpha_j^{st} E_j^{st}\left(\underline{p}\right) F_j^{st}(s) \\ \beta_j^{st} G_j^{st}\left(\underline{p}\right) H_j^{st}(s) \end{pmatrix}, \tag{9.33}$$

where the index *st* refers to the particular streamline along which the integration is performed.

If the integration is performed on a large enough number of streamlines, the solution everywhere can be obtained from the interpolation of solutions defined on the streamlines.

9.4 Macroscopic Scale: Flow Kinematics Solver

To facilitate coupling between the microscopic solver and the macroscopic one, we propose the use of the STM approach for solving the resulting macroscopic anisotropic Stokes problem.

In works reported in the previous chapters, the stream-tube method is only applied for different macroscopic non-Newtonian constitutive equations, but never coupled with kinetic theory descriptions of complex fluids.

STM can be directly and easily coupled with the separated representation of the Fokker–Planck equation on the streamlines, described in Sect. 9.3.1.

9.4.1 The Stream-Tube Method Revisited: Basic Concepts

We consider here a two-dimensional flow. In Cartesian coordinates, k denotes the mapping function, between the respective reference and physical domains Ω^* and Ω (Fig. 9.2). Under the assumption of a flow involving only open streamlines (for more complex flow conditions, one can refer to previous chapters of this book as well as to [9, 23] and the references therein), we may write the following relations:

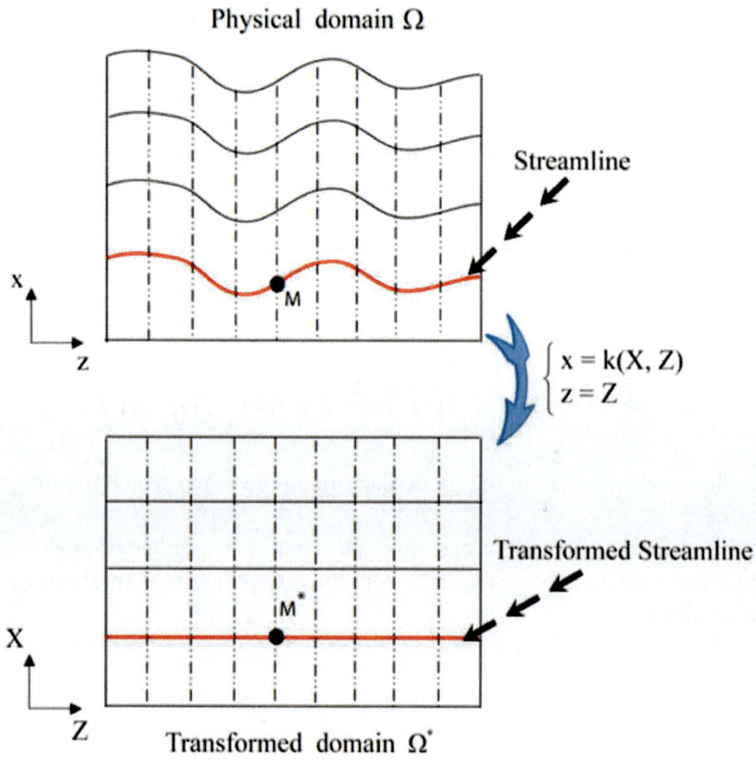

Fig. 9.2 Stream-tube method: Domain transformation*

$$\begin{cases} x = k(X, Z) \\ z = Z \end{cases}. \tag{9.34}$$

For the sake of clarity, we introduce the notation $\frac{\partial x}{\partial X} \equiv k_{,X}$ and $\frac{\partial x}{\partial Z} \equiv k_{,Z}$. The Jacobian of the transformation writes

$$\underline{\underline{J}} = \begin{pmatrix} k_{,X} & k_{,Z} \\ 0 & 1 \end{pmatrix} \tag{9.35}$$

being the Jacobian of the inverse transformation:

$$\begin{pmatrix} \frac{\partial X}{\partial x} & \frac{\partial X}{\partial z} \\ \frac{\partial Z}{\partial x} & \frac{\partial Z}{\partial z} \end{pmatrix} = \begin{pmatrix} k_{,X} & k_{,Z} \\ 0 & 1 \end{pmatrix}^{-1} = \begin{pmatrix} \frac{1}{k_{,X}} & -\frac{k_{,Z}}{k_{,X}} \\ 0 & 1 \end{pmatrix}. \tag{9.36}$$

The differential operators are transformed according to

*Reprinted from Ref. [34], Copyright 2010, with permission from Elsevier

$$\begin{cases} \frac{\partial}{\partial x} = \frac{1}{k_{,X}} \frac{\partial}{\partial X} \\ \\ \frac{\partial}{\partial z} = -\frac{k_{,Z}}{k_{,X}} \frac{\partial}{\partial X} + \frac{\partial}{\partial Z}. \end{cases} \tag{9.37}$$

In such a two-dimensional case, the components of the velocity vector $\underline{v}^T = (u, w)$ are transformed into $\underline{V}^T = (U, W)$ when the coordinate system (X, Z) is considered instead of the physical one (x, y). The just referred transformation writes:

$$\begin{pmatrix} U \\ W \end{pmatrix} = \begin{pmatrix} \frac{1}{k_{,X}} & -\frac{k_{,Z}}{k_{,X}} \\ 0 & 1 \end{pmatrix} \begin{pmatrix} u \\ w \end{pmatrix} = \begin{pmatrix} \frac{1}{k_{,X}} u - \frac{k_{,Z}}{k_{,X}} w \\ w \end{pmatrix}. \tag{9.38}$$

Now, if we enforce

$$u = \frac{\partial x}{\partial Z} w = k_{,Z} w, \tag{9.39}$$

the x-component of the velocity vector U vanishes, a fact that guarantees that the flow streamlines consist of the lines $X =$ constant.

9.4.2 Solving the Problem

Solving the problem by using STM consists of determining the mapping function that minimizes the residual of the conservation equations, by solving the resulting non-linear model using an appropriate technique (e.g. the Newton–Raphson strategy). In the two-dimensional case considered here, the momentum balance (neglecting mass and inertia terms) writes

$$\nabla \cdot \left(\underline{\underline{\tau}} - p \underline{\underline{I}} \right) = 0, \tag{9.40}$$

with $\tau = 2\eta_s \underline{\underline{D}} + \underline{\underline{\tau}}_p$. Using the just introduced transformation, the momentum balance equations read

$$\begin{cases} 0 = \frac{\partial \tau_{xx}}{\partial X} + k_{,X} \frac{\partial \tau_{xz}}{\partial Z} - k_{,Z} \frac{\partial \tau_{xz}}{\partial X} - \frac{\partial p}{\partial X} \\ \\ 0 = \frac{\partial \tau_{xz}}{\partial X} + k_{,X} \frac{\partial \tau_{zz}}{\partial Z} - k_{,Z} \frac{\partial \tau_{zz}}{\partial X} - k_{,X} \frac{\partial p}{\partial Z} + k_{,Z} \frac{\partial p}{\partial X}. \end{cases} \tag{9.41}$$

Finally, we should address the mass conservation balance for an incompressible flow. For this purpose, we consider two neighbour streamlines, closer enough to suppose that the velocity between both streamlines does not evolve significantly. Now, we consider the inflow boundary z = 0 and another generic section z within the flow domain. These two sections result in the transformed sections Z = 0 and Z. At the inflow section $dX \equiv dx$, that is, $k_{,X}|_{Z=0} = 1$ and then $dX = dx$. At section Z, $dX = k_{,x}dx$, and the mass conservation writes

$$w_{X,Z=0}dx = w(X, Z)k_{,X}|_{X,Z}dx \rightarrow w(X, Z) = w(X, Z = 0)\frac{1}{k_{,X}|_{X,Z}}. \quad (9.42)$$

The classical "velocity/pressure" formulation is then replaced by "transformation function/pressure" formulation. To solve the related equation, we propose an iterative algorithm that allows to compute the "transformation function/pressure" and the stress resulting from the microscopic calculations within a fixed point strategy. For a given value of microscopic stress at the previous iteration l − 1, the "transformation function/pressure" derived from the momentum balance equation:

$$-(\nabla p)^l + \left(2\eta_s \nabla \cdot \underline{\underline{D}}\right)^l = -\left(\nabla \cdot \underline{\underline{\tau}}_p\right)^{l-1}, \quad (9.43)$$

where η_s is the solvent viscosity and the strain rate tensor $\underline{\underline{D}}$is expressed in terms of the unknown transformation function and its derivative with respect to X and Z using the relationship (9.37) and the development of the divergence operator given in Eq. (9.41). The superscripts "l" and "$l - 1$" denote, respectively, the current and the previous iteration.

In Eq. (9.43), since the elliptic contribution related to the solvent viscosity can be negligible against the stress contribution, the introduction of an arbitrary numerical viscosity must be done. This viscosity is chosen to be in the same order as the apparent viscosity of the microscopic behaviour (that decreases for high values of shear rates). A Carreau model has been arbitrarily chosen here to represent this arbitrary viscosity. Thus, Eq. (9.43) is written as

$$-(\nabla p)^l + \left(2(\eta_s + \eta_{\text{carr}})\nabla \cdot \underline{\underline{D}}\right)^l = -\left(\nabla \cdot \underline{\underline{\tau}}_p\right)^{l-1} + \left(2\eta_{\text{carr}} \nabla \cdot \underline{\underline{D}}\right)^{l-1}, \quad (9.44)$$

with the arbitrary viscosity given by

$$\eta_{\text{carr}} = \frac{\eta_0}{\left[1 + (a\dot{\gamma})^2\right]^n}, \quad (9.45)$$

where η_0, a and n are three model parameters that are adjusted in order to describe with a Carreau model the rheological behaviour that the considered fluid exhibits a simple shear flow.

The boundary conditions are transformed using all the previous expressions, resulting in different types of conditions on the transformation function k. The interested reader can refer to [9] and [23] and the references therein.

9.5 Numerical Results

9.5.1 Transient Network Analysis in a Steady Simple Shear

First, we consider the resolution of the Fokker–Planck equation related to the fibre network model. Figure 9.3 depicts the distribution function for active and pendant fibres obtained by considering $D_{r1} = 0.5$, $D_{r2} = 0.2$, $2\eta_s N_{p1} = 1$, $2\eta_s N_{p2} = 1$, $V_c = 0.1$ and $V_d = 0.1$. An isotropic orientation of fibre distribution is assumed as the initial orientation state. A mesh of 642 nodes of the unit sphere related to the configuration space Ω_p and 100 nodes on the $1D$ domain Ω_t related to the time coordinate. After imposing the shear flow, the fibre orientation concentrates around the direction of the flow. The computed solutions for different Weissenberg numbers $\left(W_e = \max\left(\frac{\dot{\gamma}}{D_{r1}}, \frac{\dot{\gamma}}{D_{r2}}\right)\right)$ in simple $3D$ shear flows defined by $(u = \dot{\gamma} y, v = 0)$, are

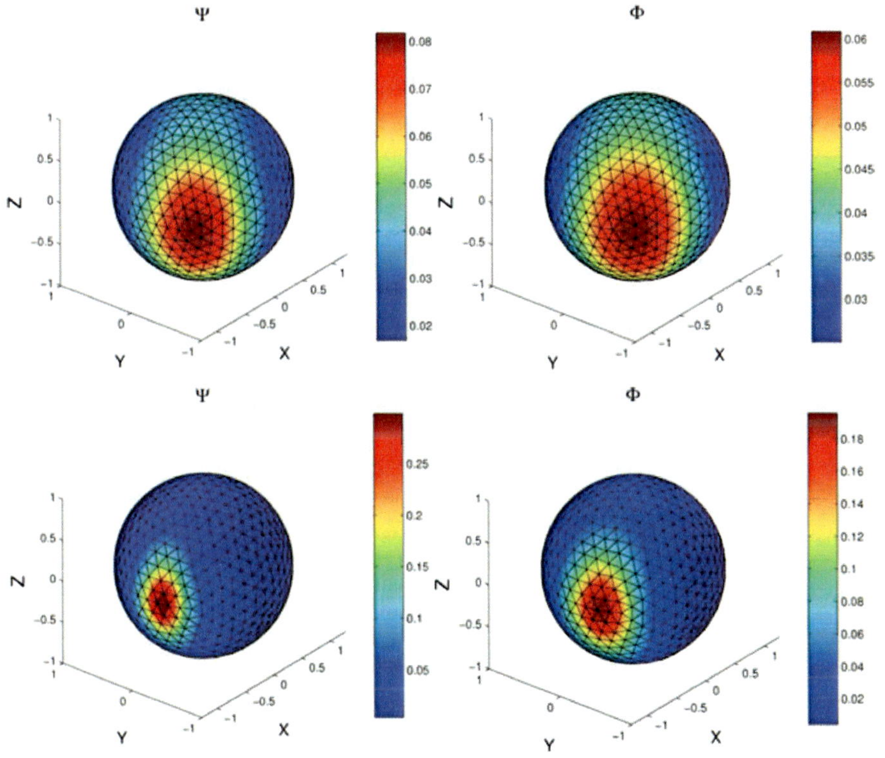

Fig. 9.3 Steady distribution function related to the fibre network model in a three-dimensional transient state shear flow with $D_{r1} = 0.5$, $D_{r2} = 0.2$, $V_c = 0.1$ and $V_d = 0.1$: (top-left) active fibres for $W_e = 1$ (top-right) pendant fibres for $W_e = 1$ (bottom-left) active fibres for $W_e = 10$ (bottom-right) pendant fibres for $W_e = 10$*

*Reprinted from Ref. [34], Copyright 2010, with permission from Elsevier

compared. Under these conditions, we can notice that the active fibres have a considerable contribution in comparison with the pendant fibres. This contribution increases with the Weissenberg number.

Figures 9.4 and 9.5 depict the most significant approximation functions, related to the conformation space and time, involved in the separated representation of the

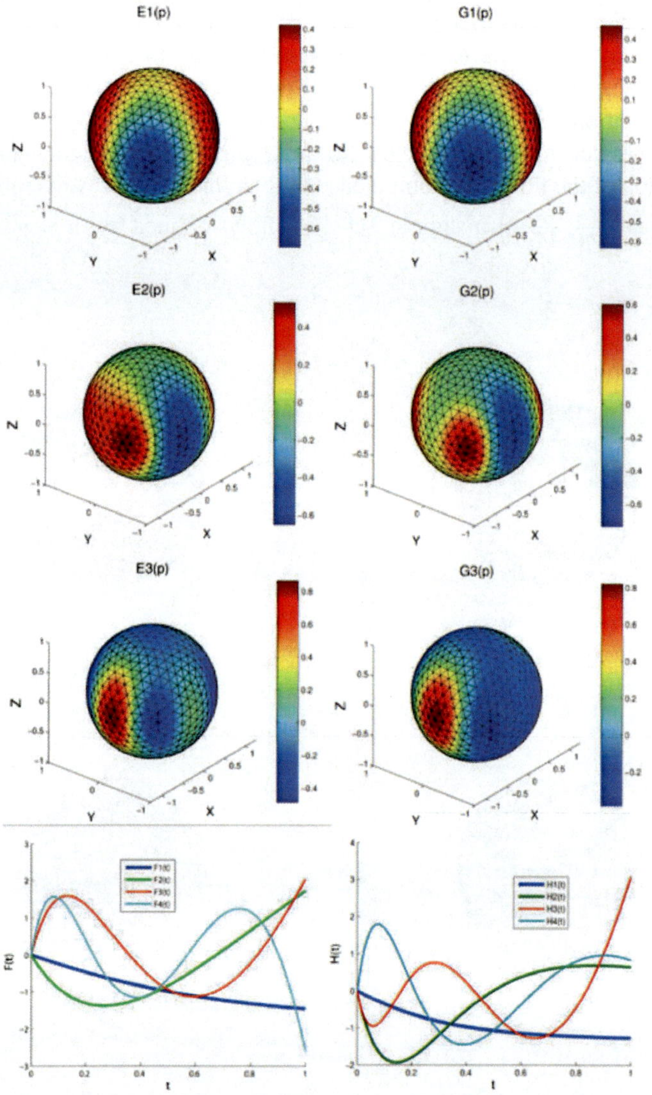

Fig. 9.4 Most significant approximation functions involved in the separated representation of the distribution function related to the fibre network model in a 3D transient shear flow characterized by $W_e = 1$, $D_{r1} = 0.5$, $D_{r2} = 0.2$, $V_c = 0.1$ and $V_d = 0.1$: (left) active fibres, (right) pendant fibres*

*Reprinted from Ref. [34], Copyright 2010, with permission from Elsevier

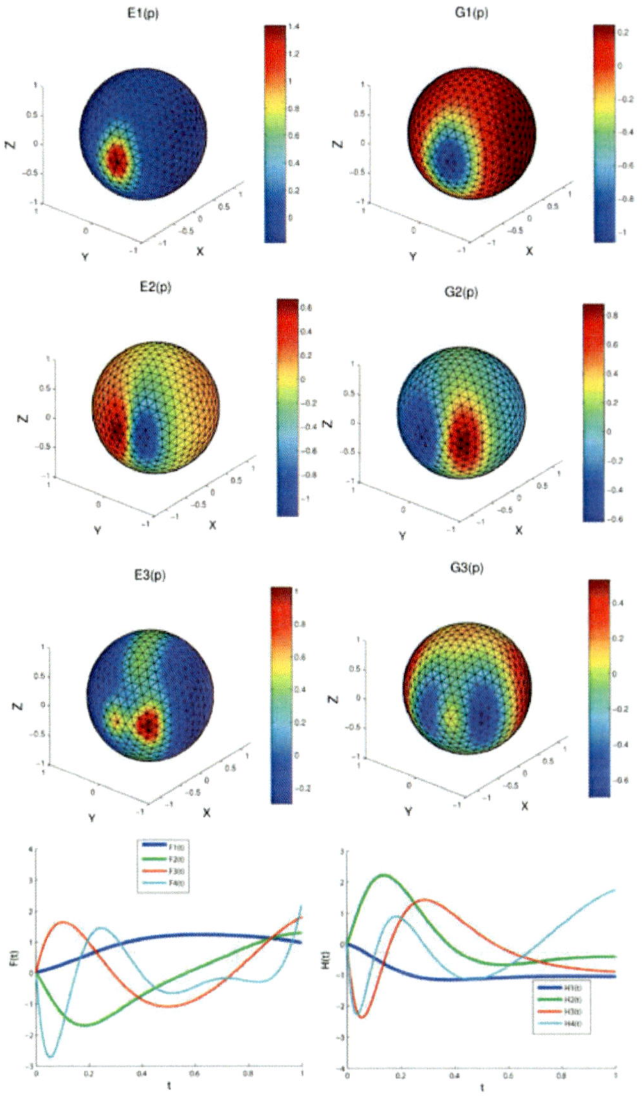

Fig. 9.5 Most significant approximation functions involved in the separated representation of the distribution function related to the fibre network model in a Three-dimensional transient shear flow characterized by $W_e = 9$, $D_{r1} = 0.5$, $D_{r2} = 0.2$, $V_c = 0.1$ and $V_d = 0.1$: (left) active fibres, (right) pendant fibres*

*Reprinted from Ref. [34], Copyright 2010, with permission from Elsevier

distribution function for both $We = 1$ and $We = 10$, respectively. The dimensionless shear stresses for fibre network model are depicted in Fig. 9.6. It was pointed out that these results confirm a classical profile of the polymer stress translating a rheological behaviour of such models.

Figure 9.7 depicts the time evolution of the active population in a simple shear flow characterized by $We = 0.1$. Obviously, the pendant population constitutes the complementary part, such that the addition of both becomes constant during the entire simulation. Moreover, we can notice that independently of the initial concentration of active fibres (40, 60 and 80% in our simulations), the same steady state is reached.

Finally, Figs. 9.7 and 9.8 compares the evolution of active populations for two different Weissenberg numbers $We = 0.1$ and $We = 1$. We can notice that the higher is the shear rate (We) the higher is the time evolution of both populations.

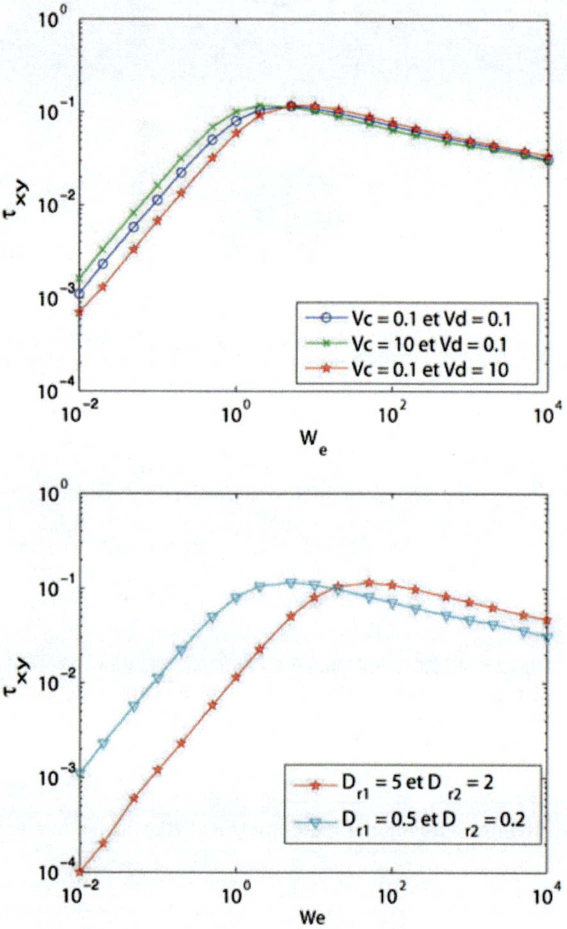

Fig. 9.6 Dimensionless microscopic shear stress evolution related to the fibre network model: (top) $D_{r1} = 0.5$ and $D_{r2} = 0.2$; (bottom) $V_c = V_d = 0.1$*

*Reprinted from Ref. [34], Copyright 2010, with permission from Elsevier

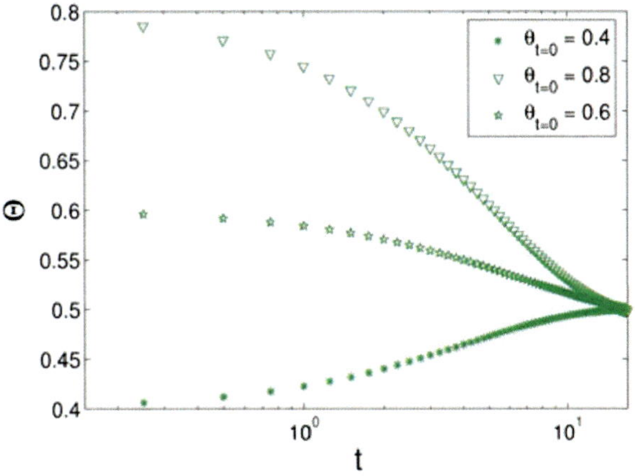

Fig. 9.7 Time evolution of the active population of fibres in a simple shear flow at $We = 0.1$*

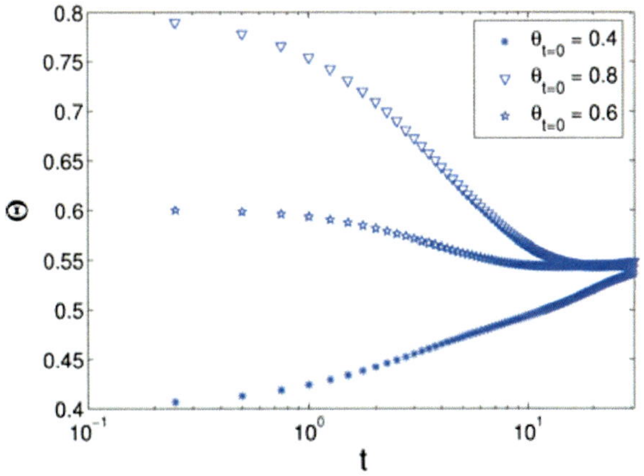

Fig. 9.8 Time evolution of active populations in a simple shear flow for $We = 1$*

9.5.2 Analysis of a Contraction Flow

The algorithm consists of repeating until convergence the following three steps:

1. Compute the microscopic stress from the orientation distributions obtained by solving the Fokker–Planck equation;
2. Compute the flow kinematics;
3. Check the convergence (of the fixed point strategy).

This algorithm will be applied to a physical domain involving a contraction. In this section, we present the macroscopic results. Figure 9.9a–d plot the axial velocity component w, the stress components τ_{xz} and τ_{zz} and the pressure field. It can be noticed the velocity overshoot just downstream of the convergent section, the peaks

---

*Reprinted from Ref. [34], Copyright 2010, with permission from Elsevier

*Reprinted from Ref. [34], Copyright 2010, with permission from Elsevier

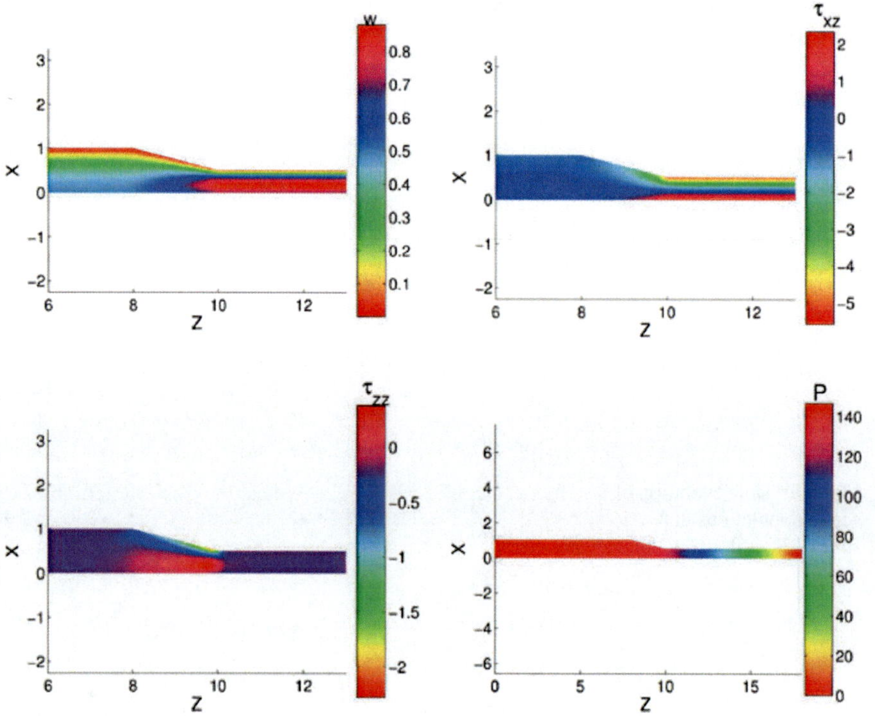

Fig. 9.9 **a** Dimensionless velocity component w; **b** dimensionless stress component τ_{xz}; **c** dimensionless stress component τ_{zz}; **d** dimensionless pressure*

of the stress near the re-entrant corner and the variation of the total pressure drop with elastic effects.

These macroscopic results reveal a variation of the analyzed field near the re-entrant corner. We can conclude that this expected macroscopic behaviour is a consequence of the fibre orientation at the microscopic level. To illustrate this dependence, we depict in Fig. 9.10 the distribution function along some streamlines. Indeed, at the central streamline, the microscopic state shows the existence of only the elongational effects is activated. For the other streamlines, the distribution functions become a compromise between elongation and shear.

9.5.3 *Convergence Analysis*

In order to check the convergence of the coupling strategy, we are considering the contraction flow and the two populations kinetic theory model defined by the set of

*Reprinted from Ref. [34], Copyright 2010, with permission from Elsevier

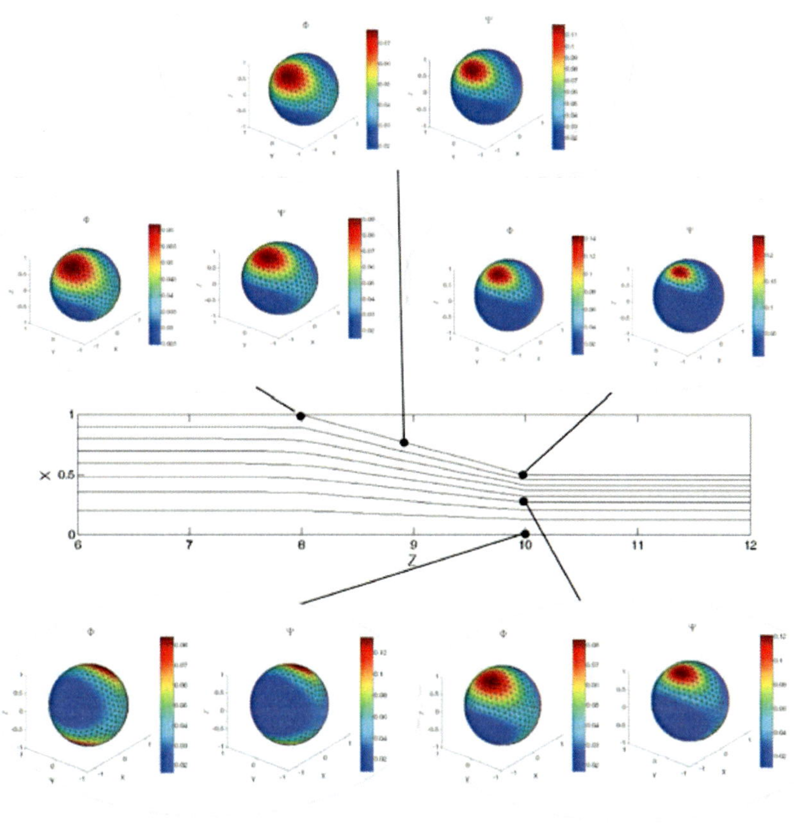

Fig. 9.10 Micro–macro coupling for $h = -1$ and $\eta_s = 1$: *distribution* functions for active, Ψ, and pendant, Φ, populations of fibres along the streamlines[*]

parameters $D_{r1} = 0.5$, $D_{r2} = 0.2$ and $V_c = V_d = 0.1$. The integration of the Fokker–Planck equation along the flow streamlines is performed by decoupling, within the separated representation framework, the conformational coordinates $\underline{p}$ and a coordinate mapping the streamline. To parameterize the streamline, we are using a pseudo-time, as is usual in the method of characteristics, where any point on the streamline can be defined from $\underline{x}(t)$, where t represents the time that a fluid particle spent for reaching position $\underline{x}$. Different time steps were considered to represent the different temporal functions involved in the separated representation: $\Delta t = 0.1, 0.05, 0.02, 0.005$ and $\Delta t = 0.001$. Figure 9.11 shows the computed a_{1212} component of the fourth-order orientation tensor associated with the first population using the different time steps on a streamline segment $\left[\underline{x}(t = 0), \underline{x}(t = 0.5)\right]$. The numerical convergence can be easily noticed in this figure.

---

[*]Reprinted from Ref. [34], Copyright 2010, with permission from Elsevier

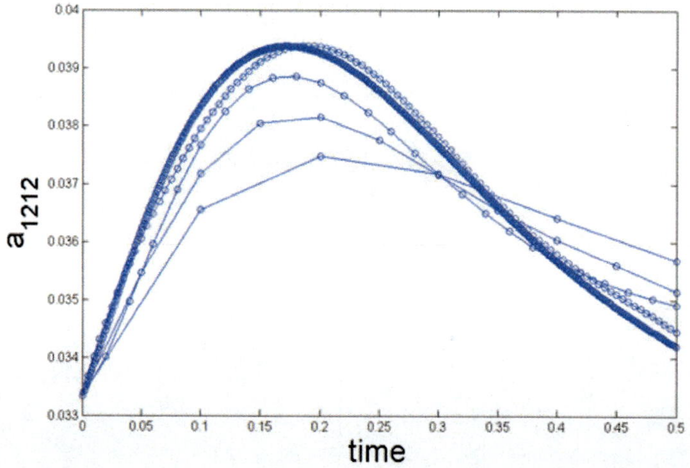

Fig. 9.11 Variations of a_{1212} along a streamline segment for different integration time steps*

The interest in using a separated $(\underline{x}(t), \underline{p})$ representation, instead of a standard backward integration along the streamline, allows for impressive CPU time savings, that can attain some orders of magnitude depending on the considered problem.

If one proceeds by applying a standard integration by the method of characteristics, a problem defined on the unit sphere surface should be solved at each time step. In the contraction flow previously addressed, standard integration schemes need the solution of thousands of problems defined on the surface of the unit sphere, along each streamline. However, when one uses separated representations if the solution consists of a finite sum of N terms, and P iterations were needed for computing each term of the sum, one must solve $N \times P$ problems defined on the surface of the unit sphere and other $N \times P$ defined on the time axis (or the curvilinear abscise related to the considered streamline). In the example here considered, $N \approx 9$ and $P \approx 9$ and then the computing time savings are of one order of magnitude. However, by decreasing the time step (needed when the Weissenberg number increases), the computing time involved in the separated representation remains more or less unchanged because only the one-dimensional integration is affected, but this integration is extremely fast. On the contrary, when one proceeds using an incremental technique the computing time increases linearly with the time step reduction. Finally, we must mention that the stability constraints are much less critical in the case of separated representations than in usual incremental techniques. We have computed solutions using larger time steps that the ones allowed by the stability constraint of incremental procedures, without stability issues. The same time steps induce stability issues when they are used within incremental strategies.

*Reprinted from Ref. [34], Copyright 2010, with permission from Elsevier

9.6 Concluding Remarks

- The stream-tube method provides possibilities to fit the solving flow problems like those presented in this chapter, in a micro–macro approach.
- The proposed deterministic micro–macro approach for simulating complex fluid flows adopts a separated representation of the fields involved in the kinetic theory. This allows circumventing the curse of dimensionality usually considered in the kinetic theory models.
- At the macroscopic level and for general complex flows, the microscopic information can be easily taken into account by integrating the kinetic theory model along the flow streamlines (in steady flows) or along the nodal pathlines in the transient case.
- For general transient and geometrically complex flows, one must integrate the microscopic model along the flow pathlines automatically defined when adopting a Lagrangian description of the flow kinematics. Obviously, this kind of description induces some numerical difficulties when one proceeds in the finite element framework because of the high element distortion. The stream-tube method allows a simple description of the flow streamlines, allowing simple microscopic integrations.
- The use of meshless techniques could be an appealing choice for addressing this kind of micro–macro models involving complex fluid models, complex geometries and transient analysis. This analysis constitutes a work in progress.

References

1. Ahn KH, Osaki K (1995) Mechanism of shear thickening investigated by a network model. J Non-Newton Fluid Mech 56:267–288
2. Ammar A, Mokdad B, Chinesta F, Keuning R (2006) A new family of solvers for some classes of multidimensional partial differential equations encountered in kinetic theory modeling of complex fluids. J Non-Newton Fluid Mech 139:153–176
3. Ammar A, Mokdad B, Chinesta F, Keunings R (2007) A new family of solvers for some classes of multidimensional partial differential equations encountered in kinetic theory modeling of complex fluids. Part II: transient simulation using space-time separated representations. J Non-Newton Fluid Mech 144:98–121 (2007)
4. Annable T, Buscall R, Ettelaie R, Whittlesdtone D (1993) The rheology of solutions of associating polymers: comparaison of experimental behavior with transient network theory. J Rheol 37:695–726
5. Anubhav T, Kam CT, Gareth HM (2006) Rheology and dynamics of associative polymers in shear and extension. Theory and experiment. Macromolecules 39:1981–1999 (2006)
6. Bereaux Y, Clermont JR (1995) Numerical simulation of non-newtonian complex flows using the Stream-Tube Method and memory integral constitutive equations. Int J Nume Method Fluids 21:371–389 (1995)
7. Bungartz HJ, Griebedl M (2004) Sparse grids. Acta Numer 13:1–123
8. Chauvière C, Loszinski AZ (2004) Simulation of dilute polymer solutions using a Fokker-Planck equation. Comput Fluids. 33:687–696

9. Chaubal CV, Strinivasan A, Egesioglu O, Leal LG (1997) Smoothed particle hydrodynamics techniques for the solution of kinetic theory problems. J Non-Newton Fluid Mech 70:125–154
10. Clermont JR (1988) Analysis of incompressible three-dimensional flows using the concept of stream tubes in relation with a transformation of the physical domain. Rheol Acta 27:357–362
11. Grecov DR (2000) Formulations théoriques et simulations numériques d'écoulements de fluides viscoélastiques entre cylindres excentriques et en conduites. PhD thesis, INPG-Grenoble (2000)
12. Grecov DR, Clermont JR (2002) Numerical study of flows of complex fluids between eccentric cylinders using transformation functions. Int J Numer Meth Fluids 40:669–695
13. Grecov DR, Normandin M, Clermont JR (2002) A numerical approach for computing flows by local transformations and domain decomposition using an optimization algorithm. Comput Methods Appl Mech Engrg 191:4401–4419
14. Grecov DR, Clermont JR (2008) Numerical simulations of non-stationnary flows od non-newtonian fluids between concentric and excentric cylinders by stream-tube method and domain decomposition. Rheol Acta 47(5–6):609–620
15. Hernandez Cifre JG, Barenbrug TMAOM, Schieber JD, van der Brule BHAA (2003) Brownian dynamics simulation of reversible polymer networks under shear using a non-interacting dumbbell model. J Non-Newton Fluid Mech 113: 73–96 (2003)
16. Hernandez Cifre JG, Pamies R, Kjoniksen AL, Knudsen KD, Nystrom B, Garcia de la Torre J (2007) Brownian dynamics simulation of reversible polymer networks using a non-interacting bead-and-spring chain model. J Non-Newton Fluid Mech. 146:3–9 (2007)
17. Keunings R (2004) Micro–macro methods for the multiscale simulation of viscoelastic flow using molecular models of kinetic theory. Rheol Rev; Binding DM, Walters K (eds) Br Soc Rheol 67–98 (2004)
18. Lodge AS (1956) A network theory of flow birefringence and stress in concentrated polymer solutions. Trans Faraday Soc 52:120–130
19. Lodge AS (1968) Constitutive equations from molecular network theories for polymer solutions. Rheol Acta 7:379–392
20. Lozinski A, Chauvière C (2003) A fast solver for Fokker-Planck equation applied to viscoelastic flows calculations: 2D FENE model. J Comput Phys 189:607–625
21. Models of shear-thickening behaviour in physically cross-linked networks (1993) G. MArrucci, S. Bhargava, S.L. Copper. Macromolecules 26:6483–6488
22. Mokdad B, Pruliere E, Ammar A, Chinesta F On the simulation of kinetic theory models of complex fluids using the Fokker-Planck approach. Appl Rheol 17:26494–1–26494–14
23. Normandin M, Clermont JR, Guillet J, Raveyre C (1999) Three-dimensional extrudate swell: experimental and numerical study of a polyethylene melt obeying a memory-integral equation. J Non-Newton Fluid Mech 87:1–25
24. Normandin M, Radu DG, Mahmoud A, Clermont JR (2002) Finite element and stream-tube formulations: two-dimensional applications. Math Comput Simul 60:129–134 (2002)
25. Öttinger HC, Lqaso M (1992) Smart polymers in finite element calculation. Int Congr Rheol, Brussel, Belgium
26. Pellens L, Ahn KH, Lee SJ, Mewiq J (2004) Evaluation of a transient network model for telechelic associative polymers J Non-Newton Fluid Mech 121:87–90
27. Pruliere E, Ammar A, El Kissi N, Chinesta F (2009) Recirculating flows involving short fiber suspensions: numerical difficulties and efficient advanced micro–macro solvers. Arch Comput Methods Eng State Art Rev 16:1–30 (2009)
28. Rincon E, Chavez AE, Herrera R, Manero O (2005) Rheological modelling of complex fluids: a transient network model with microstates. J Non-Newton Fluid Mech 131:303–334 (2005)
29. Sarvestani AS, Picu CR (2004) Network model for the viscoelastic behavior of polymer nanocomposites. Polymer 45:7779–7790
30. Tanaka F, Edwards SF (1992) Viscoelastic properties of physically cross-linked networks: transient network theory. Macromolecules 25:1516–1523 (1992)
31. Tanaka F, Edwards SF (1992) Viscoelastic properties of physically cross-linked networks, Parts 1–3. J Non-Newton Fluid Mech 43:247–271, 273–288, 289–309 (1992)

32. Vaccaro A, Marruci G (2000) A Model for the nonlinear rheology of associating polymers". J. Non-Newton Fluid Mech 92(2):261–273
33. Van Dan Brule BHAA, Hoogernrugge PJ (1995) Brownian dynamics simulation of reversible polymeric networks. J Non-Newton Fluid Mech 60:303–334
34. Mokdad B, Ammar A, Normandin M, Chinesta F, Clermont J-R (2010) A fully deterministic Micro–macro simulation of complex flows involving reversible network fluid models. Math Comput Simul 80(9):1936–1961. Elsevier

Appendix A4.1
Detailed Coefficients for Differential Equations

This appendix provides coefficients $\boldsymbol{A_1}$, $\boldsymbol{A_2}$, $\boldsymbol{A_3}$ and $\boldsymbol{B_1}$, $\boldsymbol{B_2}$, $\boldsymbol{B_3}$ considered in the non-linear differential equations (4.69)–(4.70) of Chap. 4, related to the mapping function f, in cylindrical coordinates:

$$\begin{aligned} A_1 = & -ff_R'^2 f_{R'^2 Z}''' + 3ff_R' f_{RZ}'' f_{R^2}'' + f_R'^3 f_{RZ}'' + ff_R' f_Z' f_{R^3}''' \\ & - f_Z' f_R'^2 f_{R^2}'' - 3ff_Z' f_{R^2}''^2 - 6ff_Z' f_R'^2 f_{RZ}''^2 - 3ff_R'^2 f_Z'^2 f_{R^2 Z}''' \\ & + 9ff_Z'^2 f_R' f_{R^2}'' f_{RZ}'' + 3ff_R'^3 f_Z' f_{RZ^2}''' + 3ff_R'^3 f_{RZ}'' f_{Z^2}'' \\ & - ff_R'^4 f_{Z^3}''' - 3ff_Z' f_R'^2 f_{Z^2}'' f_{R^2}'' + ff_Z'^3 f_R' f_{R'^3}''' - 3ff_Z'^3 f_{R^2}''^2 \end{aligned} \tag{A4.1.1}$$

$$\begin{aligned} A_2 = & -2ff_{RZ}' f_R'^2 + f_Z' f_R'^3 + 3ff_Z' f_{R^2}'' f_R' - 6ff_Z'^2 f_R'^2 f_{RZ}'' \\ & + 3ff_Z' f_R'^3 f_{Z^2}'' + 3ff_Z'^3 f_{R^2}'' f_R' \\ A_3 = & -ff_R'^2 f_Z' - ff_R'^2 f_Z'^3 \end{aligned} \tag{A4.1.2}$$

and

$$\begin{aligned} B_1 = & f^2 f_R' f_{R^3}''' - f_R'^4 - ff_R'^2 f_{R^2}'' - 3f^2 f_{R^2}''^2 - 2f^2 f_R'^2 f_Z' f_{R^2 Z}''' \\ & + f^2 f_Z'^2 f_R' f_{R^3}''' + 6f^2 f_Z' f_R' f_{R^2}'' f_{RZ}'' - 3f^2 f_Z'^2 f_{R^2}''^2 \\ & - 2f^2 f_R'^2 f_{R^2 Z}''^2 + f^2 f_R'^3 f_{RZ^2}''' - f^2 f_R'^2 f_{R^2}'' f_{Z^2}'' \end{aligned} \tag{A4.1.3}$$

$$\begin{aligned} B_2 = & ff_R'^3 + 3f^2 f_R' f_{R^2}'' - 4f^2 f_Z' f_R'^2 f_{RZ}'' + 3f^2 f_R' f_Z'^2 f_{R^2}'' \\ & + f^2 f_R'^3 f_{Z^2}'' \\ B_3 = & -f^2 f_R'^2 - f^2 f_R'^2 f_Z'^2. \end{aligned} \tag{A4.1.4}$$

J.-R. Clermont and A. Ammar, *Stream-Tube Method*,
https://doi.org/10.1007/978-3-030-65470-2

Appendix A9.1
Separated Representation Solver: Notation

Using the notation introduced in Sect. 9.3 Chap. 9, we start defining the different discrete fields. Thus, functions $E_j\left(\underline{p}\right)$ (resp. $F_j(t)$, $G_j(\underline{p})$, $H_j(t)$) and $R(\underline{p})$ (respectively, $S(t)$, $V(\underline{p})$ and $W(t)$) are defined using usual finite element interpolations.

The conformation vector $\underline{p}$ related to a node on the unit sphere $\underline{p}_i$ is defined as

$$\underline{p}_i^T = (x_i, y_i, z_i), \tag{A9.1.1}$$

where $x_i^2 + y_i^2 + z_i^2 = 1$.

From now on, we denote by $\underline{N}$ (respectively, $\underline{M}$) the vector containing the shape functions $N_r\left(\underline{p}\right)$ (resp. $M_s(t)$). Thus, any function depending on $\underline{p}$, e.g. $E_j\left(\underline{p}\right)$, can be written as follows:

$$E_j\left(\underline{p}\right) = \sum_{r=1}^{N_n^p} N_r\left(\underline{p}\right) \cdot E_j\left(\underline{p}_r\right), \tag{A9.1.2}$$

where N_n^p the number of nodes distributed on the surface of the unit sphere and $N_r\left(\underline{p}\right)$ the shape function related to node $\underline{p}_r$.

In our simulations, the unit sphere is substituted by polyhedra composed of a set of triangular facets whose vertices are the nodes $\underline{p}$ distributed on the unit surface. In each triangle, the shape functions are the usual piecewise linear shape functions used in the context of finite elements.

Finally, $\underline{E}_j, \underline{F}_j, \underline{G}_j, \underline{H}_j, \underline{R}, \underline{S}, \underline{V}$ and $\underline{W}$ define the vector containing the respective nodal values.

We define the following matrix:

$$\underline{\underline{N}} = \int_{\Omega_{\underline{p}}} \underline{N}\,\underline{N}^{\mathrm{T}} d\Omega_{\underline{p}}; \quad \underline{\underline{M}} = \int_{\Omega_t} \underline{M}\,\underline{M}^{\mathrm{T}} d\Omega_t, \tag{A9.1.3}$$

J.-R. Clermont and A. Ammar, *Stream-Tube Method*,
https://doi.org/10.1007/978-3-030-65470-2

$$\underline{\underline{\mathcal{N}}}_{11} = \int_{\Omega_{\underline{p}}} \left\{ (\underline{\underline{e}}_0)_{1,1} \underline{N}\,\underline{N}^{\mathrm{T}} + (\underline{\underline{e}}_1)_{1,1} \underline{N}\,\underline{dN}^T + \frac{\Gamma h_{\underline{p}}}{2} (\underline{\underline{e}}_1)_{1,1} \underline{dN}\,\underline{dN}^T - (\underline{\underline{e}}_2)_{1,1} \underline{dN}\,\underline{dN}^T \right\} d\Omega_{\underline{p}}, \tag{A9.1.4}$$

$$\underline{\underline{\mathcal{N}}}_{12} = \int_{\Omega_{\underline{p}}} \left\{ (\underline{\underline{e}}_0)_{1,2} \underline{\underline{N}}\,\underline{\underline{N}}^T \right\} d\Omega_{\underline{p}}, \tag{A9.1.5}$$

$$\underline{\underline{\mathcal{N}}}_{21} = \int_{\Omega_{\underline{p}}} \left\{ (\underline{\underline{e}}_0)_{2,1} \underline{\underline{N}}\,\underline{\underline{N}}^T \right\} d\Omega_{\underline{p}}, \tag{A9.1.6}$$

$$\underline{\underline{\mathcal{N}}}_{22} = \int_{\Omega_{\underline{p}}} \left\{ (\underline{\underline{e}}_0)_{2,2} \underline{N}\,\underline{N}^T + (\underline{\underline{e}}_1)_{2,2} \underline{N}\,\underline{dN}^T + \frac{\Gamma h_{\underline{p}}}{2} (\underline{\underline{e}}_1)_{2,2} \left(\underline{p}\right) \underline{dN}\,\underline{dN}^T - (\underline{\underline{e}}_2)_{2,2} \underline{dN}\,\underline{dN}^T d\Omega_{\underline{p}} \right\} \tag{A9.1.7}$$

and

$$\underline{\underline{\mathcal{M}}} = \int_{\Omega^T} \underline{M}\,\underline{dM}^T d\Omega_t. \tag{A9.1.8}$$

Appendix A9.2
Separated Representation Solver: Projection Step

If we assume the functions E_j, F_j, G_j and $H_j (\forall j \in [1, \ldots, n])$ to be known (verifying the boundary conditions), the coefficients α_j and β_j can be computed by introducing the approximation of $\underline{\Xi}$ into the Galerkin variational formulation associated with Eq. (9.26) of Chap. 9.

At the n-th iteration, the components of $\underline{\Xi}$ yield (using the notation introduced in the previous appendix):

$$\psi\left(\underline{p}, t\right) = \left[\underline{N}^T \underline{E}_1 \underline{M}^T \underline{F}_1 \quad \cdots \quad \underline{N}^T \underline{E}_n \underline{M}^T \underline{F}_n\right] \cdot \begin{bmatrix} \alpha_1 \\ \vdots \\ \alpha_n \end{bmatrix} \tag{A9.2.1}$$

and

$$\phi\left(\underline{p}, t\right) = \left[\underline{N}^T \underline{G}_1 \underline{M}^T \underline{H}_1 \quad \cdots \quad \underline{N}^T \underline{G}_n \underline{M}^T \underline{H}_n\right] \cdot \begin{bmatrix} \beta_1 \\ \vdots \\ \beta_n \end{bmatrix}. \tag{A9.2.2}$$

Then, the components of the test field $\underline{\Xi}^*$, which is involved in the variational formulation, are given by

$$\psi^*\left(\underline{p}, t\right) = \left[\underline{N}^T \underline{E}_1 \underline{M}^T \underline{F}_1 \quad \cdots \quad \underline{N}^T \underline{E}_n \underline{M}^T \underline{F}_n\right] \cdot \begin{bmatrix} \alpha_1^* \\ \vdots \\ \alpha_n^* \end{bmatrix} \tag{A9.2.3}$$

and

J.-R. Clermont and A. Ammar, *Stream-Tube Method*,
https://doi.org/10.1007/978-3-030-65470-2

$$\phi^*\left(\underline{p},t\right) = \left[\underline{N}^T\underline{G}_1\underline{M}^T\underline{H}_1 \quad \ldots \quad \underline{N}^T\underline{G}_n\underline{M}^T\underline{H}_n\right]\cdot\begin{bmatrix}\beta_1^*\\ \vdots \\ \beta_n\end{bmatrix}. \tag{A9.2.4}$$

Now, if all these expressions are introduced into the weak form of the Fokker–Planck equation

$$\begin{aligned}&\int_\Omega \Xi^*\frac{d\Xi}{dt}d\Omega + \int_\Omega \Xi^*\underline{\underline{e}}_0\Psi d\Omega + \int_\Omega \Xi^*\underline{\underline{e}}_1\frac{\partial\Xi}{\partial\underline{p}}d\Omega\\ &+\int_\Omega \frac{\Gamma h_{\underline{p}}}{2}\frac{\partial\Xi^*}{\partial\underline{p}}\underline{\underline{e}}_1\frac{\partial\Xi}{\partial\underline{p}}d\Omega - \int_\Omega \frac{\partial\Xi^*}{\partial\underline{p}}\underline{\underline{e}}_2\frac{\partial\Xi}{\partial\underline{p}}d\Omega\\ &= -\int_\Omega \Xi^*\underline{\underline{e}}_0\Xi_0 d\Omega - \int_\Omega \Xi^*\underline{\underline{e}}_1\frac{\partial\Xi_0}{\partial\underline{p}}d\Omega\\ &-\int_\Omega \Xi^*\underline{\underline{e}}_2\frac{\partial^2\Xi_0}{\partial\underline{p}^2}d\Omega,\end{aligned} \tag{A9.2.5}$$

the following linear system, from which we can determine the α_i and β_i coefficients $\forall i = 1, \cdots, n$, is obtained:

$$\underline{\underline{K}}\underline{\alpha} = \underline{V}, \tag{A9.2.6}$$

where

$$\underline{\alpha} = \left[\alpha_1 \ldots \alpha_n \; \beta_1 \ldots \beta_n\right]^T, \tag{A9.2.7}$$

and the components of $\underline{\underline{K}}$ and $\underline{V}$ are given by

$$\begin{aligned}\underline{\underline{K}}_{i,j} = &\left[\begin{pmatrix}\underline{E}_i^T\\ \underline{G}_i^T\end{pmatrix}\begin{pmatrix}\underline{\underline{N}}_{11}+\underline{\underline{N}}_{12}\\ \underline{\underline{N}}_{21}+\underline{\underline{N}}_{22}\end{pmatrix}\begin{pmatrix}\underline{E}_j\\ \underline{G}_j\end{pmatrix}\cdot\begin{pmatrix}\underline{F}_i^T\\ \underline{H}_i^T\end{pmatrix}\begin{pmatrix}\underline{\underline{M}}\\ \underline{\underline{M}}\end{pmatrix}\begin{pmatrix}\underline{F}_j\\ \underline{H}_j\end{pmatrix}\right]\\ &+\left[\begin{pmatrix}\underline{E}_i^T\\ \underline{G}_i^T\end{pmatrix}\begin{pmatrix}\underline{\underline{N}}\\ \underline{\underline{N}}\end{pmatrix}\begin{pmatrix}\underline{E}_j\\ \underline{G}_j\end{pmatrix}\cdot\begin{pmatrix}\underline{F}_i^T\\ \underline{H}_i^T\end{pmatrix}\begin{pmatrix}\underline{\underline{M}}\\ \underline{\underline{M}}\end{pmatrix}\begin{pmatrix}\underline{F}_j\\ \underline{H}_j\end{pmatrix}\right]\end{aligned} \tag{A9.2.8}$$

and

$$\underline{V}_i = -\begin{pmatrix}\underline{E}_i^T\\ \underline{G}_i^T\end{pmatrix}\int_{\Omega_{\underline{p}}}\left\{\begin{pmatrix}\underline{\underline{N}}_{11}+\underline{\underline{N}}_{12}\\ \underline{\underline{N}}_{21}+\underline{\underline{N}}_{22}\end{pmatrix}\begin{pmatrix}\Psi^0\\ \Phi^0\end{pmatrix}\right\}d\Omega_{\underline{p}}\begin{pmatrix}\underline{F}_i^T\\ \underline{H}_i^T\end{pmatrix}\int_{\Omega_t}\begin{pmatrix}\underline{M}\\ \underline{M}\end{pmatrix}d\Omega_t, \tag{A9.2.9}$$

where

$$\begin{cases} \Psi^0 \equiv \Psi\left(\underline{p}, t = 0\right), \\ \Phi^0 \equiv \Phi\left(\underline{p}, t = 0\right). \end{cases} \tag{A9.2.10}$$

Appendix A9.3
Separated Representation Solver: Approximation Basis Enrichment

From the alpha coefficients just computed, the approximation basis can be enriched by adding the "best" couples $E_{n+1}\left(\underline{p}\right)F_{n+1}(t)$ and $G_{n+1}\left(\underline{p}\right)H_{n+1}(t)$. To determine the involved functions, we introduce the enriched approximation (9.30) of Chap. 9:

$$\underline{\Xi} = \begin{pmatrix} \psi\left(\underline{p}, t\right) \\ \phi\left(\underline{p}, t\right) \end{pmatrix} = \sum\nolimits_{j=1}^{n} \begin{pmatrix} \alpha_j E_j\left(\underline{p}\right) F_j(t) \\ \beta_j G_j\left(\underline{p}\right) H_j(t) \end{pmatrix} + \begin{pmatrix} R\left(\underline{p}\right) S(t) \\ V\left(\underline{p}\right) W(t) \end{pmatrix} \quad \text{(A9.3.1)}$$

into the variational formulation (9.26).

As described in Sect. 9.3 Chap. 9, the functions $E_{n+1}\left(\underline{p}\right)$, $F_{n+1}(t)$, $G_{n+1}\left(\underline{p}\right)$ and $H_{n+1}(t)$ are obtained by normalizing the functions $R\left(\underline{p}\right)$, $S(t)$, $V\left(\underline{p}\right)$ and $W(t)$.

In order to proceed with the discretization of the variational formulation, we must define the expression of the weighting function $\underline{\Xi}^*$, whose separated representation is given as

$$\underline{\Xi}^* = \begin{pmatrix} R^*\left(\underline{p}\right) S(t) + R\left(\underline{p}\right) S^*(t) \\ V^*\left(\underline{p}\right) W(t) + V\left(\underline{p}\right) W^*(t) \end{pmatrix}. \quad \text{(A9.3.2)}$$

Thus, introducing Eqs. (9.64) and (9.65) into the weak form (9.66) Chap. 9, we obtain the non-linear discrete system as follows:

▼▼

$$\underline{V}_1\left(\underline{R}, \underline{S}, \underline{V}, \underline{W}\right) + \underline{\underline{K}}\left(\underline{R}, \underline{S}, \underline{V}, \underline{W}\right) \begin{bmatrix} \underline{R} \\ \underline{S} \\ \underline{V} \\ \underline{W} \end{bmatrix} = \underline{V}_2\left(\underline{R}, \underline{S}, \underline{V}, \underline{W}\right), \quad \text{(A9.3.3)}$$

J.-R. Clermont and A. Ammar, *Stream-Tube Method*,
https://doi.org/10.1007/978-3-030-65470-2

where

$$
\begin{aligned}
\underline{\underline{K}}\left(\underline{R},\underline{S},\underline{V},\underline{W}\right) &= \left[\,\underline{\underline{K}}_1\;\underline{\underline{K}}_2\,\right],\\
\underline{V}_1\left(\underline{R},\underline{S},\underline{V},\underline{W}\right) &= \left[\,\underline{V}_1^1\;\underline{V}_2^1\,\right],\\
\underline{V}_2\left(\underline{R},\underline{S},\underline{V},\underline{W}\right) &= \left[\,\underline{V}_1^2\;\underline{V}_2^2\,\right],
\end{aligned}
\tag{A9.3.4}
$$

with

$$
\underline{K}_1 = \begin{pmatrix}\underline{\underline{\mathcal{N}}}_{11}\\ \underline{\underline{\mathcal{N}}}_{22}\end{pmatrix}\left[\begin{pmatrix}\underline{S}^T\\ \underline{W}^T\end{pmatrix}\begin{pmatrix}\underline{\underline{M}}\\ \underline{\underline{M}}\end{pmatrix}\begin{pmatrix}\underline{S}\\ \underline{W}\end{pmatrix}\right] + \begin{pmatrix}\underline{\underline{N}}\\ \underline{\underline{N}}\end{pmatrix}\left[\begin{pmatrix}\underline{S}^T\\ \underline{W}^T\end{pmatrix}\begin{pmatrix}\underline{\underline{\mathcal{M}}}\\ \underline{\underline{\mathcal{M}}}\end{pmatrix}\begin{pmatrix}\underline{S}\\ \underline{W}\end{pmatrix}\right],
\tag{A9.3.5}
$$

$$
\underline{\underline{K}}_2 = \begin{pmatrix}\underline{\underline{\mathcal{M}}}\\ \underline{\underline{\mathcal{M}}}\end{pmatrix}\left[\begin{pmatrix}\underline{R}^T\\ \underline{V}^T\end{pmatrix}\begin{pmatrix}\underline{\underline{N}}\\ \underline{\underline{N}}\end{pmatrix}\begin{pmatrix}\underline{R}\\ \underline{V}\end{pmatrix}\right] + \begin{pmatrix}\underline{\underline{M}}\\ \underline{\underline{M}}\end{pmatrix}\left[\begin{pmatrix}\underline{R}^T\\ \underline{V}^T\end{pmatrix}\begin{pmatrix}\underline{\underline{\mathcal{N}}}_{11}\\ \underline{\underline{\mathcal{N}}}_{22}\end{pmatrix}\begin{pmatrix}\underline{R}\\ \underline{V}\end{pmatrix}\right],
\tag{A9.3.6}
$$

$$
\begin{aligned}
\underline{V}_1^1 = {} & \sum\nolimits_{j=1}^{n}\begin{pmatrix}\alpha_j\\ \beta_j\end{pmatrix}\left[\begin{pmatrix}\underline{\underline{\mathcal{N}}}_{11}\\ \underline{\underline{\mathcal{N}}}_{22}\end{pmatrix}\begin{pmatrix}\underline{E}_j\\ \underline{G}_j\end{pmatrix}\right]\left[\begin{pmatrix}\underline{S}^T\\ \underline{W}^T\end{pmatrix}\begin{pmatrix}\underline{\underline{M}}\\ \underline{\underline{M}}\end{pmatrix}\begin{pmatrix}\underline{F}_j\\ \underline{H}_j\end{pmatrix}\right]\\
& + \sum\nolimits_{j=1}^{n}\begin{pmatrix}\alpha_j\\ \beta_j\end{pmatrix}\left[\begin{pmatrix}\underline{\underline{N}}\\ \underline{\underline{N}}\end{pmatrix}\begin{pmatrix}\underline{E}_j\\ \underline{G}_j\end{pmatrix}\right]\left[\begin{pmatrix}\underline{S}^T\\ \underline{W}^T\end{pmatrix}\begin{pmatrix}\underline{\underline{\mathcal{M}}}\\ \underline{\underline{\mathcal{M}}}\end{pmatrix}\begin{pmatrix}\underline{F}_j\\ \underline{H}_j\end{pmatrix}\right]\\
& + \sum\nolimits_{j=1}^{n}\begin{pmatrix}\beta_j\\ \alpha_j\end{pmatrix}\left[\begin{pmatrix}\underline{\underline{\mathcal{N}}}_{21}\\ \underline{\underline{\mathcal{N}}}_{12}\end{pmatrix}\begin{pmatrix}\underline{G}_j\\ \underline{E}_j\end{pmatrix}\right]\left[\begin{pmatrix}\underline{S}^T\\ \underline{W}^T\end{pmatrix}\begin{pmatrix}\underline{\underline{\mathcal{M}}}\\ \underline{\underline{\mathcal{M}}}\end{pmatrix}\begin{pmatrix}\underline{H}_j\\ \underline{F}_j\end{pmatrix}\right]\\
& + \left[\begin{pmatrix}\underline{\underline{\mathcal{N}}}_{21}\\ \underline{\underline{\mathcal{N}}}_{12}\end{pmatrix}\begin{pmatrix}\underline{V}\\ \underline{R}\end{pmatrix}\right]\left[\begin{pmatrix}\underline{S}^T\\ \underline{W}^T\end{pmatrix}\begin{pmatrix}\underline{\underline{\mathcal{M}}}\\ \underline{\underline{\mathcal{M}}}\end{pmatrix}\begin{pmatrix}\underline{W}\\ \underline{S}\end{pmatrix}\right],
\end{aligned}
\tag{A9.3.7}
$$

$$
\begin{aligned}
\underline{V}_2^1 = {} & \sum\nolimits_{j=1}^{n}\begin{pmatrix}\alpha_j\\ \beta_j\end{pmatrix}\left[\begin{pmatrix}\underline{\underline{\mathcal{M}}}\\ \underline{\underline{\mathcal{M}}}\end{pmatrix}\begin{pmatrix}\underline{E}_j\\ \underline{H}_j\end{pmatrix}\right]\left[\begin{pmatrix}\underline{R}^T\\ \underline{V}^T\end{pmatrix}\begin{pmatrix}\underline{\underline{N}}\\ \underline{\underline{N}}\end{pmatrix}\begin{pmatrix}\underline{E}_j\\ \underline{G}_j\end{pmatrix}\right]\\
& + \sum\nolimits_{j=1}^{n}\begin{pmatrix}\alpha_j\\ \beta_j\end{pmatrix}\left[\begin{pmatrix}\underline{\underline{M}}\\ \underline{\underline{M}}\end{pmatrix}\begin{pmatrix}\underline{F}_j\\ \underline{H}_j\end{pmatrix}\right]\left[\begin{pmatrix}\underline{R}^T\\ \underline{V}^T\end{pmatrix}\begin{pmatrix}\underline{\underline{\mathcal{N}}}_{11}\\ \underline{\underline{\mathcal{N}}}_{22}\end{pmatrix}\begin{pmatrix}\underline{E}_j\\ \underline{G}_j\end{pmatrix}\right]\\
& + \sum\nolimits_{j=1}^{n}\begin{pmatrix}\beta_j\\ \alpha_j\end{pmatrix}\left[\begin{pmatrix}\underline{\underline{\mathcal{M}}}\\ \underline{\underline{\mathcal{M}}}\end{pmatrix}\begin{pmatrix}\underline{H}_j\\ \underline{F}_j\end{pmatrix}\right]\left[\begin{pmatrix}\underline{R}^T\\ \underline{V}^T\end{pmatrix}\begin{pmatrix}\underline{\underline{\mathcal{N}}}_{21}\\ \underline{\underline{\mathcal{N}}}_{12}\end{pmatrix}\begin{pmatrix}\underline{G}_j\\ \underline{E}_j\end{pmatrix}\right]\\
& + \left[\begin{pmatrix}\underline{\underline{\mathcal{M}}}\\ \underline{\underline{\mathcal{M}}}\end{pmatrix}\begin{pmatrix}\underline{W}\\ \underline{S}\end{pmatrix}\right]\left[\begin{pmatrix}\underline{R}^T\\ \underline{V}^T\end{pmatrix}\begin{pmatrix}\underline{\underline{\mathcal{N}}}_{21}\\ \underline{\underline{\mathcal{N}}}_{12}\end{pmatrix}\begin{pmatrix}\underline{V}\\ \underline{R}\end{pmatrix}\right],
\end{aligned}
\tag{A9.3.8}
$$

$$
\underline{V}_1^2 = -\left[\int_{\Omega_{\underline{p}}}\begin{pmatrix}\underline{\underline{\mathcal{N}}}_{11}\\ \underline{\underline{\mathcal{N}}}_{22}\end{pmatrix}\begin{pmatrix}\Psi\left(\underline{p},t=0\right)\\ \Phi\left(\underline{p},t=0\right)\end{pmatrix}d\Omega_{\underline{p}}\right]\left[\int_{\Omega_t}\begin{pmatrix}\underline{M}^T\\ \underline{M}^T\end{pmatrix}d\Omega_t\begin{pmatrix}\underline{S}\\ \underline{W}\end{pmatrix}\right]
$$

$$
-\left[\int_{\Omega_{\underline{p}}}\begin{pmatrix}\underline{\underline{N}}_{21}\\ \underline{\underline{N}}_{12}\end{pmatrix}\begin{pmatrix}\Phi\left(\underline{p},t=0\right)\\ \Psi\left(\underline{p},t=0\right)\end{pmatrix}d\Omega_{\underline{p}}\right]\left[\int_{\Omega_t}\begin{pmatrix}\underline{M}^T\\ \underline{M}^T\end{pmatrix}d\Omega_t\begin{pmatrix}\underline{S}\\ \underline{W}\end{pmatrix}\right] \tag{A9.3.9}
$$

and

$$
\begin{aligned}
\underline{V}_2^2 = &-\left[\begin{pmatrix}\underline{R}^T\\ \underline{V}^T\end{pmatrix}\int_{\Omega_{\underline{p}}}\begin{pmatrix}\underline{\underline{N}}_{11}\\ \underline{\underline{N}}_{22}\end{pmatrix}\begin{pmatrix}\Psi\left(\underline{p},t=0\right)\\ \Phi\left(\underline{p},t=0\right)\end{pmatrix}d\Omega_{\underline{p}}\right]\left[\int_{\Omega_t}\begin{pmatrix}\underline{M}^T\\ \underline{M}^T\end{pmatrix}d\Omega_t\right]\\
&-\left[\begin{pmatrix}\underline{R}\\ \underline{V}\end{pmatrix}\int_{\Omega_{\underline{p}}}\begin{pmatrix}\underline{\underline{N}}_{21}\\ \underline{\underline{N}}_{12}\end{pmatrix}\begin{pmatrix}\Phi\left(\underline{p},t=0\right)\\ \Psi\left(\underline{p},t=0\right)\end{pmatrix}d\Omega_{\underline{p}}\right]\left[\int_{\Omega_t}\begin{pmatrix}\underline{M}^T\\ \underline{M}^T\end{pmatrix}d\Omega_t\right].
\end{aligned} \tag{A9.3.10}
$$

According to the discussion in Sect. 9.3, the solution of this non-linear system is accomplished within an alternating direction scheme. Thus, at each iteration m of this non-linear iterative solver, assuming the couple $\left(\underline{R}^m, \underline{S}^m\right)$ to be known, the other couple $\left(\underline{V}^{m+1}, \underline{W}^{m+1}\right)$ is computed by solving a system of size $N_{\underline{p}} + N_t$. Now, with the just computed couple $\left(\underline{V}^{m+1}, \underline{W}^{m+1}\right)$, the first one is updated by solving a system of size $N_{\underline{p}} + N_t$ which leads to $\left(\underline{R}^{m+1}, \underline{S}^{m+1}\right)$. The iteration procedure stops after reaching convergence, i.e. when $\left\| \underline{R}^{m+1}, \underline{S}^{m+1} - \underline{R}^m, \underline{S}^m \right\| < \varepsilon$ and $\left\| \underline{V}^{m+1}, \underline{W}^{m+1} - \underline{V}^m, \underline{W}^m \right\| < \varepsilon$, where ε is a sufficiently small parameter.

Index

J.-R. Clermont and A. Ammar, *Stream-Tube Method*,
https://doi.org/10.1007/978-3-030-65470-2